Nursing, Caring, and Complexity Science:

For Human–Environment Well-Being

Alice Ware Davidson, RN, PhD
Marilyn A. Ray, RN, PhD, CTN-A
Marian C. Turkel, RN, PhD, NEA-BC

Springer Publishing Company, LLC
11 West 42nd Street
New York, NY 10036
www.springerpub.com

Acquisitions Editor: Allan Graubard
Senior Editor: Rose Mary Piscitelli
Cover design: Joseph DePinho
Project Manager: Vanavan Jayaraman
Composition: S4Carlisle Publishing Services

ISBN: 978-0-8261-2587-3
E-book ISBN: 978-0-8261-2588-0
11 12 13/ 5 4 3 2 1

Library of Congress Cataloging-in-Publication Data

Nursing, caring, and complexity science: For human-environment well being/[edited by] Alice Ware Davidson, Marilyn A. Ray, Marian C. Turkel.
p. ; cm.
Includes bibliographical references.
ISBN 978-0-8261-2587-3
1. Nursing models. 2. Nursing—Philosophy. 3. System theory. I. Davidson, Alice Ware, 1945–2009. II. Ray, Marilyn Anne. III. Turkel, Marian C.
[DNLM: 1. Nursing Research. 2. Models, Nursing. 3. Nursing Care—methods. 4. Nursing Theory. 5. Systems Theory. WY 20.5]
RT84.5.N853 2010
610.73—dc22

2010037914

Printed in the United States of America by Bang Printing

Nursing, Caring, and Complexity Science:

For Human–Environment Well-Being

Alice Ware Davidson, RN, PhD, was one of the first nurses to study and research the relationship between the Rogerian science of unitary human beings and complexity science in a complex technological environment. She devoted her academic life to researching complexity sciences, complex health care systems, nursing, and health care. Throughout her career, Dr. Davidson advanced complexity sciences in nursing theory, research, administration, and clinical practice. Dr. Davidson received her doctor of philosophy degree from the University of Colorado College of Nursing, Aurora, Colorado. She completed postdoctoral studies at Harvard University, studied at the Santa Fe Institute in New Mexico, and the New England Center for Complex Systems in Boston, Massachusetts. In 2005, Dr. Davidson studied with the renowned complexity scientist, Dr. F. David Peat, in Pari, Italy. Dr. Davidson was an assistant clinical professor at the University of Colorado, College of Nursing, where she taught nursing theory, research, and supervised students in clinical nursing practice. In December 2009, Dr. Davidson lost her battle with cancer. This book is dedicated to Dr. Davidson. Her legacy lives on through the scholarship on complexity sciences, nursing, and caring presented in this book.

Marilyn A. Ray, RN, PhD, CTN-A, Professor Emeritus, Florida Atlantic University, is a renowned nursing educator and researcher of caring in complex health care systems and transcultural nursing. She was a colleague and good friend of Dr. Alice W. Davidson and worked with her on articles related to complexity sciences and nursing. Dr. Ray is an advanced transcultural nurse committed to the development and progress of transcultural caring in nursing worldwide. She is well known for her theory of bureaucratic caring, which integrates knowledge of human caring within complex technological, economic, legal, and political systems in hospital organizations. Previous awards include the Christine E. Lynn Eminent Scholar Chair in Nursing, Florida Atlantic University; Yingling Visiting Scholar, Virginia Commonwealth University; Visiting Lecturer, University of Alberta, Canada, Clinical Sciences Division. Ray attended seminars at the Kennedy Institute of Ethics, Georgetown University. She also studied with Dr. F. David Peat in Pari, Italy, the distinguished physicist in the sciences of complexity. Other fellowships include the Ministry of Health of Ontario and visiting scholar positions in universities in Australia. She has held faculty positions at the University of Colorado, College of Nursing; the Union Institute, Cincinnati; McMaster University, Hamilton, Canada; the University of California, School of Nursing; and the University of San Francisco, School of Nursing. From 1967 to 1999, Dr. Ray served as an officer in the United States Air Force, beginning first with the Wyoming Air National Guard followed by the United States Air Force Reserve. She held the rank of Colonel from 1984 to her retirement in 1999. During her military career, Ray held many diverse positions—flight nurse, educator, researcher, and administrator in many USAF Commands across the United States. She attended the Marshall Space Center to support the potential role of nursing in space. During the last 8 years of her career she was a researcher at the USAF School of Aerospace Medicine, Brooks Air Force Base, Texas. Ray recently completed a book titled *Transcultural Caring Dynamics in Nursing and Health Care.* She edited one book with Dr. Jean Watson and has numerous chapters and peer-reviewed articles in many journals. Her work is translated into different languages. She has over 20 funded research grants totaling almost one million dollars, and presents nationally and internationally, the most recent at the World Universities Forum in Davos, and universities in Lausanne, Switzerland. Ray recently visited the WHO, and the International Council of Nurses in Geneva, Switzerland, sharing knowledge and her experience of transcultural nursing, theory and research, and a vision for the future of nursing within the complex global environment.

Marian C. Turkel, RN, PhD, NEA-BC, is the Director of Professional Nursing Practice at Albert Einstein Healthcare Network (AEHN) and is on the faculty of the Watson Caring Science Institute (WCSI). In her role at AEHN she is responsible for advancing Watson's theory of human caring, integrating research and evidence-based practice initiatives, creating a professional practice environment, and focusing on tenets from complexity science related to innovation and organizational transformation. As part of the faculty of the WCSI, Dr. Turkel works with various hospitals on the practical application of the theory and does presentations on caring science with an emphasis on education, leadership, practice, and research. Her commitment to advancing caring science and valuing caring being the essence of nursing practice started in 1989 when she returned to school for a master's in nursing administration at Florida Atlantic University (FAU). After graduation, Dr. Turkel enrolled in the University of Miami's PhD program and in 1997 she returned to FAU as an assistant professor and taught undergraduate and graduate theory, research, and leadership courses. In 2002 she relocated to Chicago and began consulting with various hospitals on developing practice innovations related to Magnet, creating research initiatives, implementing Watson's theory of human caring into the practice setting, and working with leaders to understand the core value of caring in nursing practice. Over the course of her career she has worked in collaboration with Dr. Marilyn Ray and was the coprincipal investigator on almost one million dollars in federal research funding to study the relationship among caring, economics, and patient outcomes. Dr. Turkel authored a textbook on strategies for obtaining Magnet Program Recognition®, published in peer-reviewed journals, contributed chapters in nursing textbooks, and presented at numerous national and international conferences. She has been actively involved with the International Association for Human Caring for approximately 20 years and assumed the role of President Elect in June 2010.

Dedications

The spirit and creativity of this book on complexity science and nursing science first began with the dedication to new knowledge by our colleague and friend, the late Dr. Alice W. Davidson. Over the past 25 years, Alice encouraged me to pursue the new science and reflect on and study how the new cosmology could transform nursing. In conjunction with the philosophy and science of caring, the science of unitary human beings, transcultural nursing, and the study of complexity sciences, I have begun my journey to understand the meaning of how we, as nurses and professors, cocreate the reality we desire from the spiritual–ethical choices we make. This new consciousness of the integration of mind, spirit, and caring energy illuminates the significance of the mutual human–environment caring relationship, our unitary interconnectedness. Through this vision, and the love and support of my late husband, Jim, and friend Alice, my family, my colleagues, and friends, especially my friends and co-editor, Dr. Marian Turkel and husband, Brooks, I am committed to continue the effort for clarity of understanding the dynamic and complex caring relationship, and how the world can be transformed by caring intention, will, knowledge, and practical wisdom.

Marilyn A. (Dee) Ray

This book is dedicated to my mentors and guides on my journey to understanding the scholarship of caring science. Dr. Carolyn Brown, Dr. Anne Boykin, Dr. Marilyn Parker, Dr. Marilyn Ray, Dr. Savina Schoenhofer, and Dr. Jean Watson provided ongoing love, nurturance, and support, and inspired me to think differently, conduct research, and publish. A special thanks to my co-editor Dr. Marilyn Ray for believing in me so many years ago and inviting me to become a part of her research and scholarship. To the faculty and Board of Directors of the Watson Caring Science Institute, I am honored to be on the journey with all of you and consider each and everyone a special friend and caring colleague. A special dedication to my family, friends, and professional colleagues who understood why I have not been sending cards or e-mails as much as I have wanted to over the past few months. I would like to dedicate this book to registered nurses in the practice setting who have made the personal and professional commitment to using tenets of caring science and complexity science to inform practice. You are the change agents. Most important is the special dedication to my husband Brooks Turkel. Brooks is truly my *bashert* (soul mate in Judaism). His love and caring energy allow my inner creativity to emerge. He honors, respects, and understands my passion for reading, reflecting, and writing. He is always there for me as I explore the scholarship of caring science and make a humble attempt to advance the discipline of nursing and transform practice.

Marian C. Turkel

This book is dedicated to our colleague and friend, Dr. Alice Ware Davidson. Alice died December 2, 2009, after a relatively short battle with malignant melanoma, actually while she was in the process of editing this book with me, Marilyn Dee Ray.

Alice was gifted with a different way of thinking; a scientist, an artist, and a technologist. Alice was one of the first nurse scholars to dedicate her academic life to the study of holistic science and the science of change—complexity sciences. Alice integrated complexity sciences, Rogers' science of unitary human beings, technology, and their application to nursing, caring, health care, and the deep meaning of healing and well-being within the mutual human–environment process. Not only did Alice achieve a PhD in complexity science and nursing at the University of Colorado College of Nursing under the leadership of the former dean, Dr. Jean Watson, and professor Dr. Marilyn Ray, but also, she continued to further her knowledge of complexity sciences and methodology through the following paths: study and postdoctoral work at Harvard University and the New England Center for Complex Systems with her mentor, Dr. Yaneer Bar-Yam and colleagues; the University of Colorado Complex Systems Department; the Santa Fe Institute; and in Italy with Dr. F. David Peat, a renowned quantum physicist originally from the United Kingdom, former Director of the National Science Foundation of Canada, a colleague of scientists, Drs. David Bohm, Rupert Sheldrake, and John Briggs, and a founder and educator of Holistic Science at the Pari Center for New Learning in Pari, Italy, to further the study of the science and art of change and holism, healing, and well-being. During this time of learning and "quiet action" to understand our mysterious universe and nursing and healing, Alice raised two wonderful children, Anne and John. Both children followed in Alice's footsteps. Anne works in Idaho as a scientist, caring for the natural environment and the animals of the earth, and John is a chef, preparing nutritious foods in his restaurant for the people of Denver, Colorado. Alice also leaves a legacy of love and caring to her long-time and dedicated partner, John Smith, and her three grandchildren.

When Alice was diagnosed with malignant melanoma, she was conceiving the ideas for this book and seeking out the many scholars to write chapters of their conceptualizations and research of complexity sciences, nursing, and human–environment well-being. Alice would want you to know for preventive health purposes that the melanoma began with a very tiny mole between her right baby and fourth toes and spread throughout her body from there. Although Alice sought treatment at M. D. Anderson Medical Center in Houston, and other treatments, the cancer progressed. However, the cancer did not stop her marvelous beliefs and creative energy to bring to life the philosophy, scientific theories, and applications to practice that you will experience from reading this book. Alice was a courageous woman, dedicated to her family and her profession. She loved life! She loved to learn about this universe and health and healing. She loved her family and friends. She reverenced the environment not only to understand it through science but also as an appreciator and protector of it through her commitment to what ought to be in the hearts of all of us, ethical agents or, as Socrates challenged us thousands of years ago, giving thought and action to how we ought to live. Alice's humility, respect for all people and creatures of the earth, and her sense of reverence for the environment was also lived out in nature at the home she shared on the Pacific Ocean with John. She enjoyed gardening and care for the fauna and flora. At her home, she was committed to preserving the environment in every way so that it can always be balanced and beautiful. Alice felt that each one of us must cultivate love for and have an ethical obligation to protect, preserve, and safeguard all that we as humans have been given. We must think with our hearts and minds, with feeling and reason. Alice stated that choice was the "conductor of the symphony" in the balance and beauty of the human–environment relationship. Let us follow in her footsteps to make the right choices and find meaning and joy in the preservation of the beauty of the Universe in this dance of life.

With gratitude to a loving friend for her gifts to us,

Marilyn A. (Dee) Ray
Marian C. Turkel

Contents

Contributors

Alan Barnard, RN, PhD, MRCNA Senior Lecturer, Queensland University of Technology, Queensland, Australia

Andrea Bartoli, PhD International Center for Cooperation and Conflict Resolution, Columbia University, New York, NY

Yaneer Bar-Yam, PhD Professor and President, New England Complex Systems Institute, Cambridge, MA

Michael Bleich, RN, PhD, FAAN Professor and Dean, Oregon Health & Science University, School of Nursing, Portland, OR

Lan Bui-Wrzosinska, PhD Warsaw School of Social Psychology, Warsaw, Poland

Sylvia Bushell, MA, CPHQ Founder and CEO, BodhiCare, Scottsdale, AZ

Aric S. Campling, RN, MSN Adjunct Instructor, The Christine E. Lynn College of Nursing, Florida Atlantic University, Boca Raton, FL, and Clinical Informatics Support Center, Children's National Medical Center, Washington, DC

Sherrilyn Coffman, RN, PhD, CS Professor and Assistant Dean, Nevada State College School of Nursing, Henderson, NV

Jeffrey Cohn, MD, MHCM Chief Quality Officer, Albert Einstein Healthcare Network, Philadelphia, PA

Peter Coleman, PhD International Center for Cooperation and Conflict Resolution, Columbia University, New York, NY

Lisa Conboy, MA, MS, ScD Clinical Instructor, Osher Research Center, Division for Research and Education in Complementary and Integrative Medical Therapies, Harvard Medical School, Boston, MA

Diana M. Crowell, RN, PhD, CNAA Independent Nursing Education and Leadership Consultant, Kittery, ME

Jim D'Alfonso, RN, MSN, CNOR Chief Operating Officer/Chief Nurse Executive, Watson Caring Science Institute, Scottsdale, AZ

Alice W. Davidson, RN, PhD Formerly Assistant Clinical Professor, The Anschutz Medical Campus, University of Colorado, College of Nursing, Aurora, CO

Patricia Welch Dittman, RN, PhD, CDE Assistant Professor, Director, Graduate Programs, Nursing Department, Nova Southeastern University, Fort Lauderdale, FL

Terry Eggenberger, RN, PhD(c) Assistant Clinical Professor, The Christine E. Lynn College of Nursing, Florida Atlantic University, Boca Raton, FL

Mary Gambino, RN, PhD Assistant Dean for Community Affairs, University of Kansas School of Nursing, Kansas City, KS

Debra Hain, RN, DNS Assistant Professor, The Christine E. Lynn College of Nursing, Florida Atlantic University, Boca Raton, FL

Jane Faith Kapustin, RN, PhD Associate Professor, Assistant Dean for Master's Programs, University of Maryland, School of Nursing, Baltimore, MD

Mary Beth Kingston, RN, MSN, NEA-BC RWJ Executive Nurse Fellow, Vice President/Chief Nurse Executive, Albert Einstein Healthcare Network, Philadelphia, PA

Larry Liebovitch, PhD Professor & Dean, Division of Mathematics and Natural Sciences, Queens College, City University of New York, New York, NY

Patricia Liehr, RN, PhD Professor and Associate Dean for Research, The Christine E. Lynn College of Nursing, Florida Atlantic University, Boca Raton, FL

Claire E. Lindberg, RN, PhD, APRN, BC Professor, The School of Nursing, The College of New Jersey, Ewing, NJ

Rozzano Locsin, RN, PhD, FAAN John F. Wymer Distinguished Professor of Nursing, The Christine E. Lynn College of Nursing, Florida Atlantic University, Boca Raton, FL

Joy Longo, RN, PhD Assistant Professor, The Christine E. Lynn College of Nursing, Florida Atlantic University, Boca Raton, FL

Jacqueline M. Lopez-Devine, RN, MSN Hospice of Palm Beach, West Palm Beach, FL

Janice M. Morse, RN, PhD (Nursing), PhD (Anthropology), FAAN Professor and the Ida May "Dotty" Barnes, RN, and D. Keith Barnes, MD, Presidential Endowed Chair, University of Utah, College of Nursing, Salt Lake City, UT

Andrzej Nowak, PhD University of Warsaw Center for Complex Systems Research, Warsaw, Poland

Kyoko Osaka, RN, PhD Assistant Professor, University of Tokushima, Japan

Nikhil S. Padhye, PhD Department of Research, University of Texas Health Science Center at Houston, Houston, TX

Barbara Penprase, RN, PhD, CNOR Associate Professor, School of Nursing, Oakland University, Rochester, MI

Joyce Perkins, RN, PhD Augsburg College, Department of Nursing and the Mayo Clinic, Rochester, MN

Marguerite Purnell, RN, PhD, AHN-BC Associate Professor, The Christine E. Lynn College of Nursing, Florida Atlantic University, Boca Raton, FL

Mary Pat Rapp, RN, PhD School of Nursing, University of Texas Health Science Center at Houston, Houston, TX

Marilyn A. Ray, RN, PhD, CTN-A Professor Emeritus, The Christine E. Lynn College of Nursing, Florida Atlantic University, Boca Raton, FL

Pamela Reed, RN, PhD, FAAN Associate Dean, University of Arizona, Tucson, AZ

Francelyn M. Reeder, CNM, PhD Associate Professor Emeritus, University of Colorado Anschutz Medical Campus, College of Nursing, Aurora, CO

Nancy Shirley, RN, PhD Associate Professor, Creighton University, Omaha, NE

Marlaine Smith, RN, PhD, FAAN Professor, Associate Dean, and Helen K. Persson Eminent Scholar, The Christine E. Lynn College of Nursing, Florida Atlantic University, Boca Raton, FL

Mary Jane Smith, RN, PhD Professor and Associate Dean Graduate Affairs, University of West Virginia, School of Nursing, Morgantown, WV

Todd Swinderman, RN, DNS, PhD Quality Coordinator, North Florida Regional Healthcare, Gainesville, FL

Tetsuya Tanioka, RN, PhD Professor, University of Tokushima, Japan

Stefan Topolski, MD Assistant Professor, University of Massachusetts, School of Medicine, Boston, MA; Clinical Instructor, University of New England, and Founder and Director of Caring in Community, Inc., Shelbourne, MA

Marian C. Turkel, RN, PhD, NEA-BC Director of Professional Nursing Practice, Albert Einstein Healthcare Network, Philadelphia, PA

Michael Brooks Turkel, MBA Chief Executive Officer, Chestnut Hill Hospital, Philadelphia, PA

Kathleen Valentine, RN, PhD Associate in Center Administration, The Christine E. Lynn College of Nursing, Florida Atlantic University, Boca Raton, FL

Robin Vallacher, PhD Professor, Center for Complex Systems and Brain Sciences, Florida Atlantic University, Boca Raton, FL

Jean Watson, PhD, RN, AHN-BC, FAAN Distinguished Professor of Nursing, Endowed Chair in Caring Science, University of Colorado, Anschutz Medical Campus, College of Nursing, Aurora, CO. Founder of Watson Caring Science Institute, Boulder, CO

Bruce J. West, PhD Chair Army ST Corps, ST/Chief Scientist Mathematics, Information Sciences Directorate, US Army Research Office, Research Triangle Park, NC

Foreword

This book asks: "What can science say about the complex task of nursing?" One of the remarkable advances in science in the last few decades is the opportunity to address questions that previously were inaccessible to scientific inquiry. Traditionally, science organized itself around questions that could be answered. Increasingly, we ask the questions we want to answer about the world around us.

I have been privileged to participate in developing and applying new methods of science that can better satisfy our desire to know and understand the world around us. Fifteen years ago, I offered a course at Boston University on the fundamental science of complex systems. Among my students was Professor Alice W. Davidson. Of this mathematical treatment of complex systems theories, Alice demanded insight and application to real-world problems of nursing. In parallel with the course activities, she performed research on the complexity of elderly living environments, eventually publishing a paper that validated the abstract theoretical understanding of complexity in this real-world context (Davidson, Teicher, & Bar-Yam, 1997). Complexity is a real world property, which we all encounter whether we use quantitative definitions or not.

Our ability to successfully engage with the world and our tasks within it depends on a reasonable matching of the complexity of our environments with our own complexity. The importance of this concept has increased as the complexity of our society has increased. Alice's work has direct implications for the environments of our elderly individuals. Designing them carefully is necessary in order to enable them to be intellectually active and yet not overwhelmed.

This demonstration has much broader implications across all domains of our existence. For this volume, the condition in which the nurse works, the complexity of his or her environment is similarly important to recognize, calibrate, and ensure. As I engaged in this concept and its applications in the 15 years since Alice's study, I have found the pervasive importance of this insight in the design of our environments, our organizations and those of our children and parents.

Over the years, Alice continued to challenge me to make practical for nursing the formal and quantitative insights of complex systems research.

Alice and her co-editors, Marilyn Ray and Marian Turkel, are pioneers in the insight that it is possible to bring science to address the most complex, personal, interpersonal, biological, and social conditions. Nurses help people under complex stresses, biologically, physically, and socially. Essential to future advances in this field is recognizing this complexity as well as the opportunities we have for deep understanding and insights. This volume illustrates the many opportunities for scientific study.

It is difficult to write this foreward knowing that Alice will not be present in any of my future classes, though her contributions are surely always represented in the slides where I describe her findings.

As the editors and authors clearly state, "nursing is about caring." It is appropriate that the effort to bring complex systems science into nursing received such early attention. After all, it is complex systems science that provides a framework for thinking about relatedness and relationships, a concept diminished in traditional science. The example I often give of the lack of perspective on relationships and their importance is the traditional dictionary definition of *mother*: "A female parent." The often-missing relational definition might be "What a child calls his or her female parent." The difference is both simple and profound, with pervasive significance for our society.

It should be clear that despite abstract concepts and formulations, underlying the effort of this book is a profound sense of relationship with those who need help, those we care for.

Yaneer Bar-Yam
New England Complex Systems Institute

REFERENCE

Davidson, A., Teicher, M. H., & Bar-Yam, Y. (1997). The role of environmental complexity in the well-being of the elderly. *Complexity and Chaos in Nursing, 3,* 5–12.

Reflection

This is a long overdue work by visionary scholars in complexity science: Dr. Alice W. Davidson (who sadly left this earth plane as this manuscript was being completed) and her beloved gifted colleagues, Dr. Marilyn (Dee) Ray and Dr. Marian Turkel, who serve as editors of a comprehensive project with outstandingly diverse nursing, physician, science, and administrator authors. This contemporary and futuristic publication offers a special gift by integrating nursing science, caring science, unitary, and complexity science into a new whole, helping to uncover the effects of caring on nursing, complexity, and human environments. This work intersects with personal/professional practice, education, research, administration, and health care systems at all levels.

It is an honor and privilege to endorse this critical work at this point in time. The developments and focus in this publication also pass the test of time. It brings us into a new era of human consciousness and the role and relevance of complexities and dynamics of human caring–healing environments in which we live and work and find our being and becoming as persons and as human systems.

This collected, edited manuscript is a magnificent exemplar of the evolving work in caring science. It is a unique honor to have this work included in the Watson Caring Science Institute Library at Springer Publishing Company.

The focus of this scholarship brings new meaning and depth of complexity science to caring science, to healing relationships and human healing systems. It brings entirely new dimensions to the phenomenon and overused mind-set of "quality."

When one falls into the depth of the scholarship in this emerging field, one is drawn into the ethic, the philosophical grounding of an emerging world-view, a cosmology that unites, connects, explains, and helps to order our very reality, our chaotic world. It offers a passionate new order of human evolution that embraces the paradox, the chaotic, the disorder, offering a new lens to understand, to comprehend, to personalize, to professionalize, and to give scientific and wisdom insights into creative emergence for what might be called Ontological Development or Ontological Design programs, projects and purposive transformative practices that inform and authenticate human existence, human caring, and healing at the human relationship and environmental level.

This manuscript combines an array of authors' talents and perspectives in this growing field. Each chapter includes a scholarly response to further inform the ontological and epistemological underpinnings of this dynamic science, integrating the science of unitary human beings, complex caring and human inquiry, and the complexity of the dynamics of human caring. The multiple authors bring together and unify deep dimensions, which invite personal and professional reflective scholarship and philosophical and theoretical integration that is combined into a new synthesis of understanding for nursing and health care.

These diverse chapters and the multiple authors' foci help one to grasp the relevance as well as the ironic complexity of disease, illness, and caring systems as well as treatment approaches projecting the reader into technological, electronic documentation, and the future of humanoid relationships in nursing and complexity science(s).

Such a comprehensive collection of work in the field of a unitary model of complexity is a testimony to the importance of this work and how it both grounds and transcends conventional views of science and reality and opens up new horizons of unitary visions. This collected scholarship in this area will inform the personal/professional evolution of caring and nursing in this century and beyond, inviting new visions of the evolved human in the world of practice, education, research, administration, and clinical care. It is truly a visionary futuristic manifesto for this time in nursing and health sciences at all levels.

Jean Watson, PhD, RN, AHN-BC, FAAN
Distinguished Professor of Nursing
Endowed Chair in Caring Science
University of Colorado, Anschutz Medical Campus
College of Nursing
Aurora, CO
Founder, Watson Caring Science Institute
Boulder, CO
www.nursing.ucdenver.edu/caring
www.watsoncaringscience.org

Prologue

Nursing's disciplinary focus is the relationship of caring within a mutual human–environment health experience for healing and well-being (Newman, Sime, & Corcorran-Perry, 1991; Newman, Smith, Pharris, & Jones, 2008; Ray, 2010a, 2010b; Ray & Turkel, 2010; Turkel, Ray, & Kornblatt, in press; Watson, 2005, 2008). With over 150 years of caring science and concentrated research over the past 30 years in the scholarship of caring, we have much to build on. As such, two central perspectives are highlighted in this book: Nightingale's (1859/1969, 1992, 2010b) conceptual system along with Watson (1985, 2005, 2008) and Ray's (1981a, 1984, 1989, 2001, 2006, 2010b) caring theories. Also described is a comparison of the caring theories, the science of unitary human beings (SUHB), and the philosophy of complexity sciences. Such views on caring, the SUHB, and complexity depict a new form of trans-theoretical convergence—with each philosophy, theory of nursing, and sciences of complexity considering the mutual processes of human–environment interaction. In nursing, of course, it was Nightingale (1859/1969) who first identified this perspective.

This prologue thus underscores Nightingale's (1859/1969, 1992) theory of nursing as a reparative process that facilitates knowledge of the integrality of the human–environment relationship. The SUHB of Rogers (1970, 1990), emphasizing the continuous and emergent nature of the simultaneous human and environmental fields at any given point in space and time, is illuminated. Although there are a number of caring theories, for the purposes of this prologue, the caring sciences of Watson and Ray will be highlighted.

Indeed, Watson's (1985, 2005, 2008) transpersonal theory of caring, which centers on caring as a moral ideal, love, and harmony of body, mind, and spirit, is emphasized. Equally so is Ray's theory of bureaucratic caring (1981a, 1984, 1989, 2006, 2010b) (Coffman, 2010; Ray & Turkel, 2010; Turkel, 2007), which describes caring as the relationship between human and spiritual caring dimensions (spiritual, ethical, humanistic, social), and the organizational context of hospital health care systems (with its economic, political, technological, and legal caring dimensions).

The trans-theoretical identification recognizes the commonalities of nursing's philosophy and conceptual frameworks. The commonalities within nursing also show how the theories relate to the philosophy of the sciences of complexity, specifically paralleling the central tenets—unitary nature, connectedness, belongingness, relationship, mutual human–environment process, energy, pattern, increasing complexity, nonrepeatable phenomena, self-organization, choice, and transformation (Anderson, Crabtree, Steele, & McDaniel, 2005; Davidson & Ray, 1991; Davidson, Ray, Cortes, Conboy, & Norman, 2006; Davidson, Teicher, & Bar-Yam, 1997; Hamilton, Pollack, Mitchell, Vicenzi, & West, 1997; Lindberg, Nash, & Lindberg, 2008; Ray, 1994, 1998, 2006; Smith, 1999; Vicenzi, White, & Begun, 1997; Watson, 2005, 2008; Watson & Smith, 2002). Unlike complexity sciences, caring sciences focus on the uniqueness of the mutual human–environment relational caring process, which articulates the *depth* of meaning of caring within the human health experience from empirical, ethical, aesthetic, personal, and sociocultural patterning (Carper, 1978; Ray, 2010a; White, 1995). The following is a presentation of the nature of caring in nursing, Rogers' *Science of Unitary Human Beings*, and complexity sciences and nursing.

THE NATURE OF CARING

Caring is holistic and is the essence of nursing. Nurses and health care administrators in complex organizations recognize that caring, in the human health experience, facilitates excellence in nursing care, health care delivery, and patient outcomes. Caring is not only humanistic, spiritual, and an ethical phenomenon, but it integrates knowledge of the sociocultural environment, the technological, economic, political, and legal dimensions into its meaning structure and conceptual foundation. Caring, as complex, captures the *genuine* science of quality because its science is also the art of practice, an aesthetic which illuminates the beauty of the dynamic nurse–patient relationship, that makes possible authentic spiritual–ethical choices for transformation—healing, health, well-being, and a peaceful death. Thus, caring is universal and particular, ubiquitous in its appeal as the core of nursing philosophy, and particular in its diversity of expression in nursing practice.

Contemporary nursing practice focuses on creating caring environments for nurses, patients, and families within today's complex health care organizations. With the emergence of the American Nurses Credentialing Center's (ANCC) Magnet Recognition Program® (Magnet), nursing theory has moved from its central place in academia and research to practice. The majority of Magnet hospitals have implemented a theoretical framework grounded in caring science. The theme for the 2010 ANCC National Magnet Conference is "Magnet: A Culture of Caring." Watson's (1985, 2005, 2008) theory of human caring has become the theory of choice as direct care registered nurses (RNs) return to caring values. Theory-guided practice advances both the discipline and profession of nursing. Practice outcomes demonstrate the creation of a caring-healing environment, at all levels, and facilitate both human and environmental well-being (Ray 1981a, 2010b; Turkel & Ray, 2004; Watson, 2008).

THEORETICAL EXEMPLARS OF CARING SCIENCE AND THE MUTUAL HUMAN–ENVIRONMENT PROCESS

Nightingale (1859/1969, 1992) is credited as the "mother" of professional nursing. She laid the foundation for caring as central to nursing. Early in her life, Nightingale was concerned about others' suffering, especially from her vantage point of considerable means. She was a deeply religious woman and believed that God called her to help others and to practice the art of charity, love of one's fellow man (human), and a faith in God (Calabria & Macrae, 1994). Nightingale's vision of nursing was seeing nurse caring actions as improving health through the reparative process of disease (dis-ease), which nature instituted (whom she articulated as God) within the human–environment relationship. Her work was not only influenced by her deep faith, but also by what she had experienced while traveling, training, and caring for wounded soldiers during the Crimean War. Nightingale was concerned about the human–environment relationship for enhancing healing and well being. Nursing was not only the administration of medications and dressings, but also the proper use of fresh air, light, warmth, cleanliness, quiet, punctuality, and care, and the extreme importance of nursing in determining the issue of disease . . . "all at the least expense of vital power of the patient" (Nightingale, 1992, p. 6). Many of Nightingale's ideas parallel the ways of thought and action present in nursing today. As such, Nightingale (Ray, 2010a) demonstrated the importance of nursing as nature and nurture in caring and health; the relationship among theology, spirituality, science, nature, environment, human action, and morals (ethics); the dynamic process of the nurse–patient caring relationship; the nurse as epidemiologist; employing evidence-based practice (from a knowledge of empirics and quantitative statistics); and nursing as open to interdisciplinary and cross-cultural practices. Thus, the importance of the human–environment caring relationship for health and well-being was born.

Rogers' (1970) nursing science is a unique conceptual system with its origin in the unitary nature of the human–environment mutual process for healing and health. Rogers' (1970) conceptual system is scientific, a synthesis of ideas and facts, and creative—as it describes an irreducible whole, the SUHB. Over many years, Rogers (1970, 1990) identified homeodynamic principles and finally established the concepts of resonancy, helicy, and integrality to facilitate understanding and study of the continuous mutual human–environment process. *Resonancy* refers to continuous change in wave patterns in human and environmental fields, *helicy* is the continuous innovative, unpredictable, and increasing diversity of human and environmental field patterns, and *integrality* refers to the continuous mutual human field and environmental field process (Rogers, 1990).

While Rogers' theory did not identify directly with complexity sciences, for example, the subsets of quantum theory, chaos theory, or the theory of complex adaptive systems, her vision was synchronous with many concepts common to complexity sciences (Briggs & Peat, 1989; Peat, 2002): evolving dynamic irreducible, nonrepeatable and nonlinear processes, wave patterning, energy fields, increasing complexity, greater diversity, timelessness, facets (suggests fractal patterns or ubiquitous wholes that are self-similar), and emergence (Reeder, 1984; Rogers, 1970, 1990). The implicit SUHB philosophy always begins with the *unitary* nature of the human and environment *in* mutual process, which is continually open and emergent. Thus, locality and space and time properties are not absolute and emerge anywhere. This idea shows the unpredictable, increasingly diverse, and emergent patterning that becomes manifest as humans and the environment dynamically evolve. This patterning unfolds in nursing within the nurse–patient relationship, the unitary nature, and increasing complexity of the mutual human and environment field process.

Although Rogers viewed caring as important within nursing, she did not embrace it as a substantive area (Smith, 1999; Watson & Smith, 2002). She did, however, embrace the notion of unconditional love, not as a substantive area of study, but as critical to nursing practice. In the Rogerian sense, love is considered a unitary, irreducible mutual human-environmental energy field process (Rogers, 1990; Smith, 1999). As Rogers and others engaged with the study of the SUHB were advancing their ideas of the unitary nature of the mutual human-environment process, scholars in the caring sciences were philosophizing about the nature of caring as love and a human and spiritual connectedness. The religious or divine interconnectedness of caring, the caring consciousness of nurses (caritas or charity and compassion and loving kindness) can be understood as a higher human-environmental field process (Ray, 1981b, 1997, 2010a, 2010b; Smith, 1999; Watson, 2005, 2008; Watson & Smith, 2002).

Smith (1999) wrote an extensive comparison and contrasting of the SUHB and caring, and identified patterns of caring meanings that are both implicit and explicit in the SUHB, such as (a) manifesting intentions (creating, holding, and expressing thoughts and will for caring–healing and well-being) (Purnell, 2006; Reeder, 1984); (b) appreciating pattern (the discovery of knowing wholeness and essence) (Cowling & Repede, 2010); (c) placing value on the other as lovable or worthy of being loved; (d) acknowledging the emergence of pattern (Ray, 1997, 2010a); (e) attuning to dynamic flow, for example, attuning to the rhythmic dance within the continuous mutual process—being present in the moment (Boykin & Schoenhofer, 2001) or in the caring science of Watson (1985, 2005, 2008), the caring moment; (f) experiencing the infinite or the pandimensional awareness of the coextensiveness of the universe within the context of human relating (Ray, 1997; Watson, 2005, 2008); and (g) inviting creative emergence or reflection of the transformative potential of caring for self and other and the belief in the continuing innovation of emergent patterning and the panorama of possibilities (Davidson & Ray, 1991; Davidson et al., 1997, 2006; Ray, 1998; Smith, 1999). From a trans-theoretical viewpoint, a human being is a caring energy field with information rooted in the body, interacting with the caring energy and information of others and the universe (Cannato, 2010).

Watson's (1979, 1985, 2005, 2008) philosophy brought to light caring science as the essence of nursing and as the foundational core of the discipline. Caring is a dynamic transpersonal relationship between the nurse and the patient that involves ethical choice and action within the present moment (past, future, and present all at once), which manifests the potential for harmony of body, mind, and soul (spirit) (Watson, 1985, 2005, 2008). Thus, the process of caring is a moral ideal committed to a specific end, ". . . the protection, enhancement, and preservation of the person's humanity which helps to restore inner harmony and potential healing" (Watson, 1985, p. 58). Caring consists of the 10 caritas processes, or caring practices as they have been referred to by RNs in the practice setting, and are known to facilitate healing. Caritas nursing practice involves working from a human-to-human connection, a practice that is "heart-centered." Watson's (2008) caritas processes are exemplars of caring science. Watson's theory continues to be advanced in the practice setting as RNs transform from a focus on the tasks of nursing to "the practice of loving kindness, authentic presence, cultivation of one's spiritual practices, and being in the caring–healing environment and allowing for miracles" (Watson, 2008, p. 34). Watson's philosophy and research reveals that we must become increasingly aware of who we are, the nature of and mystery of caring, and how we influence others and the environment in terms of the choices made for caring, which is life giving for all.

WATSON'S CARITAS PROCESSES

The 10 caritas processes grounded in the tenets of philosophy and ethics potentiate the creation of a caring–healing environment for nurses, patients, and families. RNs from various practice settings developed the caritas literacy, where various attributes of the caritas processes were identified and inform the profession of nursing. A brief overview follows.

Caritas Process 1: *Cultivating the Practice of Loving Kindness and Equanimity Toward Self and Others* involves listening and respecting others, honoring human dignity, treating self and others with loving kindness, recognizing vulnerabilities in self and others, and accepting self and others as they are.

Caritas Process 2: *Being Authentically Present: Enabling, Sustaining, and Honoring the Faith, Hope and Deep Belief System and the Inner-Subjective Life World of Self and Others* means creating opportunities for reflection, silence, and pause; promoting intentionality and human connections with others viewing life as a mystery to be explored rather than a problem to be solved; and interacting with caring arts and sciences to promote healing and wholeness.

Caritas Process 3: *Cultivation of One's Own Spiritual Practices and Transpersonal Self, Going Beyond Ego-Self* means practicing self-reflection, transforming tasks into caring–healing interactions, demonstrating genuine interest in others, and valuing the goodness of self and others as human beings.

Caritas Process 4: *Developing and Sustaining a Helping-Trusting Caring Relationship* includes the concept of a caring moment where the experience transforms both nurse and patient. This involves the practice of authentic presence, holding a sacred space for healing in others' time of need, and entering into the experience to explore the possibilities in the moment and in the relationship.

Caritas Process 5: *Being Present To and Supporting the Expressions of Positive and Negative Feelings* means creating and holding sacred space (a safe space for unfolding and emerging); encouraging story telling as a way to express understanding, allowing the story to emerge, change, and grow; and encouraging reflections of experiences and feelings.

Caritas Process 6: *Creative Use of Self and All Ways of Knowing as Part of the Caritas Process: Engage in the Art of Caritas Nursing.* In practice, a caritas nurse uses self to create healing environments via intentional touch, artistic expression, journaling, music, and play; integrates aesthetic, empirical, ethical, personal, and metaphysical ways of knowing into practice; and helps others to find new meaning in their journey.

Caritas Process 7: *Engage in Genuine Teaching-Learning Experience that Attends to the Unity of Being and Subjective Meaning Attempting to Stay Within the Other's Frame of Reference.* In nursing practice, these calls for a focus on honoring the wholeness of persons, giving information to someone in a way they can receive it, seeking to learn from others and understand their world view, and helping others understand how they are thinking about their health.

Caritas Process 8: *Creating a Healing Environment at All Levels.* A caritas nurse creates space for: human connections to occur, caring intentions, and a healing environment by attending to nursing as environment, light, art, water noise, hand washing, and comfort measures.

Caritas Process 9: *Administering Sacred Nursing Acts of Caring-Healing by Tending to Basic Human Needs.* In practice this means seeing the wholeness of the patient, respecting patients' special needs, involving the family or significant other in the plan of care, and transforming the tasks of nursing into sacred acts of love and caring.

Caritas Process 10: *Opening and Attending to Spiritual/Mysterious and Existential Unknowns of Life-Death or Allowing for Miracles.* This is often difficult to grasp or explain. The caritas nurse is comfortable dealing with the unknown, being open to possibilities, and nurturing and supporting.

As RNs are returning to acknowledging the humanity of nursing through caring practices, health care is transforming and new visions for the future are being cocreated. As Watson's caring science continues to transform practice and the system, traditional outcome measures will not capture the essence of caring moments or allow for miracles. However, the emergence of new and innovative approaches to measure success or outcomes, such as human flourishing, purposeful involvement, and building capacity will occur.

Watson (2008) remarked, "A Caring Science/Caritas orientation to nursing education intersects with the arts and humanities and related fields of study, beyond the conventional clinicalized and medicalized views of human and health-healing" (p. 255). Therefore, to facilitate successful health and healing outcomes, and to advance our knowledge of what it means to be human, Watson (2008) spoke about the importance of advancing caritas education in all arenas. Watson emphasized that a caring science "engages with the diversity of the sciences and humanities and the notions of personal growth, of transformative learning by which the terms in which people think and the words they speak can actually be changed in educative situations" (p. 258). By engaging Watson's caring science as caritas (love, ethics, mystery, transcendence) with, for example, the SUHB, and Ray's theory of bureaucratic caring (highlighting organizations and caring), and the philosophy and research of the sciences of complexity (interconnectedness, self-organization, emergence), a greater understanding of science and the arts will emerge that will, in effect, change the world from disengagement, competition, and violence to compassion, justice, and peace.

RAY'S THEORY OF BUREAUCRATIC CARING

Ray's (1981a, 1984, 1989, 2006, 2010a, 2010b; Coffman, 2006, 2010; Ray & Turkel, 2010; Turkel, 2001, 2006, 2007) theory of bureaucratic caring helps us to see how we can look at complex health care, community, and global systems so that the caritas processes that Watson (2008) articulates can be more understood and better implemented in practice. Health care organizations are hierarchical and exhibit leadership and management system processes that manifest some degree of power, authority, and control for effective functioning. Hospitals therefore tend to be bureaucratic; that is, they are not only places for the care of the sick, but they also are integrated technical-politico-economic and legal organizations. These contextual patterns serve to facilitate efficient and effective functioning of the system. Although new approaches to leadership have been proposed, such as self-governing systems more often than not, workplace communities survive because of knowledge of their economic, technical, legal and political integrative patterning. Thus, organizations are complex and dynamic cultural systems that people consciously developed for the purposes of coordinating activities for specific ends (Wheatley, 2006). In health care, these ends are an efficient organization and healthy patient outcomes.

Organizations are small cultures. Cultures are co-created by people as they interact and construct meaning from diverse or common values, beliefs, attitudes, and behaviors that are communicated and transmitted from one group to another over time. Organizational cultures deal with values and beliefs about what they are there for, products they may produce, how they govern and manage, how they use technology, and how they deal with human relationships—employer–employee ideologies and social interactions (Bar-Yam, 2004; Perrow, 1986). As such, organizations as small cultures include most social structural elements that are visible in society, such as human systems, political, economic, technical, legal, ethical, religious/spiritual, and educational systems. Thus, highly concentrated organizational cultures are woven into meaningful dimensions of the social structure in order to make sense out of what occurs among people and professions in a given system. Today, organizational cultures comprises patterns of behavior, social norms, complex adaptive systems, creative ideas, complex problem-solving processes, explanatory models, and action, including spiritual–ethical caring action (Bar-Yam, 2004; Ray, 2010).

To more fully understand the theory of bureaucratic caring, it is necessary to understand the nature of bureaucracy as it relates to the theory. Developed over many centuries of Western civilization, complex organizations were recognized as rational–legal forms of bureaucracy, with a bureaucratic model finally identified by Max Weber in 1947 (Boone & Bowen, 1980). The process acknowledged key elements of a rational–legal system that included the major areas and tools of economic, political, social, technological, and legal dimensions that we recognize in complex systems today. Within a bureaucracy (organizational culture), diverse people cocreate and construct the meaning of their complex and dynamic social reality through social interaction. Within a bureaucracy, the interplay of communication and ethical choice making within relationships form a moral community within the workplace community; this social interaction sets the moral tone of the work life environment (Turkel & Ray, 2000, 2001, 2004, 2009). Bureaucracy, thus, is a *social* tool that identifies the dynamics of relational authority, power, control, and ethical behavior that has an influence on what we value and how we think and reason in work life interactions (Perrow, 1986).

Ray's (1981a, 1984, 1989, 2006, 2010a, 2010b; Coffman, 2010; Turkel, 2001, 2006, 2007) theory of bureaucratic caring emerged from studying the meaning of caring in a hospital's the complex institutional culture. As a central focus of the discipline of nursing (Newman et al., 2008; Ray, 1981a, 1981b; Watson, 1979, 1985, 2005, 2008), caring "unfolds within a participative process formed by the informational patterns of the nurse and the patient" (Newman et al., 2008, p. 4). Thus, the sphere of knowledge in nursing captures views of caring, copresence, wholeness (health and well-being—integration of body, mind, spirit), choice, and emergence in the human–environment relationship.

Through her research, Ray (1981a, 1984, 1989, 2006, 2010a, 2010b; Coffman, 2010; Ray & Turkel, 2010) also discovered how the context itself (the bureaucracy) played a role in the meaning structure of caring, identifying both substantive and formal theories, differential caring and a synthesis, bureaucratic caring. The theory is holographic (Coffman, 2010; Ray, 2006, 2010a; Ray & Turkel, 2010). Caring is a whole, yet is a part of the complexity of a hospital, and in essence, the society at large. Within the differential caring theory, caring in the complex organization of the hospital differentiated itself in terms of meaning by its unit context—dominant caring dimensions related to areas of practice or units wherein professionals worked and patients resided.

Differential caring theory also showed that different units espoused different (dominant) caring meanings based on their organizational goals and values, such as technological caring in intensive care units or surgical suites, humanistic and spiritual caring in the oncology unit, and legal, economic, and political caring in the administrative environments (Ray, 2006; Ray & Turkel, 2010).

In this light, differential caring was organized into a hierarchical structure (psychological, practical, interactional, philosophic) from data on the meaning of caring (Ray, 1981a, 1984, 2010b). Here, the data revealed that differential caring theory is an expression of beliefs and behaviors relating to competing technological, ethical, religious, political, legal, economic, educational, humanistic, and social factors of the organization and dominant culture.

For its part, the formal theory of bureaucratic caring emerged by integrating the qualitative data from interviews, participant–observations of the hospital culture, and the interpretation of meaning of caring using an Hegelian philosophy of thesis, antithesis, synthesis: illuminating the *thesis* of caring (a meaning system) in relation to the potential *antithesis* of caring (the hospital as a bureaucracy) to a *synthesis*, the theory of bureaucratic caring. The theory illuminates the nature of caring as the *integration* of humanistic, social, ethical, religious/spiritual, political, economic, technological, and legal caring. By researching and understanding the meaning of caring, nurses and other professionals showed the uniqueness of caring as an expression of the complex human-environment mutual process. To capture this uniqueness of caring within contemporary nursing practice in complex organizations, the nursing process has been reconceptualized with a new mnemonic: recognizing, connecting, partnering, and reflecting (RCPR) (Turkel & Ray, in press).

COMPLEXITY SCIENCES AND NURSING

Complexity sciences, called the science of quality, look to networks of relationship as the underpinning for choices in the continual and emerging mutual human–environment process (Bar-Yam, 2004; Briggs & Peat, 1989; Peat, 2002). Complexity sciences also fit the ontology (what we know about being) and epistemology (the way we know) of nursing science (Anderson et al., 2005; Davidson & Ray, 1991; Davidson et al., 1997; Lindberg, Nash, & Lindberg, 2008; Ray, 1994, 1998). The mutual human–environment process for both complexity sciences and nursing science involves phenomena that are complex, dynamic, relational, nonlinear, structurally similar, integral, pandimensional, holonomic, difficult to study, qualitative, self-organizing, and open to emergence. The human–environment mutual process recognizes the unitary nature of the human and the environment—everything in the universe is interdependent; is sensitive to initial conditions (hysteresis); is patterned; evolves over space and time, is similar through constant change (homeodynamic) but never again exactly the same (irreversible); has "properties" that arise in the act of observation itself; has phenomena that are capable of filtration (choice) by involved agents-in-relationship; and has phenomena that self-organize (change or transform). Thus, there is continuous emergence (Bar-Yam, 2004; Briggs & Peat, 1989; Davidson & Ray, 1991; Peat, 2002; Ray, 1994, 1998). In nursing science, the mutual process of humans and the environment evolves toward health, healing, or well-being through the caring relationship. This transformation is relational self-organization (Ray, 1994).

Complexity sciences and nursing science have the power to promote a deeper understanding of human beings as they evolve with the environment. It is a symphony of mutual continuous change or transformation: "The complexity and the interdependency of agents who interact in a self-organizing manner make it impossible for any single agent to control the processes and outcomes of care for any patient" (Wiggins, 2008, p. 11). However, relational caring self-organization in nursing exposes the power and energy of the caring relationship and the caring moment (Watson, 2008). Choice within networks of relationship or honoring the patient and family through caring cocreates a living organization (Nirenberg, 1993) where "seeing" with the body, mind, and soul (spirit and heart) symbolizes loving kindness, caring communication, and mystery. There is a deep spiritual meaning to caring in the mutual human–environment process (Watson, 2008). In a world of many complex dimensions, the available information exceeds more than what our senses perceive.

Choosing the best ethical and spiritual action is based on information gathered from the multidimensionality of patients' and families' clinical caring experiences by nurses and other professionals who have moral commitment and caring knowledge, and use their intuition (intention and practical wisdom) to cocreate change and relational self-organization. Intuition, which reflects the art of nursing, is a dimension of Watson's (2008) theory of human caring as described in Caritas Process 6 (which calls for *Creative Use of Self and All Ways of Knowing as Part of the Caritas Process: Engage in the Art of Caritas Nursing*). This concept of intuition is nonlinear and requires the nurse to be open to innovation and creativity in nursing practice. As a complex human concept, intuition is one of the multiple ways of knowing that is acknowledged by a caring nurse as he/she comes to know and understand self and patient as a caring person.

AN INVITATION

The authors invite all who read this book to reflect on each chapter and the responses that follow each chapter. Here, reading requires an open mind, letting go of conventional ways of thinking, and a willingness to move beyond one's comfort zone as one opens up to emerging possibilities. This book, a unique contribution to the discipline, is the first to focus on both caring science and complexity sciences within the realm of nursing science, practice, and health care organizations. The commonality between caring science and complexity sciences is the human connection, the relationship, and the pattern recognition that facilitates understanding of the mutual human and environment process for emergence of health, healing, well-being, or a peaceful death.

The vision for this book grew out of the commitment, passion, and dedication of the co-editors, the late Dr. Alice W. Davidson, one of the first nursing scholars to advance nursing and complexity sciences; Dr. Marilyn Ray, who integrated caring science and complexity sciences in contemporary nursing theory and health care practices; and Dr. Marian Turkel, who is committed to the teaching and implementation of caring science in complex health care systems and research initiatives.

The chapters and responses with a variety of practice exemplars in this book describe, explicate, and reflect unique trans-theoretical approaches that converge tenets from the philosophies of caring sciences, the SUHB, and complexity sciences. Scholars with knowledge related to caring science, the

SUHB, and complexity sciences contributed to this book. The chapters and responses are diverse, illuminating the diversity of complexity itself, but similar in the underlying articulation of and need for human caring and concern for the delivery of quality health care. There are chapters more focused on complexity sciences, highlighting, for example, entropy, methods, organizational paradoxes, and conflict relationships from more theoretical, quantitative, and/or mathematical research approaches. There is a chapter focused on the disease process of diabetes that shows the complexity of diabetes from the cellular to policy levels. There are chapters focused on theoretical and qualitative research methods or newer research methods capturing the science of complexity, such as the comparing and contrasting of complexity sciences and the SUHB, complex caring dynamics, and story theory and method. There are chapters related to leadership, caring in complex health care organizations, and nursing education that address both complexity and caring sciences. And finally there are chapters that challenge our ethical thinking with informatics applications in practice, and the future of nursing and caring within the realm of the human–humanoid relationship.

Each chapter begins with an introductory preface describing the relevance of the content to contemporary complexity science, nursing, and caring sciences for human–environment well-being. Each chapter has a *response,* and sometimes two responses, that highlight what the particular chapter means to nursing education, research, leadership, administration, and practice. Responses provide the reader with a practical application or understanding of the content presented. Three types of responses were offered to the response authors: (a) a structured contextual template, (b) a response to open-ended questions, and (c) an open emergent and reflective process. Response authors chose what illuminated or critiqued ideas in the chapter and many offered new approaches based upon nursing research or clinical practice applications.

Readers are invited to engage in personal reflection on the relevance of the writings as they relate to theory, research, education, leadership, and practice. The Epilogue and Addendum at the end of the book provide challenging questions and practical definitions of terms from nursing and complexity sciences to further demonstrate how nurses, physicians, and health care leaders can use or continue to embrace the use of caring in the mutual human–environment relationship in education, research, administration, leadership, and practice.

REFERENCES

Anderson, R., Crabtree, B., Steele, D., & McDaniel, R., Jr. (2005). Case study research: The view from complexity science. *Qualitative Health Research, 20*(10), 1–17.

Bar-Yam, Y. (2004). *Making things work: Solving complex problems in a complex world*. Boston: NECSI Knowledge Press.

Boone, L., & Bowen, D. (1980). *The great writings in management and organizational behavior*. Tulsa, OK: Penn Well.

Boykin, A., & Schoenhofer, S. (2001). *Nursing as caring: A model for transforming practice* (2nd ed.). Sudbury, MA: Jones & Bartlett.

Briggs, J., & Peat, F. D. (1989). *Turbulent mirror: An illustrated guide to chaos theory and the science of wholeness*. New York: Harper & Row.

Calabria, M., & Macrae, J. (Eds.). (1994). *Suggestions for thought by Florence Nightingale: Selections and commentaries*. Philadelphia: University of Pennsylvania Press.

Cannato, J. (2010). *Field of compassion: How the new cosmology is transforming spiritual life*. Notre Dame, IN: Sorin Books.

Carper, B. (1978). Fundamental patterns of knowing in nursing. *Advances in Nursing Science, 1*(1), 13–23.

Coffman, S. (2006). Marilyn Anne Ray's theory of bureaucratic caring. In A. Marriner Tomey & M. Alligood (Eds.), *Nursing theorists and their work* (6th ed., pp. 116–139). St. Louis, MO: Mosby/Elsevier.

Coffman, S. (2010). Marilyn Anne Ray's theory of bureaucratic caring. In M. Alligood & A. Marriner Tomey (Eds.), *Nursing theorists and their work* (7th ed., pp. 113–136). St. Louis, MO: Mosby/Elsevier.

Cowling, R., & Repede, E. (2010). Unitary appreciative inquiry: Evolution and refinement. *Advances in Nursing Science, 33*(1), 64–77.

Davidson, A., & Ray, M. (1991). Studying the human-environment phenomenon using the science of complexity. *Advances in Nursing Science, 4*(2), 73–87.

Davidson, A., Ray, M., Cortes, S., Conboy, L., & Norman, M. (2006). *Complexity for human-environment well-being*. Retrieved December 2, 2009, from http://necs.org/events/iccs6/viewpaper:id=216

Davidson, A., Teicher, M., & Bar-Yam, Y. (1997). The role of environmental complexity in the well-being of the elderly. *Complexity and Chaos in Nursing, 3*(1), 5–12.

Hamilton, P., Pollack, J., Mitchell, D., Vicenzi, A., & West, B. (1997). The application of nonlinear dynamics in nursing research. *Nonlinear Dynamics, Psychology, and Life Sciences, 1*(4), 237–261.

Lindberg, C., Nash, S., & Lindberg, C. (2008). *On the edge: Nursing in the age of complexity*. Bordentown, NJ: Plexus Press.

Newman, M., Sime, A., & Corcoran-Perry, S. (1991). The focus of the discipline of nursing. *Advances in Nursing Science, 14*(1), 1–6.

Newman, M., Smith, M., Dexheimer-Pharris, M., & Jones, D. (2008). The focus of the discipline revisited. *Advances in Nursing Science, 31*(1), E16–E27. doi:10.1097/01.ans. 0000311533.65941.f1

Nightingale, F. (1859/1969). *Notes on nursing: What it is and what it is not*. New York: Dover.

Nightingale, F. (1992). *Notes on nursing: What it is and what it is not*. Philadelphia: J. B. Lippincott.

Nirenberg, J. (1993). *The living organization*. San Diego, CA: Pfeiffer & Company.

Peat, F. D. (2002). *From certainty to uncertainty: The story of science and ideas in the twentieth century*. Washington, DC: Joseph Henry Press.

Perrow, C. (1986). *Complex organizations: A critical essay*. New York: McGraw-Hill.

Purnell, M. (2006). Development of a caring model for nursing education. *International Journal for Human Caring, 10*(3), 8–16.

Ray, M. (1981a). *A study of caring within the institutional culture*. Unpublished doctoral dissertation, University of Utah, Salt Lake City, UT.

Ray, M. (1981b). A philosophical analysis of caring within nursing. In M. Leininger (Ed.), *Caring: An essential human need* (pp. 25–36). Thorofare, NJ: Charles B. Slack.

Ray, M. (1984). The development of a nursing classification system of caring. In M. Leininger (Ed.), *Care, the essence of nursing and health* (pp. 93–112). Thorofare, NJ: Charles B. Slack.

Ray, M. (1989). The theory of bureaucratic caring for nursing practice in the organizational culture. *Nursing Administration Quarterly, 13*(2), 31–42.

Ray, M. (1994). Complex caring dynamics: A unifying model of nursing inquiry. *Theoretic and Applied Chaos in Nursing, 1*(1), 23–32.

Ray, M. (1997). Illuminating the meaning of caring: Unfolding the sacred art of divine love. In M. Roach (Ed.), *Caring from the heart: The convergence of caring and spirituality* (pp. 163–178). New York: Paulist Press.

Ray, M. (1998). Complexity and nursing science. *Nursing Science Quarterly, 11,* 91–93.

Ray, M. (2001). The theory of bureaucratic caring. In M. Parker (Ed.), *Nursing theories and nursing practice* (pp. 421–430). Philadelphia: F. A. Davis.

Ray, M. (2006). The theory of bureaucratic caring. In M. Parker (Ed.), *Nursing theories and nursing practice* (2nd ed., pp. 360–368). Philadelphia: F. A. Davis.

Ray, M. (2010a). *Transcultural caring dynamics in nursing and health care*. Philadelphia: F. A. Davis.

Ray, M. (2010b). *A study of caring within an institutional culture: The discovery of the theory of bureaucratic caring*. Saabrücken, Germany: LAP Lambert Academic Publishing.

Ray, M., & Turkel, M. (2010). The theory of bureaucratic caring. In M. Parker & M. Smith (Eds.), *Nursing theories and nursing practice* (3rd ed.). Philadelphia: F. A. Davis.

Reeder, F. (1984). Philosophical issues in the Rogerian science of unitary human beings. *Advances in Nursing Science, 8*(1), 14–23.

Rogers, M. (1970). *An introduction to the theoretical basis of nursing*. Philadelphia: F. A. Davis.

Rogers, M. (1990). Nursing: Science of unitary, irreducible, human beings: Update 1990. In E. Barrett (Ed.), *Visions of Rogers' science-based nursing* (Publication No. 15-2285). New York: National League for Nursing.

Smith, M. (1999). Caring and science of unitary human beings. *Advances in Nursing Science, 21,* 14–28.

Turkel, M. (2001). Applicability of bureaucratic caring theory to contemporary nursing practice: The political and economic dimensions. In M. Parker (Ed.), *Nursing theories and nursing practice* (pp. 433–444). Philadelphia: F. A. Davis.

Turkel, M. (2006). Applicability of bureaucratic caring theory to contemporary nursing practice: The political and economic dimensions. In M. Parker (Ed.), *Nursing theories and nursing practice* (2nd ed., pp. 369–379). Philadelphia: F. A. Davis.

Turkel, M. (2007). Dr. Marilyn Ray's theory of bureaucratic caring. *International Journal for Human Caring, 11*(4), 57–74.

Turkel, M., & Ray, M. (2000). Relational complexity: A theory of the nurse-patient relationship within an economic context. *Nursing Science Quarterly, 13*(4), 307–313.

Turkel, M., & Ray, M. (2001). Relational complexity: From grounded theory to instrument development and theoretical testing. *Nursing Science Quarterly, 14*(4), 281–287.

Turkel, M., & Ray, M. (2004). Creating a caring practice environment through self-renewal. *Nursing Administration Quarterly, 28*(4), 249–254.

Turkel, M., & Ray, M. (2009). Caring for "not-so-picture perfect patients": Ethical caring in the moral community of nursing. In R. Locsin & M. Purnell (Eds.), *A contemporary nursing process: The (un)bearable weight of knowing in nursing* (pp. 225–249). New York: Springer.

Turkel, M., Ray, M., & Kornblatt, L. (2011). Instead of reconceptualizing the nursing process, let's rename it. *Nursing Science Quarterly*.

Vicenzi, A., White, K., & Begun, J. (1997). Chaos in nursing: Make it work for you. *American Journal of Nursing, 97*(10), 26–31.

Watson, J. (1979). *Nursing: The philosophy and science of caring*. Boulder, CO: University Press of Colorado.

Watson, J. (1985). *Nursing: Human science and human care*. Norwalk, CT: Appleton-Century-Crofts.

Watson, J. (2005). *Caring science as sacred science*. Philadelphia: F. A. Davis Company.

Watson, J. (2008). *Nursing: The philosophy and science of caring* (2nd Rev. ed.). Boulder, CO: University Press of Colorado.

Watson, J., & Smith, M. (2002). Caring science and the science of unitary human beings: A trans-theoretical discourse for nursing knowledge development. *Journal of Advanced Nursing, 37*(5), 452–461.

Wheatley, M. (2006). *Leadership and the new science* (3rd ed.). San Francisco: Berrett-Koehler.

White, J. (1995). Patterns of knowing: Review, critique, and update. *Advances in Nursing Science, 17*(4), 73–86.

Wiggins, M. (2008). The challenge of change. In C. Lindberg, S. Nash, & C. Lindberg (Eds.), *On the edge: Nursing in the age of complexity* (pp. 1–21). Bordentown, NJ: Plexus Press.

Acknowledgments

The completion of this book was a labor of love. The work first was conceived by Dr. Alice W. Davidson who, during the process of trying to achieve the goal of advancing scholarship to integrate nursing and complexity sciences for human–environment well-being, lost her life to cancer. As a coeditor with Alice, after her death, Dr. Marian Turkel stepped to, and together we have fulfilled Alice's dream, the publication of this book. We learned that the foundation of complexity sciences is to illuminate patterns of wholeness, how all organisms are linked together coherently to reveal the artistic beauty and harmony of the universe. We know too that nursing is unitary and holistic, linked together in networks of relationship, principally the complex and beautiful patterns of caring. We know by knowledge and experience that the ways of being and knowing as caring in nursing through the quality of the nurse–patient relationship facilitates healing, health, and well-being or a peaceful death. Through the scholarship expressed in this book by examining ideas advanced in the sciences of complexity, the science of unitary human beings, caring in nursing education, research, leadership, and practice, we recognize that the organic whole is not only diverse and unique but also, dynamic and self-organizing. The unique expressions are *caritas processes*, loving kindness, to use the words of Dr. Jean Watson. We are all unique expressions of the mutual human-environment caring relationship. The philosophies, theories, and scientific information expressed in this book help us to understand, more than just in an intellectual way, how nursing, health care, and the social world are continuously unfolding in tandem, continuously emerging toward an authenticity of awareness of the meaning of holism, a developing awareness that everything is a *living relationship*.

We would like to thank the following professionals who have made this work possible:

- We are grateful to all the contributors to this book, our scientists, physicians, administrators and nursing colleagues from many universities, hospitals and military organizations who made this work a living treatise to understanding patterns of wholeness for health, healing and well-being. Our contributors are the finest scholars in their fields, and areas of scholarship and practice.

- We are grateful to our colleague and friend, Dr. Jean Watson, Founder of the Watson Caring Science Institute, Boulder, Colorado, which is also connected with the Library at Springer Publishing Company.
- We are grateful to M. Brooks Turkel, without whose diligence and attention to detail in computing and editing and his love and caring, the timeliness and accuracy of this work would not have been possible.
- We are grateful to Drs. Madeleine Leininger and Jean Watson and Sister M. Simone Roach, who advanced the scholarship of caring and gave us a deep appreciation for and knowledge of human caring as the essence of nursing.
- We are grateful to Drs. Anne Boykin and Savina Schoenhofer for the initial integration of caring into the nursing curriculum at Florida Atlantic University and to all our other professional colleagues at Florida Atlantic University who are dedicated to advancing caring science through education, practice, and research.
- We are grateful to the faculty and Board of Directors of the Watson Caring Science Institute; the professionals from the International Association for Human Caring, and the scholars from the Plexus Institute, the FAU Center for Complex Systems and Brain Sciences, and the New England Center for Complex Systems who gave us support and encouragement to fulfill the Alice's dream.
- We are grateful to Allan Graubard, Executive Editor, Springer Publishing Company, who trusted the process of the book's unfolding as a "creative emergence" from the time when Dr. Davidson first approached him to the final submission of the work. We thank Assistant Editor, Elizabeth Stump, and the Production Editor, Rose Mary Piscitelli, for their work on our behalf.

As Maturana and Varela remarked in their book *The Tree of Knowledge* (1992):

> *We have only the world we can bring forth with others, and only love helps bring it forth* . . . . (from Goodwin, B. [2003]. Patterns of wholeness: Holistic science. *Resurgence*, 1[216], p. 14)

Marilyn A. (Dee) Ray
Marian C. Turkel

ONE

Philosophical and Theoretical Perspectives Related to Complexity Science in Nursing

Marlaine Smith

Nursing science has focused on complex systems in the study of nursing phenomena. The discipline of nursing is the study of relatiohip, human–environment patterning, and the complex dynamics related to caring and healing. Rogers' science of unitary human beings (SUHB) has embedded within it many concepts consistent with complexity science. Complexity science is the transdisciplinary study of complex adaptive systems (CAS); it encompasses multiple theoretical perspectives and methods of inquiry and comprises a large number of entities displaying a high level of nonlinear (or noncausal) interactivity. Nurse scholars have recognized the importance of the unitary nature of the human and environment in caring for others, the difficulty of isolating parts of the unitary human–environment process, and the complexity or choices within continual emergence and change. In 1970, Rogers defined the unitary human–environment process as continuously changing, irreversible, and evolving toward increased innovation, diversity and higher frequency wave patterning. Building upon this foundation, the purpose ofthis chapter is to describe the interrelationships among select philosophic and theoretic perspectives in nursing and concepts in complexity science. Both Perkins and Reeder illuminate how there is a conceptual integration of complexity sciences, the SUHB, and caring in nursing. In this way they point out the uniqueness of nursing as a unitary caring science.

INTRODUCTION

The ideas embedded in complexity science are receiving increasing attention in many disciplines struggling to describe and explain phenomena in their fields of study. There is growing appreciation of the inadequacy of conventional models for understanding complex interrelated systems, and this accounts for the emerging applications of complexity science in the physical, behavioral, biological, and human sciences. Nursing science has been one of the early adopters of ideas related to complexity science. With the discipline's focus on the study of human health and healing through caring (Smith, 1994) or person–environment relationships that facilitate health (Fawcett, 2000), relationship emerges as central to the ontology and epistemology of nursing (Newman, Smith, Dexheimer-Pharris, & Jones, 2008); therefore, the study of relationships and approaches that account for interconnectedness and dynamism is essential. Multiple paradigms have emerged in nursing, with the unitary–transformative paradigm (Newman, Sime, & Corcoran-Perry, 1991) reflecting some of the distinguishing tenets of complexity science. This paradigm had its origins in the conceptual system for nursing developed by Martha Rogers (1970), the SUHB. Rogers introduced a worldview that posited humans and the environment as patterned energy fields without boundaries that were continuously changing and creatively emerging. The concepts within other nursing theories correspond with complexity thinking. The purpose of this chapter is to describe the interrelationships among selected philosophic and theoretic perspectives in nursing and concepts in complexity science. An overview and definition of the concepts in complexity science will be presented first, followed by an analysis of the fit with selected nursing theories. The chapter ends with some speculation about ways in which these ideas will continue to shape the direction of nursing inquiry, practice, and education for the future.

Salient Concepts in Complexity Science

The literature is rife with multiple meanings of complexity science. In illuminating the "perplexity in complexity science," Horgan (1995) reported that there were 31 definitions of complexity; most likely, there are many more in existence now. Fundamentally, complexity science is the transdisciplinary study of CAS; it encompasses multiple theoretical perspectives and methods of inquiry. Richardson, Cilliers, and Lissack (2000) provide a simple definition of a complex system as one "that is comprised of a large number of entities displaying a high level of non-linear interactivity" (p. 8). "Complexity science examines systems comprised of multiple and diverse interacting agents and seeks to uncover the principles and dynamics that affect how such systems evolve and maintain order" (Lindberg & Lindberg, 2008, p. 32).

Those espousing reductionistic perspectives of complexity science assert that the simple rules underlying complex patterns can be sorted out through mathematical models; others disagree. Some assert that the concepts of complexity science are not really being applied authentically in some disciplines but instead are used as metaphorical tools to explain observed phenomena (Haigh, 2002). For example, concepts such as the *edge of chaos*, *emergence*, and *far from equilibrium* may be used metaphorically to describe an observed dynamic of a system. When applied in this way, the tenets of complexity are not ontological, or definitive of the nature of reality, but rather are adopted as an epistemological tool (metaphor) to understand an aspect of reality through comparison (Richardson & Cilliers, 2001). This might be considered as a pseudoscientific application of the thinking within complexity science and not the actual science itself. Some complexity scientists approach the study of complex systems through the methods of logical positivism, whereas others embrace methodological pluralism characteristic of a postmodern view. Phelan (2001) differentiates complexity science from complexity pseudoscience. Science involves the development and testing of theory through research. In contrast, pseudoscience uses resemblance thinking, neglects empirical matters, and is oblivious to alternative theories. Both complexity science and pseudoscience exist in nursing. Phelan argues that complexity theory and complexity thinking should be replaced with complexity science in some cases when the meaning reflects multiple ways of knowing, which, he argues, are not scientific. The dialogue related to complexity science is, in itself, complex. Choosing and defining one's own perspective is necessary when clarifying one's position among these multiple points of view.

Perhaps the most encompassing notion of complexity science situates it as a worldview that underpins a particular philosophy of science. The purpose of any science is to create organizing frameworks to explain some aspect of reality and then to examine those frameworks for their empirical honesty (Smith, 1994). Dent (2000) asserts that "complexity science is an approach to research, study and perspective that . . . [embraces] the philosophical assumptions of the emerging worldview" (p. 5). A variety of authors (Bohm, 1980; Briggs & Peat, 1989; Capra, 1982; Ferguson, 1980; Harman, 1998; Prigogine & Stengers, 1984; Wilber, 1998) have described this emerging worldview and argue that its appeal is related to the recognition that the traditional or empirical–analytic–mechanistic worldview underpinning science from a logical positivist perspective is inadequate. This traditional worldview provides some guidance for understanding phenomena within certain conditions, but beyond those conditions, its explanatory power is limited. Therefore, the emerging worldview encompasses the traditional within the limitations of its applicability while moving beyond it to explain the unexplainable within its perspective. This shift from the traditional to the emerging worldview is described as a time of dislocation (Ackoff, 1981), a period between "stories" (Schwartz & Ogilvy, 1979), a turning point (Capra, 1982), or new paradigm thinking (Ferguson, 1980).

Table 1.1 is adapted from Dent (2000) and contrasts features of the traditional and emerging worldviews. The traditional or empirical–analytic–mechanistic worldview is consistent with Newtonian physics and its consequents of reductionism, discrete parts, local relationships, linear causality, objective reality, dualism, and a reliance on logic, prediction, averages,

TABLE 1.1 Comparison of the Traditional and Emerging Worldviews

TRADITIONAL WORLDVIEW	EMERGING WORLDVIEW
Reductionism	Holism
Focus on parts	Focus on patterns
Focus on discrete entities	Focus on relationships
Linear causality	Mutual causality
Local and linear relationships	Nonlocal and nonlinear relationships
Determinism	Indeterminism, unpredictability
Objective reality	Perspectival reality; cocreated reality
Observer outside the observation	Observer influences the observation
Systems adapt to stimuli	Systems self-organize
Logic	Paradox
Either/or thinking	Polarity thinking
Founded on Newtonian physics	Founded on quantum physics
Prediction	Understanding/sensitivity; analysis/explanation
Averages; homogeneity	Diversity, variation
Focus on outcomes	Focus on emergence
Matter creates mind	Presence of consciousness in mind matter

(Adapted from Dent, 2000)

and homogeneity in the scientific method. In contrast, a quantum reality can acknowledge the Newtonian worldview in some circumstances, but its perspective also can embrace holism, patterns, relationships, nonlocal and nonlinear processes, indeterminism, perspectival and cocreated reality, paradox, polarity, diversity, and emergence. Gleick (1987) predicted that the three scientific theories from the 20th century most relevant for the 21st century would be the relativity, quantum, and chaos theories. The emerging worldview reflects these theories. Any theory is a map of the territory (reality); the real territory is always more complex than any map can depict. Each theory explains a range of phenomena; however, it may be totally inadequate for others. The emerging theories of complexity science address some of the unexplained gaps inherent in the traditional worldview (Capra, 1982; Coppa, 1993).

The structure of traditional and emergent worldviews corresponds to the worldviews underpinning paradigms within nursing science. These ontological paradigms contrast the different views of the nature of human beings, human–environment relationships, and health that shape approaches to nursing epistemology, methodology, and practice. Parse (1987) describes two paradigms in nursing: the simultaneity and the totality. Those with the lens of the totality paradigm perceive persons as composed of interrelated and interacting dimensions or parts; person and environment are separate but interactive, and health is biopsychosocial well-being, often depicted on a continuum from wellness to illness (or death). Parse (1987) asserted that *Rogers' Science of Unitary Man (sic)*, now the *Science of Unitary Human Beings* (SUHB) published in 1970, ushered in the simultaneity paradigm. In this paradigm person-environment is viewed as an irreducible whole and health as a pattern of the whole. In 1991 Newman, Sime, & Corcoran-Perry named three paradigms in nursing as the particularistic–deterministic, the interactive–integrative, and the unitary–transformative. The deterministic perspective corresponds closely with the traditional worldview; the interactive–integrative worldview offers a systems perspective that acknowledges human adaptation to the environment and the expression of this adaptation in subsystems. The unitary–transformative paradigm in nursing corresponds to the emerging worldview. Coming from this perspective, science is conducted differently. There is an attempt to understand "the evolving pattern of the whole" (Newman, 2008, p. 14). Nursing theories are clustered under these worldviews. The simultaneity and unitary–transformative paradigms have close similarities with the emerging worldview and complexity theory. This paradigm asserts the existence of human–environment patterning that reflects a fundamental wholeness, a dynamic process of change, and creative emergence based on participation and growing complexity.

As stated earlier, complexity science is known also as a theory of complex adaptive systems (CAS). In Table 1.2, the defining elements of CAS are identified. CAS are characterized by qualities such as embeddedness, nonlinearity, unpredictability, self-organization, diversity, porous boundaries, and emergent behavior (Chaffee & McNeill, 2007; Lindberg & Lindberg, 2008). Chaos theory is a subset of complexity science; its central tenet is that order underlies and emerges from apparent disorder. Although turbulent systems seem to exhibit chaotic behavior, order emerges from a communication point or phase space (Ray, 1998).

CAS are composed of interdependent and adaptive elements. There is a fundamental assertion that there is a simple rule that when discovered will show the unity of the life sciences (Lewin, 1992). The internal structures of complex systems are not reducible to a mechanical system (Allen, 2001, p. 30). These systems coevolve with their environment, open to flows of energy, matter, and information (p. 39). The evolution is creative and uncertain. There is a changing system embedded in a landscape of potential attractors that influence change. This is "transformational teleology" (Stacey, Griffen, & Shaw, 2000) as the potential futures (patterns of attractors) are transformed in the present (Allen, 2001, p. 40).

TABLE 1.2 Correspondence of Concepts Related to CAS to Extant Nursing Theories

CONCEPT	DEFINITION	CORRESPONDENCE TO NURSING THEORIES
Embeddedness	CAS exist within larger systems that provide context for understanding each.	Rogers' postulate of open systems and principle of integrality: Humans and environment are energy fields, coextensive with the universe and in continuous, mutual process.
		Newman describes individual patterns of consciousness embedded in family and community patterns. Roy's philosophical assumptions related to cosmic unity.
Dynamism	CAS are in continuous change; stability is not expected or desired.	Rogers' postulates and principles of helicy and resonancy: Change in human–environmental energy field is continuous, innovative, and unpredictable with greater diversity of field patterning.
		Roy describes persons as holistic adaptive systems with coping processes that maintain adaptation and promote person and environment transformations.
Patterning	CAS have unique configurations of movement and flow that identify each.	Rogers' postulates of energy field, open systems, and pattern and organization: Each energy field has a distinguishing pattern, a single wave that differentiates it from others.
		Newman's concept of pattern of the whole with movement, time, and space as patterns of evolving consciousness.
		Davidson stated that patterning could shift from stability to radical changes and that patterning over time should be studied through a perceptual dance between the parts and the whole.
Coevolution	CAS simultaneously shape and are shaped by their interactions with other systems.	Rogers' postulates of open systems: Human and environment energy fields are in continuous mutual process.
		Newman asserts that nurse and client evolve together in the mutual relationship.
		Parse's principle that humans are co-transcending with the possibles.

TABLE 1.2 ***(Continued)***

CONCEPT	DEFINITION	CORRESPONDENCE TO NURSING THEORIES
Emergence	Creative and innovative change occurs at the edge of the system where there is the most disorganization and disorder.	Rogers' principles of helicy. Change in human–environmental field patterning is continuous, innovative, and unpredictable. Humans and environment are evolving toward increasing complexity and diversity of field patterning. Rogers' assumption of negentropy, that the life process is evolving toward greater complexity and order, was not consistent with the concept of order emerging from chaos.
		Newman's ideas that disruption and disorder (disease) can lead to a choice point and expanding consciousness (health).
		Parse's principle that humanbecoming as negentropic unfolding.
		Turkel and Ray describe the emergence of relational complexity as a creative change in the system when economics and caring are viewed in a dialectic whole.
		Reed asserts that growing self-transcendence occurs with aging and crisis.
Inherent integrity	CAS possess unity and structure.	Rogers' postulate of pattern: Pattern provides integrity to the energy field; it identifies it and distinguishes it.
		Roy's person as adaptive system emphasizes the importance of system integrity through stability and change. Adaptive responses promote integrity.
Nonlinearity	The direction of dynamic change in CAS is unpredictable. A small action by an agent may produce a large change in the CAS and vice versa.	Rogers' postulate of open systems and principle of helicy: Human–environmental energy fields are in continuous mutual process. Change in human–environmental energy field patterning is continuous, innovative, and unpredictable. Attunement and resonance are nonlinear processes of acquiring knowledge.
Porous boundaries	CAS have open boundaries that permit continuous interaction with other systems.	Rogers' postulate of energy fields and open systems: Energy fields, by their nature, have no boundaries; they are in continuous, mutual process.

(continued)

TABLE 1.2 ***(Continued)***

CONCEPT	DEFINITION	CORRESPONDENCE TO NURSING THEORIES
Self-organization	CAS have the ability to influence their structure and direction by creating new patterns through interactions with other systems.	Rogers' principle of helicy and Barrett's theory of power as knowing participation in change: Change is continuous, dynamic, and creative, and humans participate knowingly in the process of change; humans participate knowingly in change through awareness, choice, freedom to act intentionally, and involvement in creating change.
		Newman's ideas of transformation through pattern recognition; shifts occur as client recognizes own patterning.
		Davidson found that in complex human–environment interrelationships choice is important for well-being.
		Turkel and Ray describe the self-organizing characteristics of the work environment occurring as forces of economic accountability and need for relationship converge.
		Roy's scientific assumption: Systems progress to a higher level of complexity and self-organization.
Simple rules	Underlying CAS are rules that govern the behavior of interacting agents over time using feedback and algorithms. These rules, when discovered, will lead to a unity of the life sciences. The CAS evolves creative possibilities through learning as a whole.	Turkel and Ray describe a search for unity in relational complexity, a dialectic synthesis that resolves the tension between economics and caring.
		Ray defines caring as the energy by which choice is facilitated to bring order (healing or well-being) out of chaos (disease, need, pain, or crisis).
Diversity	Agents or elements in the CAS are varied and unique, and these differences are critical for system health.	Rogers described increasing diversity as a manifestation of patterning.

TABLE 1.2 ***(Continued)***

CONCEPT	DEFINITION	CORRESPONDENCE TO NURSING THEORIES
Polarity and paradox: order–disorder	CAS are characterized by a wholeness that embraces dialectic rhythms. The fluctuations of these opposing forces are preferred over stability.	Rogers' principle of resonancy asserts that human patterning evolves toward more diverse manifestations evident in opposing rhythms that evolve toward a unity (dialectic synthesis); for example, the pattern of sleeping, waking, and beyond waking. Newman asserts that the concept of health encompasses both disease and nondisease. Davidson states that chaotic patterns give rise to ordered patterns in human life. In Parse's humanbecoming school of thought, the paradoxical rhythms of revealing–concealing, connecting–separating, and enabling–limiting describe the patterns of relating. The paradoxical rhythms reflect the whole. In their theory, Turkel and Ray focus on the paradox of economics and caring and its resolution in relational complexity.
Adaptable elements	CAS are composed of elements that can evolve with environmental changes.	Roy's theory of adaptive system states that there are internal processes that act to maintain the integrity of the individual or group.
Distributed control	CAS do not direct change centrally; the agents participate in direction of change and resultant outcomes.	Rogers' principle of helicy states that the nature and direction of change are a function of knowing participation in the human–environment process.
Attractors	System catalysts that promote the emergence of new behaviors.	Ray identifies caring as an attractor that moves the system toward order in chaos.

This table was constructed from a synthesis of Lindberg, C. (2008). Nurses take note: A primer on complexity science. In C. Lindberg, S. Nash, & C. Lindberg (Eds.), *On the edge: Nursing in the age of complexity*. Bordentown, NJ: Plexus Press.
Chaffee, M., & McNeil, M. (2007). A model of nursing as a complex adaptive system. *Nursing Outlook*, *55*(5), 232–241e. DOI: 10.1016/J. outlook; 2007.04.003

Relationship of Complexity Thinking to Nursing Theories

In 1970, Martha Rogers published her groundbreaking book, *An Introduction to the Theoretical Basis of Nursing*. In this book, Rogers provided the foundations for her conceptual system, now called the science of unitary human beings (SUHB), offering a radical alternative to existing conceptualizations about the relationships between human beings, their environment, and the nature of the life process.

Rogers recognized the limitations of the existing models within science and medicine that reduced, categorized, and classified human phenomena and sought to predict human–environment relationships through hypothesizing the direction of reactions to environmental stimuli. This model led to prescriptions to change human behavior toward desired outcomes. Rogers proposed a nursing lens focused on the wholeness, interconnectedness, and complexity of the life process as the foundation for nursing science, with a goal of "better health for mankind [*sic*]" (p. xii). She drew from tenets of general systems theory, evolutionary biology and cosmology, relativity theory, and quantum mechanics in the creation of a wholly new conceptual system that, in many ways, preceded or paralleled the emerging foundations of complexity science. She argued against the application of systems theories to understand human life. Instead, she called for the life process to be understood in its wholeness as reflected in pattern and organization.

Rogers asserted that the life process is inherently complex and dynamic. "An energy field underwrites the unity of man [*sic*] and provides the conceptual boundaries which identify his [*sic*] oneness. A field transcends its component parts . . . and possesses its own integrity. Human beings are more than and different from the sum of their parts" (p. 46). The notion of both humans and environments as energy fields sets the stage for a boundaryless integral relationship between the two. Human beings coevolve with their environment. This perspective contradicts the notions of orderly or predictable adaptation to environmental stimuli. Likewise, without boundaries between humans and the environment, there can be no predictable outcomes, only dynamic change cocreated through participation.

Rogers (1990, 1994) viewed living systems as continually becoming, that is, growing, more diverse and complex. This contradicted the second law of thermodynamics, entropy, setting the stage for its opposite, the concept of *negentropy*, the continuous, dynamic change toward greater complexity that characterizes human life. Emergence and unpredictability are natural consequences of this assumption. "The capacity of life to transcend itself, for new forms to emerge, for new levels of complexity to evolve, predicates a future that cannot be foretold" (Rogers, 1970, p. 57).

Rogers' postulate of pattern and organization is founded on the premise that complex whole systems are known by their patterns. The energy field imposes pattern; pattern reflects the wholeness of human life (Rogers, 1970, p. 61). Pattern is dynamic and observable through its manifestations. Self-regulatory or self-organizing characteristics of pattern, included in her 1970 book, were later rejected because of their violation of the coevolving, integral nature of the person–environment mutual process. Human beings participate knowingly in change (1990, 1994) through making choices, but they do not regulate or organize themselves in isolation. Rogers also argued that chaos theory was not consistent with a unitary perspective. She considered the premise of chaos theory, that if the conditions and context of any event are known the outcomes can be predicted, inconsistent with an open systems perspective. Furthermore, she found the idea that order emerges at the edge of chaos as contrary to a worldview where emergence is continually unpredictable, creative, and diverse.

Theories have been derived from Rogers' conceptual system. One is the theory of accelerating evolution (Rogers, 1970, 1990), that more complex life forms evolve more rapidly and that change is accelerating. She suggested that we would witness more complex and diverse patterning manifestations such as hyperactivity, extrasensory perception, and changes in sleep–wake patterns over time. Perhaps the precipitous change in communication through technology is another example of this theory of accelerating evolution. Barrett's (1986) theory of power as participating knowingly in change is another theory derived from the SUHB. Barrett asserts that humans have the power to participate in change through awareness, choice, freedom to act intentionally, and involvement in creating change. Her theory provides a blueprint for promoting participation in change within complex situations.

Rosemarie Parse's (1998) *The Humanbecoming School of Thought* and Margaret Newman's (1986, 1994) *Theory of Health as Expanding Consciousness* (HEC) were developed, in part, from Rogers' SUHB; therefore, both of these theories fall within the simultaneity or unitary–transformative paradigms and resonate with the complexity theory concepts. The theory of humanbecoming (Parse, 1998) (formerly man-living health [Parse, 1981]), by its name, suggests that a process of emergence characterizes health. Parse asserts that human beings cocreate patterns of relating with their environments. These patterns manifest in the paradoxical rhythms of revealing–concealing, connecting–separating, and enabling–limiting. From Parse's perspective, these rhythms constitute the pattern of a whole. Humanbecoming is co-transcending with the possibles. These themes, embodied in the theory's principles, are consistent with complexity thinking related to interconnectedness, paradox, and emergence.

Margaret Newman's theory of *Health as Expanding Consciousness* (HEC) (1986, 1994) draws from Rogers' (1970) SUHB, Young's (1976) theory of the evolution of consciousness, Bohm's (1980) theory of the holographic universe, and Prigogine's (1984) theory of dissipative structures, all of which have correspondence with selected ideas in complexity science. "The focus is on the person, the pattern of the evolving whole, of transformations within transformations, including the unpredictability of chaotic systems" (Newman, 2008, p. 8). According to Newman, health is a higher unity encompassing both having disease and being free from it. Having a disease may be an impetus for movement to a choice point and a higher level of organization. This relates to Prigogine's (1984) theory in which a period of disorganization can be the impetus for movement to a higher level of organization. Persons experiencing the disruptive patterning associated with disease may arrive at a choice point where they develop awareness, recognize their pattern, make decisions, and take actions that become life transforming. This is expanding consciousness. According to HEC, knowing through attunement and resonance is a more complex way of knowing the whole that may result in intuitive insights and revelation (Newman, 2008, p. 37). This process of seeing the whole in parts is a key to pattern recognition. These ideas are consistent with the concepts of creative emergence, hidden unity, and self-organization related to complexity theory.

Sr. Callista Roy's (2008) conceptual model of person as adaptive system has evolved over time, incorporating some qualities related to complexity science and complex adaptive systems. Her model evolved from general system theory and adaptation theory, and later, she incorporated from Young (1976) the ideas of unity and meaningfulness in the created universe. Her recent work (Roy, 2009; Roy & Jones, 2007) places greater emphasis on the complexity of the times in which we live and how this compels the need for greater unity and spiritual vision. Roy's philosophical assumption related to cosmic unity reflects the interconnectedness of persons to a larger whole. In her scientific assumptions, Roy (2010) states that systems of matter and energy progress to higher levels of complex self-organization; that consciousness and meaning are consistent with person and environment integration; that choice results in the integration of creative processes; that there are integral relationships between persons and the earth; and that person and environment transformations have created human consciousness (p. 170). These scientific assumptions clearly reflect a movement toward a unitary–transformative worldview consistent with the fundamental premises of complexity science.

Pamela Reed's (1991) middle range self-transcendence theory was built on assumptions from Rogers' SUHB. Reed defines living systems as open, self-organizing, and developing through periods of disequilibrium. She asserts that in the life process humans develop in the direction of greater complexity. One of the indicators of this is growing self-transcendence, and this quality is often seen in the aging process. However, self-transcendence can be accelerated during times of vulnerability or crisis, when persons experience deeper connections to self and others and an increasing focus on contributing to a greater good.

Alice Davidson, under the guidance of her doctoral mentor, Marilyn Ray (Davidson & Ray, 1991), was one of the first nursing scholars to draw explicitly from the tenets of complexity science in her research related to human–environment interrelationships. She built on Rogers' (1970, 1990) principle of integrality and asserted that patterns may appear stable or can shift dramatically from previous patterning. Davidson hypothesized that apparent chaotic patterns may give rise to greater order in human life and that humans come to know the environment through a perceptual dance of focusing on parts and the whole. This dialectical movement is essential to understanding the complex, relational, human–environment process. In order to understand this, Davidson designed a study using multiple methods to examine complex relationships. Her process of studying the human–environment relationship evolved from complexity theory and included the following: (1) apprehension of the whole to determine data needed; (2) examination of the manifestations of the whole collected as data; (3) analysis of data according to the appropriate traditional paradigm; (4) dialectical movement among the findings; (5) pattern identification across paradigms; (6) application of boundaries of existing understandings from data, literature review, and so on; (7) boundary crossing by reversing and questioning; (8) disengagement and enabling ideas to be transformed into new patterns or creative insights of meaning making; (9) intuiting an understanding of the phenomenon (unity of meaning), and; (10) communicating meaning to others through language and practice (p. 77).

Davidson's research question was: How does the integral human–environment relationship facilitate human beings' well-being as manifest in production and creativity? Participants were workers within an organization. Data generation included the following: participant observation and action research; causal modeling; phenomenology and hermeneutic analysis of the totality of the data. Based on her research, she developed a theory of choice patterns, asserting the importance of choice for human beings as they promote their own unique well-being. Later Davidson and colleagues (Davidson, Teicher, & Bar-Yam, 1997) tested this theory and found that for the elderly to remain active and productive they needed sufficient complexity in their human–environment relationships to stimulate and challenge them.

Several nurse scholars have written about the relationship of chaos theory to nursing theories (Mishel, 1990; Vicenzi, 1994; Ray, 1994). Gleick (1987) defines chaos as the apparent irregular, unpredictable behavior of deterministic, nonlinear dynamic systems. It is referred to as "deterministic randomness" (Vicenzi, 1994, p. 37). Mishel (1990) used chaos theory as a way to reconceptualize her middle range theory of uncertainty in illness to reflect the changes that occur in people when uncertainty is prolonged. Disorder, instability, diversity, and disequilibrium characterize systems in chaos; these characteristics correspond to the experiences of persons confronting the continuing and prolonged crises associated with the uncertainty during a chronic illness. Mishel draws the comparison of prolonged uncertainty with far-from-equilibrium dynamics in which very small external changes can affect profound internal changes within the system and the system can dramatically self-organize. Continuing this analogy Mishel theorizes that entropy, or the degree of disorder and disorganization in life, increases in persons who are in the midst of unabating uncertainty. Persons cannot withstand this mounting uncertainty and are pushed to a bifurcation point when there is either death or a change toward a more complex level of self-organization.

Vicenzi (1994) suggested that chaos theory could provide a new synthesis for nursing and community health; she was one of the first to relate the concepts of chaos theory to nursing theory and health-related phenomena. For example, she suggests that perhaps nursing intuition is a prereflective sensing of chaos and that heart rate variability, neutrophil counts, epilepsy, and schizophrenia exhibit chaotic dynamics. Sensitive dependence is a concept in chaos theory sometimes called the "butterfly effect," referring to the phenomenon that small change can lead to extreme effects in the long term (p. 38). She relates these concepts to Rogers' (1990) postulates of unpredictability and irreducibility, and suggests that nurse scientists might increase their use of longitudinal research designs and computer modeling to come to understand the wholeness and complexity of human patterning.

Ray (1994) situates her theory of complex caring dynamics within the context of complexity science and chaos theory. The foundations of her theory are in a relational ontology patterned by love. She states that "self-organization in chaos theory, although seemingly without a goal, is influenced by creativity—a vision of a purposeful love, a purposeful God, a force or spirit of life expressing itself intelligently in the universe" (Ray, 1994, p. 25). Ray argues effectively that the concepts of relationality, belongingness, holism, self-organization, human–environment integrality, patterning, stages of organization–disorganization stages, unidirectionality of change, and pattern recognition are contained within nursing and consistent with complexity theory (Ray, 1994, p. 25). Caring is defined as "the energy by which choice is facilitated to bring order (healing or well-being) out of chaos (disease, disease, need, pain, or crisis)" (Ray, 1994, p. 26). Ray describes the caring relationship as an attractor that draws the system away from chaos and toward well-being and healing. In this relationship, the nurse instills hope, leading to choice making that changes the pattern so that both nurse and client are "more alive and authentic" than before (Ray, 1994, p. 26). She identifies four life pattern forms related to pattern seeing: technical, practical, critical, and creative caring dynamics. Each has its corresponding methodological frameworks for pattern mapping and integrative synthesis or pattern recognizing. The theory of complex caring dynamics highlights the primacy of relationships or love as the pull of the universe toward creative emergence or healing.

Turkel and Ray (2000, 2001) developed a theory of relational complexity that describes and explains the importance of caring in the complexity of the current health care environment. The theory's assumptions flow from a unitary–transformative worldview embracing complexity, expanding consciousness and transformation, and ideas of paradoxical patterning and living changing value priorities within Parse's theory of human becoming (Turkel & Ray, 2001). They assert that the current health care environment is driven by the economic factors of regulation, reimbursement, and costs. This environment can influence nursing practice so that it becomes shaped by the demands of increasing patient acuity and time constraints imposed by increasing documentation and decreased staffing. This can lead to compromised quality and safety outcomes for patients, including well-being and quality of life. On the other hand, nursing places caring and relationship prominently within its discipline and profession. The seeming paradoxical forces of economics and caring must be reconciled through a codetermining relationship (interidentification), a transformation of quantity into quality (qualitative difference), the negation of negation (dialectic resolution), and a spiral form of development (transformation and change) (Turkel & Ray, 2001, p. 282).

Turkel and Ray (2000) propose that a shift in the self-organizing pattern of interrelationships among the nurse, administrator, and patient can change this dynamic. This shift occurs when caring and relationship become the organizing force for patterning. When this occurs, the nurse–patient–administrator relationship becomes an economic resource for the organization. They propose a synthesis of caring and economics that will result in fulfillment of the organization's mission, economic enhancement, actualization of the nurse–patient relationship to its full potential, and patient outcomes of enhanced healing and well-being. In Turkel and Ray's theory, human caring and economics are synthesized into relational complexity in an economic context in which the result is "ethical economic caring" (Ray, 1994, p. 50). This reflects the tenets of complexity science related to the edge of chaos, where the emergence of order from disorder can be found in a choice point and hidden unity becomes revealed.

Complexity Thinking and the Future of Nursing Inquiry, Practice, and Education

Complexity thinking is taking root and growing across disciplines; it can only continue to spread within the discipline of nursing, affecting nursing inquiry, practice, and education. In the past 30 years, nursing has recognized the importance of multiple epistemologies and philosophies of science. Complexity science is founded in an emerging worldview that contrasts with the traditional worldview that guides scientific inquiry. This emerging worldview (Table 1.1) will continue to guide inquiry within the discipline allowing for, perhaps even compelling, the use of multiple methods of inquiry in understanding the whole of complex phenomena. Several authors (Koithan, 2006; Resnicow & Page, 2008) point to the advantage of chaos and complexity in offering a valuable explanatory model for public health research and behavior change, and the study of whole systems approaches to healing. The limitations of randomized clinical trials in understanding the depths of the human experience, the interrelationship of power and knowledge, and the participatory nature of knowledge generation are evident. However, randomized clinical trials and other methods of empirical inquiry have their place in answering important questions related to phenomena of concern to nursing. In the future, nursing will transcend the boundaries of qualitative/quantitative and traditional science/human science dichotomies to embrace methods that require multiple perspectives on a single phenomenon in order to understand it fully. This is a hermeneutic process used by Davidson (1988) and described by Wilber in his AQUAL or all-quadrant thinking as "the eye of the spirit" (Wilber, 1997). Through this process, the multiple perspectives are synthesized through the lens of the observer; meaning is cocreated between the observer and multiple sources of knowledge, and wisdom replaces knowledge development.

Nursing practice will incorporate complexity thinking in the development of practice models and through the use of technologies for practice. The complexity of the health care environment is at a critical point, exhibiting the characteristics of a system in chaos that is at or approaching a bifurcation point. At this choice point, nursing, the profession that holds the energy of love, relationship, and compassion, may present a path in which both caring, technology, and the realities of economics may by synthesized in harmony (Ray, 1994). This can be combined with attention to a high-touch/high-tech world of practice for nursing. For example, new technology at the bedside may provide a holistic narrative of what is most important to the person, including his/her preferred name, interests, culture, family, community, and what is meaningful. At the same bedside, there will be the integrated patient record, incorporating relevant and current data, and technology to enable the ability to access information and knowledge for providing safe, effective care. This attention to caring may be a stimulus for the higher order of caring in nurses (Clark, 2003).

The American Nurses Credentialing Center requires hospitals aspiring to or achieving Magnet™ status to have a professional practice model that reflects nursing theory. In this way, the practice environment is creating an alignment of nursing knowledge for practice. Nursing theories in the unitary–transformation paradigm reflect complexity thinking, and there is accelerated movement of these theories into the practice environment. For example, Watson's Caritas Consortium is a group of health care organizations that are advancing caring-based practice models that reflect the tenets of complexity, such as interconnectedness, creative emergence, and dynamism. Nurses in practice will be generating and testing the concepts within chaos and complexity thinking (Haigh, 2002).

Finally, nursing education will incorporate complexity science in its approaches to teaching and learning. The Carnegie Foundation report on the nursing education (Benner, Sutphen, Leonard, & Day, 2010) includes commentary on the chaotic, dysfunctional U.S. health care system and how nursing students must learn to function within it. The study found that nurses are undereducated for practicing in this chaotic health care system. For this reason, nursing educators will be challenged to prepare students for practice in complex environments. This will require abandoning a focus on delivering content and instead teaching students how to access and use knowledge in the context of dynamic patient care situations within complex systems. Nursing students need practice in clinical reasoning and clinical imagination, thinking through what can happen as changes occur in patient care situations, and creatively responding to these situations. In addition, nurse educators need to attend to the formation of the student as person. This formation requires the development of personhood and may involve the use of reflective and spiritual practices and aesthetic engagement. Such practices will expand awareness in order to better apprehend patterns of the whole and to live a compassionate practice. Benner et al. (2010) describe the importance of teaching ethical comportment, day-to-day living, and the essential values of nursing in practice. The necessary content and practice competencies must be embedded in patient narratives or stories. In this way, the nursing content currently taught is viewed within the context for promoting the health, healing, and quality of life for the person, family, and the community. The complexity of the nursing situation can be revealed through the story's details, and knowledge related to nursing science as well as pathophysiology and disease management can be integrated and explored within it. Students need to learn complexity thinking and how to understand and negotiate the complex environments in which they work. Skills for pattern seeing, systems thinking, and facilitating patient transitions will be essential for the future. Students will learn through simulations that can embed complexity within a safe environment. In a health care environment characterized by interrelationships and diversity, learning needs to occur with other professional students and with diverse communities.

Complexity thinking is penetrating the consciousness of many disciplines, including nursing. Martha Rogers described in 1970 a unitary worldview consistent with complexity theory, making nursing the pioneer in advancing these ideas. Other nursing theorists have followed, expanding complexity thinking in nursing. The concepts within the science of complexity will shape the future of nursing inquiry, practice, and education.

REFERENCES

Ackoff, R. (1981). *Creating the corporate future: Plan or be planned for*. New York: Wiley.

Allen, P. (2001). What is complexity science? Knowledge of the limits to knowledge. *Emergence, 3*(1), 24–42.

Barrett, E. A. M. (1986). The principle of helicy: The relationship of human field motion and power. In V. M. Malinski (Ed.), *Explorations on Martha Rogers' science of unitary human beings* (pp. 173–188). Norwalk, CT: Appleton-Century-Crofts.

Benner, P., Sutphen, M., Leonard, V., & Day, L. (2010). *Educating nurses: A call for radical transformation*. New York: Jossey-Bass.

Bohm, D. (1980). *Wholeness and the implicate order*. London: Routledge.

Briggs, J., & Peat, F. D. (1989). *Turbulent mirror: An illustrated guide to chaos theory and the science of wholeness*. New York: Harper & Row.

Capra, F. (1982). *The turning point: Science, society & and the rising culture*. New York: Simon & Schuster.

Chaffee, M. W., & McNeill, M. M. (2007). A model of nursing as complex adaptive system. *Nursing Outlook, 55,* 232–241.

Clark, C. (2003). The transpersonal caring moment: Evolution of higher ordered beings. *International Journal of Human Caring, 7*(3), 30–39.

Cooksley, R. W. (2001). What is complexity science? A contextually grounded tapestry of systemic dynamism, paradigm diversity, theoretical eclecticism, and organizational learning. *Emergence, 3*(1), 77–103.

Coppa, D. F. (1993). Chaos theory suggests a new paradigm for nursing science. *Journal of Advanced Nursing, 18*(6), 985–991.

Davidson, A. W. (1988). *Choice patterns: A theory of human-environment relationship*. Unpublished doctoral dissertation, University of Colorado, Boulder, Colorado.

Davidson, A. W., & Ray, M. A. (1991). Studying the human-environment phenomenon using the science of complexity. *Advances in Nursing Science, 14*(2), 73–87.

Davidson, A. W., Teicher, M., & Bar-Yam, Y. (1997). The role of environmental complexity in the well-being of the elderly. *Complexity and Chaos in Nursing, 3,* 5–12.

Dent, E. B. (2000). Complexity science: A worldview shift. *Emergence, 11*(4), 5–19.

Fawcett, J. (2000). *Analysis and evaluation of contemporary nursing knowledge: Nursing models and theories*. Philadelphia: F. A. Davis.

Ferguson, M. (1980). *The aquarian conspiracy: Personal and social transformation in the1980s*. Los Angeles: J. P. Tarchers.

Gleick, J. (1987). *Chaos: Making a new science*. New York: Viking.

Haigh, C. (2002). Using chaos theory: Implications for nursing. *Journal of Advanced Nursing, 37*(5), 462–468.

Harman, W. (1998). *Global mind change: the promise of the last years of the twentieth century*. Indianapolis, IN: Knowledge Systems.

Horgan, J. (1995). From complexity to perplexity. *Scientific American, 5,* 104–109.

Koithan, M. (2006). Models for the study of whole systems. *Integrative Cancer Therapies, 5*(4), 293–307.

Lewin, R. (1992). *Complexity: Life at the edge of chaos*. New York: Macmillan.

Lindberg, C., Nash, S., & Lindberg, C. (2008). *On the edge: Nursing in the age of complexity*. Bordentown, NJ: Plexus Press.

Mishel, M. H. (1990). Reconceptualization of the uncertainty in illness theory. *Image, 22*(4), 256–262.

Newman, M. (1986). *Health as expanding consciousness*. St. Louis, MO: Mosby.

Newman, M. (1994). *Health as expanding consciousness*. St. Louis, MO: Mosby.

Newman, M. (2008). *Transforming presence: The difference nursing makes*. Philadelphia: F. A. Davis.

Newman, M. A., Sime, M., & Corcoran-Perry, S. (1991). The focus of the discipline of nursing. *Advances in Nursing Science, 14*(1), 1–6.

Newman, M. A., Smith, M. C., Dexheimer-Pharris, M., & Jones, D. (2008). The focus of the discipline of nursing revisited. *Advances in Nursing Science, 31*(1), E16–E27.

Parse, R. R. (1981). *Man-living-health: A theory of nursing*. New York: Wiley.

Parse, R. R. (1998). *The human becoming school of thought*. Philadelphia, Thousand Oaks, CA: Sage.

Parse, R. R., Coyne, A. B., & Smith, M. J. (1987). *Nursing science: Paradigms, theories and critiques*. Philadelphia: W. B. Saunders.

Phelan, S. E. (2001). What is complexity science? *Emergence, 3*(1), 120–136.

Prigogine, E., & Stengers, I. (1984). *Order out of chaos: Man's new dialogue with nature*. Toronto, ON: Bantam Books.

Ray, M. (1994). Complex caring dynamics: A unifying model of nursing inquiry. *Theoretic and Applied Chaos in Nursing, 1*(1), 23–32.

Ray, M. (1998). Complexity and nursing science. *Nursing Science Quarterly, 11*(3), 91–93.

Reed, P. (1991). Toward a nursing theory of self-transcendence: Deductive reformulation using developmental theories. *Advances in Nursing Science, 13*(4), 64–77.

Resnicow, K., & Page, S. E. (2008). Embracing chaos and complexity: A quantum change for public health. *American Journal of Public Health, 98*(8), 1382–1389.

Richardson, K. A., & Cilliers, P. (2001) What is complexity science? A view from different directions. *Emergence, 3*(1), 5–23.

Richardson, K. A., Cilliers, P., & Lissack, M. (2000). Complexity science: A grey science for the stuff in between. *Proceedings of the first International Conference on Systems Thinking in Management*. Geelong, Australia, 532–537.

Rogers, M. E. (1970). *An introduction to the theoretical basis of nursing*. Philadelphia: F. A. Davis.

Rogers, M. E. (1990). Nursing: Science of unitary, irreducible human beings: Update 1990. In E. A. M. Barrett (Ed.), *Visions of Rogers' Science-based Nursing* (pp. 5–11). New York: National League for Nursing Press.

Rogers, M. E. (1994). The science of unitary human beings: Current perspectives. *Nursing Science Quarterly, 7*(1), 33–35.

Roy, C. (2008). *The Roy Adaptation Model* (3rd ed.). Upper Saddle River, NJ: Prentice Hall Health.

Roy, C., & Jones, D. (Eds.). (2007). *Nursing knowledge development and clinical practice*. New York: Springer.

Roy, C. & Zhan, L. (2010). Sister Callista Roy's adaptation model. In M. Parker & M. Smith (Eds.), *Nursing theories & nursing practice* (3rd ed., pp. 167–181). Philadelphia: F. A. Davis Company.

Schwartz, P., & Ogilvy, J. A. (1979). *The emergent paradigm. Changing patterns of thought and belief.* Menlo Park, CA: SRI International.

Smith, M. C. (1994). Arriving at a philosophy of nursing. In J. F. Kikuchi & H. Simmons (Eds.), *Developing a philosophy of nursing* (pp. 43–60). Thousand Oaks, CA: Sage.

Stacey, R., Griffen, D., & Shaw, P. (2000). *Complexity and management*. London and New York: Routledge.

Turkel, M. C., & Ray, M. A. (2000). Relational complexity: A theory of the nurse-patient relationship within an economic context. *Nursing Science Quarterly, 13*(4), 307–313.

Turkel, M. C., & Ray, M. A. (2001). Relational complexity: From grounded theory to instrument development and theoretical testing. *Nursing Science Quarterly, 14*(4), 281–287.

Vicenzi, A. (1994). Chaos theory and some nursing considerations. *Nursing Science Quarterly, 7*(1), 36–42.

Wilber, K. (1997). *The eye of the spirit*. New York: Random House.

Wilber, K. (1998). *The marriage of sense and soul: Integrating science and religion*. New York: Random House.

Young, A. (1976). *The reflexive universe: Evolution of consciousness*. San Francisco: Robert Briggs.

Response to Chapter 1

REFLECTIVE QUESTIONS

Joyce Perkins

1. What is the relationship between theory and practice (clinical, administrative or educational) highlighted in this chapter?

The most important relationship between theory and practice highlighted in this chapter is that found in the unitary-transformative or simultaneity paradigm as described in the nursing literature. This world view notes the human-environment as a unitary whole. As such, the theory is the practice, and may be said to happen or unfold via the direct perception of the nurse. As nurses, theory is the cognitive or conceptual framework that maps the terrain of our work and world. Practice is the lived experience of a certain perspective or theoretical world view embodied within the nurse. Theory is our way of organizing and bringing clarity to the process of living life and working with our patients. Practice is the direct experience of what works or does not work in the "field" of patient care. This constant interplay or daily feedback expands the awareness of the nurse exponentially, expanding consciousness with the emergence of new insights in a consistent, even rhythmic way, in relation to the intensity of the situations presented. Nurses enter into relationship with patients in times of crisis or "bifurcation points". They are poised to help initiate shifts in the lives of persons who enter into a caring relationship with them.

In this chapter, Smith explicates the complexity of this relationship between nurse, patient, and environment. She traces the origins of what is fast becoming the "map" of choice across multiple venues including clinical practice, administration, and educational institutions.

In clinical practice, an example of this model of nursing care is displayed by nurses on a depression, mood disorder, in-patient hospital unit in a large tertiary care hospital in the Midwest. These nurses practice what is referred to as "relationship centered care" (Suchman, 2006) by putting the patient's needs first. They also practice what is called "unitary-caring science" in nursing. Unitary caring science is described as a melding of Roger's *Science of Unitary Human Beings* and Watson's *Transpersonal Caring Theory*. They go about their work with patients modeling the complexity descriptions noted by Butcher (2002) as: 1) cultivating creativity, 2) using butterfly power, 3) flowing with turbulence,

4) exploring integrality, 5) seeing the beauty and art of nursing, 7) living in pan dimensionality, and 8) participating with the whole. Each of these themes describes a quality specific to complexity science but also to unitary caring science.

In administrative relationships, a complexity model of operating may be seen in the literature and in the style of management that encourages a non-hierarchal approach to problem solving and transformational leadership. Margaret Wheatley (2001) has been the spokes person for this creative venue that brings all participants to the table, giving voice to potential and emergent possibilities for resolving complex problems. Knowing the difference between simple, complicated and complex problems (Zimmerman, 1999) allows nurses the opportunity to choose the most efficient avenue for resolution, choosing to focus their efforts in paradigm I, II, or III as articulated by Newman et al. (1991).

In education, teaching complexity principles to student nurses brings light bulb moments in the class room as they grasp the implications of functioning with these essential guidelines alongside them. Complexity principles are "essential" in that all of nature follows the process mapped by complexity science. It inherently makes sense, and at the same time lends a refreshing and inspiring vision that uplifts the spirit. In this chapter Smith has managed to take an often overwhelmingly complex subject and revealed the path by giving us navigational tools to help make sense of our world and our individual roles in the global scheme of things. From the microscopic to the macroscopic or cosmic environment, from the quantum wave/particle to the path of stars and planets, all may be better understood as fractal elements of the universal design of our cosmos. The term "fractal" refers to the shapes and patterns found in nature that are iterative or repeat at all levels of design or dimension, each embedded within the one before. Information is shared, distributed in this way via a non local process as stated in complexity theory.

Nurses are constantly placed in situations that facilitate an accelerated learning curve as they walk 24/7 alongside their patients. As they choose to step into caring relationships with each other and with their patients, our world becomes a better place and we all become healthier people. In complexity science, nurses now have a map which affirms our historical nursing theories, yet also inspires a new and dynamic approach. As each nurse individually grasps the concepts and makes them his/her own, a new life and energy is brought to the fore. In complexity science, we have the patterns and dynamics that express how one can both express individually yet be globally connected and conscious in a deeply caring and inspired way.

2. **How does a nurse or a physician or health care administrator learn about complexity science/s and caring through the presentation of ideas or research?**

   People expand their horizons of meaning by being in dialogue with one another. Truly listening or hearing what each other is saying is imperative. By learning something of the various cultural perspectives we enrich the dialogue and ourselves by being able to envision new possibilities or other ways of knowing and doing. The natural world demonstrates the unfolding of complexity principles, so paying attention and having the intention to learn is particularly helpful. There are many short videos available on line now to assist one in grasping graphically or visually principles that may be difficult if just the verbal description is offered. We are becoming a multi sensory population and there are tools that can be shared to open up these doorways for exploration. Indigenous people who think in images instead of words are showing us the way in many respects. Learning to pay attention to intuition, "feeling" dynamics, and subtle impressions brings the world to life in a way that objective rational thinking has left void.

3. **What meanings are illuminated in this chapter for nursing or other health care professions? Direct your answer to education, administration, research or practice based upon the focus of the chapter.**

   The meanings illuminated in this chapter for nursing and other health care professionals are as follows:

   a. Nurse educators must now teach about the world and nature as it really is, not as isolated phenomena but in context with many influences and dimensions to consider. We are to enjoy the process. In other words, we are to come from a place of love in the heart instead of fear. Surprising and amazing happenings unfold as we work in harmony and synchronicity with all that is. A resonance pervades that unifies our efforts in mighty ways. Native peoples would say that we follow the "beauty way."
   b. Nurse administrators need to avoid "top down" hierarchical models. They need to obtain input from the "bottom-up," bringing all players to the table to make decisions, especially those who will be doing the work. More insight into problem situations comes into play when one includes outliers with unique perspectives. Outliers are seen as valuable players rather than irritations to be ignored as in the traditional paradigm of medical science or paradigm I in nursing. In Complexity dynamics we have a "both/and" philosophy in which each perspective plays its part. In all, workers take more responsibility for decisions that they have helped to create.

c. Nurse researchers must now resolve the paradox of duality perspectives with an expanded vision. Quantitative and Qualitative, reductionist and pluralistic measures, and many other qualities listed in this chapter co-exist each with their purpose, but we are rising to a new level of operating within ourselves. As consciousness expands individually, the collective consciousness also takes a leap. New research methods must be devised that actually speak to this new reality. We live in a quantum world that "leaps" to surprising solutions, no longer just using rational linear logic to explore our world.

d. Practicing nurses are embedded in the mix of patient dilemmas. Solutions emerge from constant feedback as new information is digested and recycled through creative applications to problematic areas. Nurses need to get used to change, be flexible, and become comfortable with uncertainty, learning to tap into "group mind" where solutions lay waiting for one who asks. Life takes on a vitality that can only come from needing to stay awake and on ones toes. Not at all the boring and tedious actions called forth under a predictable conceptual model of the universe. All becomes important and each is needed to do his/her part for the whole to radiate a mature persona of love and kindness.

4. What is the relevance of this work to the future of the discipline and profession of nursing, health care professions and health care in general?

Nursing, at its best, has always worked with the idea of compassion, which means to join with passion or enthusiasm into the endeavor at hand. Nurses facilitate well being or the healing process for and with others. Nurses have always been in full participation in that they walk along side their patients rather than act as hierarchical dictators of orders. Nurses are thus, ready to move forward with complexity thinking. As a group, nurses do not have energy invested in hierarchical power structures. Complexity Science coupled with caring dynamics seems to be our true language, our way of being in the world. We have always considered multiple options and avenues for exploration to help take us toward the focus of our attention and intention. That focus is to care for our patients, ourselves, and experience a harmony of wholeness and health of body, mind and spirit.

Complexity thinking is the venue that will allow us the quantum leap forward in our ability to solve previously inexplicable problems. Other health professions as they come to understand the dynamics of a complexity model will be more willing to engage with one another for mutual benefit, each bringing the wisdom of their way of doing things. We will learn to work together and share the responsibility for our own health and healing along with that of our earth.

REFERENCES

Butcher, H. (2002). Living in the heart of helicy: An inquiry into the meaning of compassion and unpredictability in rogers' nursing science. *Visions, 10*(1) 6–22.

Newman, M., Sime, A., Corcoran-Perry, S. (1991). The focus of the discipline of nursing. *Advances in Nursing Science, 14*(1), 1–6.

Suchman, A. L. (2006). A new theoretical foundation for relationship-centered care: Complex responsive processes of relating. *Journal of General Internal Medicine, 21*, S40.

Wheatley, M. J. (2001). Restoring hope to the future through critical education of leaders. *The Journal for Quality and Participation, 24*(3), 46.

Zimmerman, B. (1999). Complexity science: A route through hard times and uncertainty. *Health Forum Journal, 42*(2), 42.

Response to Chapter 1

REFLECTIONS FROM A ROGERIAN SCIENCE PERSPECTIVE

Francelyn M. Reeder

THOUGHTS ON THE SCIENCE OF UNITARY HUMAN BEINGS

The Science of Unitary Human Beings (SUHB) (Rogers, 1970), the conceptual system of the nursing theorist Martha Rogers, is its own unique science with its origin in the unitary nature of the human–environment mutual process. Nursing is a science and an art. "A science is an organized abstract system. It is a synthesis of facts and ideas, a new product" (Rogers, 1990, p. 6). A science has many theories. The mutual human–environment process is the focus of nursing; it is complex and often difficult to study. The abstract system of the SUHB exists as an irreducible whole. Definitions, principles, and theories derive from this irreducible whole. "The evolution of unitary human beings is a dynamic, irreducible, nonlinear process characterized by increasing diversity of energy field patterning" (Rogers 1990, p. 9). Thus, developing nursing's abstract system demands a new worldview, a language that has specificity, clarity, precision, and communication (Rogers, 1990). Although the *Science of Unitary Human Beings* is often directly compared with complexity sciences, I believe that the SUHB can be classified as different because of the unique principles that are espoused and the wholeness expressed in the life process as reflected in pattern manifestation and organization in the energy fields of humans and environment. Rogers did not use the language of complexity, such as chaos theory, complex adaptive systems, or from a physiological view, normal variability, or from a methods perspective, fractal analysis to identify or analyze fractal patterns and the interactive process that creates them. Rogers, on the other hand, advocated multiple naturalistic field methods or forms of inquiry to study phenomena within context, recognizing the unitary nature between the observer and the observed or that the knower and the known are one (Reeder, 1984). Rogers (1990) used the revised homeodynamic principles of resonancy, helicy, and integrality to understand and study the continuous human and environmental process (p. 97). "*Resonancy* refers to continuous change from lower to higher frequency wave patterns in human and environmental fields. *Helicy* refers to the continuous innovative, unpredictable, increasing diversity of human and environmental field patterns. *Integrality* refers to continuous mutual human field and environmental field process" (Rogers, 1990, p. 8). These principles provide a fundamental guide to the study and practice of nursing. The SUHB *begins* with the unitary nature of humans in the environment, which is in mutual process and continuously moving forward. The increasing complexification of energy field patterning shows that there is no generalization from parts to whole (Rogers, 1990). Outcome subsequently is not the origin, nor is the science a stimulus–response system, nor is it divisible into parts. Consequences or conclusions although may be *captured* as pattern outcomes in research, however, are not applicable in the strict sense of the words. Thus complexity science, although illuminating the relationship networks or the tenet of interconnectedness, is different from Rogerian science.

Philosophically, Rogers' SUHB is essentially unitary. Its essence of meaning always begins with the unitary nature of the human and environment in mutual process, which is continually open to pattern change. As a unitary mutual human–environment process, *complex patterns become manifest* as humans and their environment dynamically evolve; they are unpredictable and increasingly diverse.

From an inquiry perspective, complex ways of knowing are manifest and unfold in the real world. The SUHB always encompasses the ever-changing human–environment integrality, including the environment, for example, of space. Rogers reinforced the phrase advanced by Robinson and White (1986), home spatialis, to illuminate the ever-changing human–environment field patterning. Complexity, thus, is interpreted through a lens—a telescope to look at humans in a real-world mutual process. These manifestations are experienced in practice, studied in research, and communicated in education.

APPLICATION OF THE SUHB IN NURSING RESEARCH, EDUCATION, AND PRACTICE

Unitary field pattern methods or the study of pattern manifestations have been advanced by Rogerian scholars, such as Rawnsley, Butcher, and Cowling (Madrid & Barrett, 1994; Locsin & Purnell, 2009). Complex field patterning manifestations of the human–environment mutual process in practice can be interpreted by nurses through education, research, and practice. Nurses are not separate from patients; they are in an integral relationship manifest as integral evidence. Persons intended are ever changing, never static, transcendent as well as immanent, and potential as well as actual; as such, preference is given to wave-seeing (whole) rather than particle-seeing (parts) consideration or judgment in the world. The benefit of understanding integral evidence ". . . holds promise for the development of pattern seeing of wave phenomena" (Reeder, 1984, p. 21) in Rogers' *Science of Unitary Human Beings*. *The integral evidence*, observational statements or communicative phenomena of "consciousing the world" or manifesting intentional consciousness, can be, for example, energy as felt, healing as imagined, health as remembered or health as desired, pain as avoided, and action as judged.

The integral evidence includes not only five sense perceptible things but also phenomena such as willing, loving, judging, imagining, remembering, intuiting, feeling, and anticipating (Reeder, 1984). *Learned pattern seeing* in nursing adopting the SUHB for research, education, and practice must come from a *centered* perspective of knowledgeable compassion. "art, wisdom and compassion, underwritten by transcendent imaginative conceptual skill index the ways of knowing that are integral to this unique science of nursing" (Reeder, 1984, p. 23). Knowledge and wisdom thus are lived uniquely by the knower and the known (Mitchell, 2009). Rogers taught us to appreciate that "as individuals attune to a unity consciousness [the unity of the knower and the known], infinite possibilities are born, and the status of finite consciousness disappears in the beauty and flow, revealing a kaleidoscope of patterns that are ever-changing" (Cowling & Repede, 2009, p. 78). As Einstein revealed, the most beautiful thing that we can experience is the mysterious; the source of all true art and science is the mysterious (Madrid & Barrett, 1994, xix). In nursing, we hold this beauty and mystery in our hands, hearts, and minds through knowledge and the wisdom of Rogers' incomparable *Science of Unitary Human Beings*.

REFERENCES

Cowling, R., & Repede, E. (2009). Consciousness and knowing: The patterning of the whole. In R. Locsin & M. Purnell (Eds.), *A contemporary nursing process: The (un) bearable weight of knowing in nursing* (pp. 73–121). New York: Springer.

Madrid, M., & Barrett, E. (Eds.). (1994). *Rogers' scientific art of nursing practice*. New York: National League for Nursing Press.

Mitchell, G. (2009). Evidence, knowledge, and wisdom: Nursing practice in a universe of complexity and mystery. In R. Locsin & M. Purnell (Eds.), *A contemporary nursing process: The (un)bearable weight of knowing in nursing* (pp. 99–121). New York: Springer.

Reeder, F. (1984). Philosophical issues in the Rogerian science of unitary human beings. *Advances in Nursing Science, 8*(1), 14–23.

Robinson, G., & White, H. (1986). *Envoys of mankind*. Washington, DC: Smithsonian Institute Press.

Rogers, M. (1970). An *introduction to the theoretical basis of nursing*. Philadelphia: F. A. Davis.

Rogers, M. (1990). Nursing: Science of unitary, irreducible, human beings: Update 1990. In E. Barrett (Ed.), *Visions of Rogers' science-based nursing* (pp. 5–11). New York: National League for Nursing.

CHAPTER 1
PHILOSOPHICAL AND THEORETICAL PERSPECTIVES RELATED TO COMPLEXITY SCIENCE IN NURSING

- *How do you see the work of Martha Rogers' science (the Science of Unitary Human Beings) influencing contemporary health care practices?*
- *Do you see Martha Rogers' work more congruent with caring science or complexity science?*
- *How do you see caring science and complexity sciences guiding your professional practice? Discuss some concepts from Watson's "Caritas" theory and the Science of Unitary Human Beings. Are they comparable?*
- *In the response, Dr. Perkins defines "fractal"; what shape or pattern found in nature best describes your professional work environment?*
- *In the response, Dr. Perkins talks about the importance of integrating concepts from complexity into education. What education would your peers need to make the connection between the theory of complexity sciences and professional practice?*
- *In the response, Dr. Reeder speaks of Rogers' Science of Unitary Human Beings. What does unitary mean to you both from a Rogerian point of view and how you have developed the idea?*

TWO

Complex Caring Dynamics: A Unifying Model of Nursing Inquiry*

Marilyn A. Ray

Changing perspectives on the nature of nursing within the emerging sciences of complexity, Rogers' Science of Unitary Human Beings (SUHB), and the nature of nursing as relational caring as the focus of inquiry promise to transform nursing. This chapter provides an explanation of the epistemological basis of complexity sciences and complex caring inquiry in nursing. It speaks of the complex and dynamic nature of nursing as facilitating choice patterning with a relational caring perspective. Further, a unifying model of complex caring dynamics is presented to illuminate the complexity of inquiry. Complexity sciences show us that the philosophical approaches of the hypothetical deductive quantitative approach and the qualitative inductive approach, although seemingly paradoxical, operate together. The unifying model illustrates four dynamical processes that can be selected as research approaches, either separately or in combination: technical, practical critical, and creative caring dynamics. They are used to reflectively see, map, and recognize the different choice patterns in the human health and caring experience. These research processes will contribute to a new understanding of nursing philosophy, the debate about mixed research methods in nursing, and nursing practice. Respondent Morse addresses ideas of caring as consciousness or consciousness as caring and presents questions about the nature of reciprocal caring. Morse presents a view of mixed-method design to resolve some issues in holistic inquiry. Respondent Reed presents the ontological (ways of being) and epistemological (way of knowing) in terms of the transtheoretical evolution of complexity science and nursing science, and offers new ideas on how the sciences are integrated.

*An original publication of this chapter appeared as an article in 1994 in the now out-of-print and circulation journal *Theoretic and Applied Chaos in Nursing*. Permission has been granted by the former editor, Angela Vicenzi, RN, EdD, to republish portions of the paper as a chapter in this book.

INTRODUCTION

Scientists during the 20th century and now into the 21st century from different disciplines are seriously studying what knits the nature of the universe together. A new understanding of the concepts of wholeness, wave patterns, fractals, networks of relationship, the discovery of nonlinear dynamical or chaotic (unpredictable) systems, ideas such as self-organization, and emergence are at the center of the changes in science. This unified science is called the sciences of complexity (Bar-Yam, 2004; Briggs & Peat, 1989, 1999; Davidson & Ray, 1991; Lewin, 1992; Lindberg, Nash, & Lindberg, 2008; Mitchell, 2009; Peat, 2002; Ray, 1998; Vicenzi, 1994; Vicenzi, White, & Begun, 1997; Waldrop, 1992; West, 2006), and has made it possible for scientists studying biophysical, chemical, economic, environmental, sociocultural, and nursing phenomena to arrive at similar conclusions about their holistic and relational nature. Their findings include the way patterns are established in living and nonliving systems, how these patterns are interconnected, and how they are linked to corresponding principles of order.

Revisiting nursing's fundamental nature shows the influence of complexity sciences on nursing science. At the same time, nursing paradigms—which include the conceptual system of Rogers' (1970; Barrett, 1990a, 1990b) SUHB illuminating the mutual human–environment process and human and environmental field patterning; the Caring Sciences that illuminate relationships, compassion, loving kindness, justice, and harmony of body, mind, and spirit (see Ray, 2010; Watson, 1985, 2005, 2008, 2009; Watson & Smith, 2002) and how the organizational culture influences caring, meaning systems in practice and relational self-organization (see Ray, 2006, 2010; Turkel & Ray, 2000, 2001, 2009a)—now appear as central phenomena of the discipline (Newman, Smith, Dexheimer-Pharris, & Jones, 2008).

Caring in nursing has been conceptualized in various ways: an all-encompassing view, such as the being and becoming through caring (Boykin & Schoenhofer, 2002; the six Cs of Roach [2002], commitment, compassion, conscience, confidence, competence, and comportment); a human trait; a moral imperative; an affect; an interpersonal interaction and an intervention (Morse, Bottorff, Neander, & Solberg, 1991); evidence-based nursing (Turkel, 2006a, 2006b); and a spiritual and ethical relationship (Ray, 1981, 1997, 2006, 2010; Watson, 1985, 2005, 2008). Within the context of the diversity, the primary focus of these conceptualizations is caring's relational character; a sense of belongingness and interconnectedness that manifests itself in actions that evoke, direct, or facilitate understanding and change in terms of health, illness, healing, dying, death, and well-being.

This chapter examines the sciences of complexity movement, compares them to Rogers' SUHB, discusses the dynamics of the phenomenon of caring in the human health experience, examines the relationship between caring and choice, and proposes a unifying model of complex caring dynamics for nursing inquiry.

THE SCIENCES OF COMPLEXITY MOVEMENT

Complexity science is the science of change (Briggs & Peat, 1989, 1999; Peat, 2002). There are a number of theories, such as quantum theory, chaos theory, and string theory, that fall under the science of complexity, thus named the sciences of complexity (Arntz, Chasse, & Vicente, 2005). The fundamental idea in the sciences of complexity is that all things in nature are interrelated and organized into patterns. Over time, these patterns exhibit no rational predictable behavior, but produce new, more complex structures known as nonlinear emergent properties (Arntz et al., 2005; Waldrop, 1992; West, 1993). Complexity scientists state that the future is open, and all must participate in a new dialogue with nature (Nicolis & Prigogine, 1989): a dialogue where all phenomena are "an integral part of the time-bound, spontaneously organized movement of nature, not a low probability accident" (Briggs & Peat, 1989, p. 151). Rather than maintaining a view of the universe as mechanistic, controllable, objective, and predictable, there is acknowledgment of a new perspective that links the notions of spontaneous activity, uncertainty, and unpredictability (Briggs & Peat, 1984). The sciences of complexity provide a foundation for a clearer understanding of the universe as holistic, complex, and dynamic: a universe that is interdependent and relational, where the observer cannot be separated from the observed (Heisenberg's Principle of Uncertainty), and the future is open and always changing (emergent) (Peat, 2002). Although phenomena are spontaneous, changing, and nonlinear, they are, however, dependent on their earlier historical past states (histeresis). That is, there is an initial condition with an unpredictable possible future.

The essential picture of complexity or dynamic relational activity is seen at the boundary between evolving systems. At the boundary, there are forces of reciprocity between order and disorder (chaos) and between systems that are considered settled, rigid, and orderly and systems that are turbulent—what complexity theorists call the edge of chaos (Briggs & Peat, 1984, 1989, 1999; Prigogine & Stenger, 1984; Waldrop, 1992). Therefore, "[c]haos theory explains the ways in which natural and social systems organize themselves into stable entities that have the ability to resist small disturbances or perturbations." Give a system a slight nudge and "it will move into a radically new form of behavior [nonlinear behavior] such as chaos" (Peat, 2002, p. 125).

The tension between disorder and order (chaos theory) drives change and a creative reordering or self-organization in systems (Capra, Steindl-Rast, & Matus, 1991). For example, chemical and biological systems, or even social or economic systems, are driven to change and creative reordering. The edge of chaos is, thus, a communication and information process that feeds back on itself and coordinates system behavior through continual, mutual interaction (Capra et al., 1991). "If amplifications have increased to the level where the system is at maximum instability (a crossroads between death and transformation known technically as a bifurcation point), the system encounters a future that is wide open" (Wheatley, 1992, p. 96).

> The role of information is revealed in the word itself: information . . . In a constantly evolving, dynamic universe, information is a fundamental yet invisible player, one we can't see until it takes physical form. Something we cannot see, touch, or get our hands on is out there, influencing life. . . The source of life is new information-novelty-ordered into new structures . . . The greatest generator of information is the freedom of chaos [disorder and order], where every moment is new. (Wheatley, 2006, pp. 96–97)

As such, one fluctuation becomes dominant, and a new pattern forms with greater order. The system then remains stable until at some point, too much change again creates a disordered system, and new arrangements for order are activated (Briggs & Peat, 1989).

What is most interesting in these systems is the creative or autocatalytic (self-organizing) process. At the potential decay point (entropy or equilibrium state), there is a phase space, "which exerts a 'magnetic' appeal for a system, seemingly pulling the system toward it" (Briggs & Peat, 1989, p. 36). The system seems to hesitate in this phase space, to be offered a choice among various possible directions of evolution. Through the interweaving of iterations or feedback loops, the turbulent state containing the possibilities for self-organization—structure, pattern, and process—is free to seek out its own solution to the current environment (Prigogine & Stenger, 1984). The unpredictable system "chooses" one possible future while leaving others behind (Briggs & Peat, 1989). Chaos, rather than merely a mindless jiggling, has awareness; it seeks out and chooses its own subtle form of order (Briggs & Peat, 1989, 1999; Capra et al., 1991).

Enfolded and etched in all patterns and processes of a living organism or sociocultural system are a number of these bifurcation points. Because it is assumed that time is irreversible and that all phenomena in the universe are interconnected, these bifurcation points constitute a living history of the choices by which we evolved—from primordial beginnings to the complex cellular and social forms of today. Therefore, "every complex system is a changing part of a greater whole, a nesting of larger and larger wholes [holonomy] leading eventually to the most complex dynamical system of all"—the universe (Briggs & Peat, 1989, p. 148).

THE PHILOSOPHY, EPISTEMOLOGY, AND METAPHYSICS OF COMPLEXITY SCIENCES

Capra et al. (1991) identified belongingness or belonging to the universe as the central metaphor of the new sciences of complexity. Belongingness represents a vision of reality that is relational; the self can be understood in the context of belonging because we first belong to, from a spiritual sense, the transcendent mystery of God (Ray, 1997, 2010; Watson, 2005, 2008), and second, the self. Our personhood is defined through relationships. The complexity sciences paradigm is embedded in the social paradigm, which is also a pattern of relationship (in the Rogerian sense, the mutual human–environment process [Rogers, 1970, 1990]). This is congruent with a philosophy that holds that the kind of society we live in, to a large extent, determines the kind of science we have. If we believe that individuals can be understood only from the dynamics of the whole (society) and that a complex, global society with different coexisting cultures illuminates our social interrelatedness, then the only way to understand any one culture is to understand its relationship to the whole (Capra et al., 1991).

The new epistemology recognizes that the wholeness of the universe is brought forth as a process of "knowledge-in-process" (Harman, 1991, p. 27). Complexity sciences reinforce the perspective that science cannot provide any complete and definitive understanding of reality. Because all phenomena in the universe are interconnected and dynamic, all knowledge (including scientific concepts, theories, and research results) is limited and approximate.

In this scientific view/theory of living systems, the processes of life are seen as mental processes. There is a relationship between mind and matter that differs from the dualistic philosophy of Descartes, which advocated the separation of mind and matter. The new view of unitive consciousness or interconnectedness between mind and matter reveals an unfolding of value, purpose, and meaning. Capra et al. (1991) claim that "Logos [the word] is the pattern that makes a cosmos out of chaos" (p. 126) and further, that the relational pattern of the cosmos is drawn by love—"we love what attracts us" (p. 118).

Self-organization in chaos theory, although seemingly without a goal, is influenced by creativity—a vision of a purposeful love, a purposeful God, a force or spirit of life expressing itself intelligently in the universe (Cannato, 2010). As participants in the universe we, thus, say "yes" to belonging. We choose to belong not just at the level of the intellect, but within a mode of consciousness that is moral, where the choice to act is a reflection of knowledge and understanding of how to act when people and things belong together (Capra et al., 1991)—in essence, within the realm of caring, the action of love (Ray, 1981, 1989, 1997, 2006, 2010; Watson, 1985, 2005, 2008).

In this philosophical treatise, there is recognition that, again, the notion of consciousness is not an individual thing but is continually formed in and by relationships. It is intrinsically apprehended through a spiritual or loving relationship and extrinsically understood through shared relationships with persons integral with their environment (Ray, 2010; Webb, 1988). Relational subjectivity is coupled with the cognitive objectivity of intentional moral operations—attention, understanding, and knowing within relationships and communitarian ethics (Lonergan, cited in Webb, 1988; Ray, 2010). The experiential data of relational knowing now are considered the mark of exploration or inquiry.

THE PHENOMENON OF RELATIONAL CARING IN THE MUTUAL HUMAN–ENVIRONMENT PROCESS AND THE SCIENCES OF COMPLEXITY

Rogers's conceptual system of the SUHB is a unique science in nursing with its origin in the *unitary* nature of the mutual human–environment process. What distinguishes the SUHB is the idea of its unitary nature that has no beginning and no end, the whole that illuminates the notion of the science as dynamic, irreducible, nonlinear characterized by increasing diversity of energy-field patterning and increasing complexity (Barrett, 1990a, 1990b; Rogers, 1970, 1990). The field is infinite and inclusive of all creation. Within this field according to the focus of consciousness (attention and intention), wave phenomena form and become attractor patterns upon which patterns of physical and behavioral reality become manifest (Perkins, 2010). Although the SUHB is compared and contrasted with the sciences of complexity, Rogers did not use the language of complexity. She used homeodynamic principles to articulate the continuous mutual human and environment process, such as resonancy, helicy, and integrality. Resonancy highlights the continuous change from lower to higher frequency wave patterns in human and environmental fields. Helicy refers to continuous innovative, unpredictable, and increasing diversity of human and environmental field patterns. Integrality refers to the continuous mutual human and environmental field process (Rogers, 1990, p. 8). Thus, Rogers always expressed the unitary nature or the mutuality of the human–environment process, which is continually open to pattern change. From this unitary whole, complex patterns are manifested as humans and environment continually evolve. As such, research is focused on the study of pattern manifestations in education, administration, and practice. Four concepts (energy fields, openness, pattern, and pandimensionality) and the three principles (resonancy, helicy, and integrality) provide the abstract framework for seeing the pattern manifestations in nursing (Barrett, 1990a, 1990b). The energy field is "a means of perceiving people and their respective environments as irreducible wholes" (Rogers, 1990, p. 5). Learned pattern seeing is centered on knowing participation in change (Barrett, 1990a, 1990b). Rogers did not perceive caring to be a substantive area for study; however, she believed in unconditional love (Rogers, 1990; Smith, 1999). Butcher (2002) believed that Rogers' concept of helicy also paralleled compassion in nursing.

Perkins (2010) remarked that caring and the science of unitary human beings must be synonymous because if unconditional love prevails, healing happens at some level of information, whether it be in a "spirit or sacred" energy form or manifested in the physical form. Therefore, caring is the key that unlocks healing potential and makes the human and environmental unitary field better (Perkins, 2010).

In her writings, Rogers manifested her commitment to health promotion as a pattern manifestation; nurses were in the service of others, mutually facilitating the actualization of patients' health potential. Thus, "in Rogerian science, practice modalities concern human life patterning and reflect the wholeness of the unitary person in continuous innovative change with the universe" (Barrett, 1990a, 1990b, p. 35), and as Rawnsley pointed out "harmony of the human and environmental fields" (in Barrett, 1990a, 1990b, p.37).

The rise of the caring movement in nursing has paralleled the rise of Rogerian science and complexity sciences (Benner & Wrubel, 1989; Briggs & Peat, 1989; Davidson & Ray, 1991; Leininger, 1981, 1984, 1991; Leininger & McFarland, 2006; Morse et al., 1991; Newman, Sime, & Corcoran-Perry, 1991; Ray, 1981, 1984, 1989, 1998, 2006, 2010; Roach, 1987/2002; Rogers, 1970; Smith, 1999; Turkel, 2006a, 2006b, 2007; Turkel & Ray, 2000, 2001; Watson, 1979, 1985, 2005, 2008, 2009). Concepts advanced in complexity sciences are well represented in the nursing literature, especially the caring literature, as we have seen. They include the shared concepts of human–environment integrality, relationality, belongingness, holism value, purpose, choice, love, and qualitative measures for seeing the movement of disorder and order (chaos) and continuous change in a nonlinear world. This overlap is also evidenced by Newman et al. (1991, 2008) who wrote about the focus of the discipline of nursing. Nursing, thus, focuses on the unitary human being, the unitary transformative paradigm that places emphasis on consciousness and complexity and relationship. Relationship and caring in the human health experience are the unifying foci of nursing and nursing inquiry. The following unitive characteristics are identified: relationship, presence, caring, human–environment integrality, human field patterning, expanded consciousness, increasing complexity, human becoming, stages of disorganization and organization (chaos theory), change that is unidirectional and unpredictable, pattern manifestation and pattern recognition, and self-organization (Fawcett, 1993; Ray, 1994). These views ground nursing firmly within the caring sciences, the sciences of complexity, the science of unitary human beings, and the philosophy of consciousness. The new metaphysics states that the mind gives rise to matter rather than matter giving rise to matter—"mind and matter one and same" (Arntz et al., 2005, p. 77)—and patterns emerge in consciousness from the cellular to networks of relationship to collective memory. Human beings, the world, and everything in it are complex, and our obligation now is to become more aware of what it means to be human in the world. From a loving–caring–healing perspective, we are not just in a process of understanding the ascent to consciousness, but we are engaged in the ascent to superconsciousness where "the laws of love are directed not merely to the fulfillment of his [humans'] own will but rather to the transcendent and mysterious purposes of the Spirit, that is, the good of all men [human beings and animals]" (Merton, 1985, p. 193). People are at the heart of the choice to change, to transform. "Because of the new reality of belongingness and interconnectedness [mind and matter are one] in the global world today, there is an ethical demand (transcultural spiritual-ethical caring communication) for increasing *awareness*, dialogue, and *understanding* to facilitate *choices* that will help heal and transform person in the environments and cultures" (Ray, 2010, p. 98).

Expanding on the ideas of Rogers (1970) and Newman (1986), but incorporating additional concepts from the sciences of complexity, specifically chaos theory, Davidson (1988; Davidson & Ray, 1991) recognized that *choice* was central to the integral patterning of the mutual human–environment relational process. Davidson's research demonstrated that "choice is felt capacity within the multidimensional [dimensional] human field that is manifested as knowingness influencing environmental pattern selection" (Davidson & Ray, 1991, p. 81). In like manner, the life pattern of the human field (what has happened) and the self-pattern (what will happen) are mediated by choice in the continuous present at the "edge of chaos" (the paradox of far-from-equilibrium and order). Choice mediates "self"-organizing phenomena (structure, process, and pattern) at the critical bifurcation point in the complex dynamic of the mutual human–environment relationship.

THE NATURE OF CARING AND CHOICE

Nursing is a relational caring process. Patients need nursing at "critical bifurcation points" in their lives (Vicenzi, 1994; Vicenzi, White, & Begun, 1997). "Caring is the coherent 'radiation' of nursing and the energy of nursing research inquiry" (Davidson & Ray, 1991, p. 83). Caring is the energy by which choice is facilitated to bring order (health, healing, or well-being) out of disorder (the processes of the theory of chaos) (disease, dis-ease, need, pain, or crises).

The caring relationship (Figure 2.1, Caring Attractor) is the attractor of the magnetic appeal in the human system of the mutual human–environmental field process, which pulls the system away from disorder. Caring portrays all the ways of knowing. Caring is a complex phenomenon with philosophical and sociocultural views and aesthetic levels of expression (Carper, 1978; Leininger & McFarland, 2006). It is a personal knowing (Polanyi, 1958), an intellectual and emotional commitment to others for whom one has responsibility; it is both individual and wholly communal, thus it is ethical (Ray, 2010; Roach, 2002). At the same time, caring encapsulates knowledge of the available

FIGURE 2.1 Caring Attractor

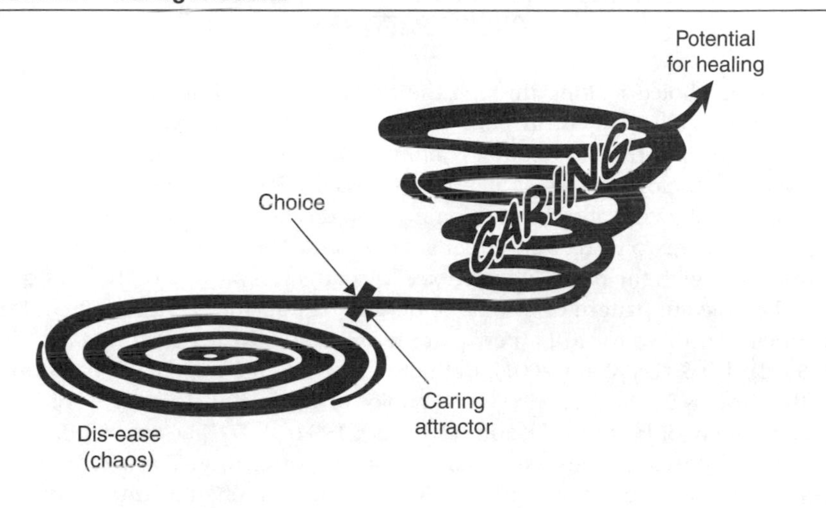

research, and thus, caring captures the empirical evidence for care with confidence and competence (Roach, 2002; Turkel, 2006a, 2006b).

In my philosophical analysis of caring, I identified the magnetic appeal of caring as love and co-presence (Ray, 1981, 1989, 1991, 1997, 2010). Caring in this sense is a spiritual–ethical knowing—a process understood within the virtues of faith, hope, love, and technical caring competency (Ray, 1987, 1997, 2007). Caring as love and co-presence is an existential quality of felt realness, a relational sign of "one with the other" (Levinas, 1949). It is characterized in consciousness as an ontologic mystery (Marcel, 1949) and the "divine grace within" (Capra et al., 1991, p. 59; Ray, 1997).

The magnetic appeal of caring (the action of love) "pulls" the other in such a way that the power of the relationship brings forth faith and trust—"a courageous trust in belonging" (Capra et al., 1991, p. 24; Swinderman, 2005). Within this view of relationship, the nurse instills hope. By anticipating and working through a knowledgeable, caring presence, and using technical caring competency, the nurse evokes, directs, or facilitates reasoned moral choice (Ray, 1987, 2007). These intentional operations, at the edge of chaos (the bifurcation point for potential disintegration or dis-ease, or transformation and change in well-being) cocreate choice-making opportunities toward the best possible future to enhance well-being, health, healing, or a peaceful death. This complex choice-making patterning is a caring moment (Watson, 1985, 2008) of insight, a spiritual transformation. The quality of the magnetic appeal enables both the nurse and the patient to be "more alive" or "more authentic" than before. The process occurs simultaneously within the will to act or choose, and it is an emergent quality (Swinderman, 2005).

A RELATIONAL CARING EXEMPLAR ILLUSTRATING CHOICE PATTERNING

Facilitating choice making through the professional and authentic caring relationship can be clarified, in part, with complexity sciences. Chaos or unpredictability in "far-from-equilibrium" human and environmental systems can be viewed in human systems as disease, dis-ease, dilemmas, crises, or need. For example, at the critical or bifurcation point in disease, the patient is the specific agent of change to effect a healthier state of self-organization or well-being and participates with the nurse in the choice to change. As a result of the caring relationship, the life pattern of disease or dis-ease is transformed through choice or a series of choices toward a new pattern of "*relational self-organization*" (Ray, 1994; Turkel & Ray, 2000, 2001). Both the nurse and the patient are transformed in the process by the reasoned moral choice toward a transformative spirituality and harmony of body, mind, and spirit (Ray, 1994, 1997; Watson, 2008).

The following is an exemplar of choice patterning. The exemplar illuminates the caring teams' participation in a pattern change, and shows how the patterns can be the foci of nursing and multidisciplinary caring inquiry.

A PERSON WITH SEVERE CHRONIC PAIN*

A person with severe chronic pain, after years of "doctor shopping," has taken medications with no lasting relief and is in physical, emotional, and spiritual chaos. The person arrives at a critical or bifurcation point where a choice must be made either to continue in the same life pattern leading to deterioration or death or to a new beginning in pain management and healing.

By a deep call "within, and by encouragement of significant others," the patient is motivated to seek pain rehabilitation from a community of caregivers. The pain center is the last hope. The person, deeply connected to the pain itself, is in a crisis of spirit brought upon by past abuse or loss of a relationship that is collectively physical, emotional, sexual, and spiritual.

The multidisciplinary team of caregivers offers different views and treatment modalities toward healing. The nurse caregivers, in particular, encourage faith in the self, instill hope, and, through loving compassion (even against the patient's own self-sabotage), motivate and facilitate choice toward a new future possibility at the bifurcation point or edge of chaos. This new pattern is one of participation in life and living: a renewal of the quality of a life. At the edge of chaos is hope, anticipation of the future possibility of controlling pain. This process of self-patterning, through the caring relationship of the nurses and the patient, results in a transformation, an emergent life pattern (new order), through the process of "relational caring self-organization."

---

*Exemplar presented by Julia Howell, MSN, ARNP (personal communication, December 1993).

UNIFYING MODEL OF COMPLEX CARING DYNAMICS FOR NURSING INQUIRY

The focus of nursing as caring in the mutual human–environment relationship suggests the need for a new model of inquiry that can deal with the complexity of nursing. Clearly, caring relationality is the ground of nursing, and knowledge-in-process, understanding, and choice within a relational reflective consciousness are critical components (Ray, 2010). As such, nursing inquiry is a reflective journey toward possibilities.

A unifying model of complex caring dynamics assumes that nursing is a holistic, complex, and dynamic discipline and a part of the unfolding of the complex phenomena of the universe (Marcel, 1949; Ray, 1981). Seeing complex life patterns (the neuro-biophysicial, sociocultural, ethical, personal, aesthetic, and spiritual) and mapping and recognizing those patterns that potentiate choices toward health, healing, well-being, or a peaceful death become the foci of nursing inquiry.

ORGANIZATION OF THE MODEL OF COMPLEX CARING DYNAMICS FOR NURSING INQUIRY

Figure 2.2 is a unifying model of complex caring dynamics. It represents an overarching metaparadigm that integrates the structural components of nursing inquiry and shows the complexity and dynamism of nursing. The model incorporates the views of the sciences of complexity, caring, and choice patterning (Davidson & Ray, 1991) including the Rogerian perspective of the mutuality of the human–environment integral relationship (Rogers, 1970). It is consistent with Newman's tenets of the unitary transformative paradigm for nursing (Newman et al., 2008), which identifies features of complexity sciences and nursing as caring in the human health experience. Moreover, it includes components of other paradigms identified in nursing—the simultaneity paradigm of human becoming, and, in part, the more technically oriented particularistic–deterministic paradigm (Fawcett, 1993; Newman, 1986; Newman et al., 1991). The model also reflects Habermas' comprehensive approach to inquiry, which integrates competing social paradigms, technical, practical, and emancipatory (Bernstein, 1983; Habermas, 1995; Sumner, 2008). The final element of the model illuminates the creative potentialities in pattern recognizing and integrative synthesis, which, according to the philosophy of Hegel (in Ray, 2006), forms the foundation to meet the challenge of paradox in all knowledge development (inquiry and research) for the generation of new ideas, new theses, potential antitheses, and new syntheses.

FIGURE 2.2 Complex Caring Dynamics

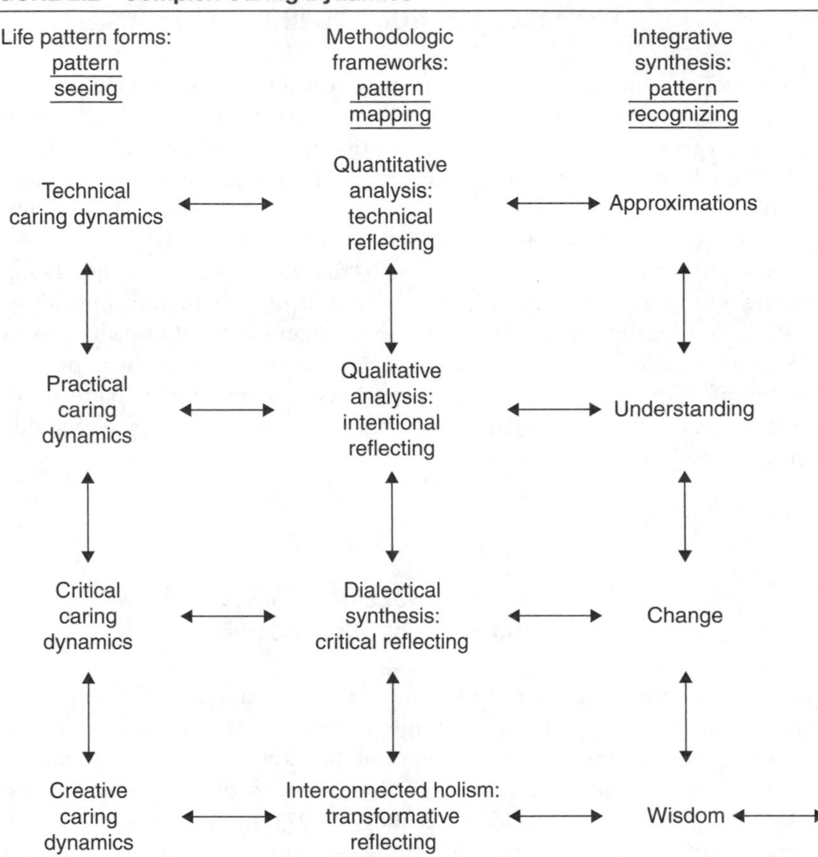

ASSUMPTIONS

In the complex caring dynamical model of inquiry, there are three fundamental assumptions: first, that caring is the essence of nursing; second, that the nurse researcher engages in a caring relationship when in the process of research, a "participatory we" (Marcel, cited in Speigelberg, 1982); or a "compassionate we" (Ray, 1991, p. 182) in the process of inquiry; and third that the caring consciousness of the researcher is central to the inquiry both at the levels of data generation and data analysis. The caring consciousness of the researcher offers significant challenges to the research process and reinforces Heisenberg's principle of uncertainty and the point in Husserlian philosophy and complexity sciences that the observer and the observed are one (Peat, 2002).

EXPLANATION OF THE MODEL

All components of caring inquiry (see Figure 2.2) are considered relational and reflect a process of engagement or mind–matter connectedness on the part of the researcher. All knowledge generated for research is considered "approximate, that is, all concepts, theories, and findings are limited and approximate" (Capra et al., 1991, p. xv) rather than being considered final, predictable, linear, or causal. The technical caring dynamics' component should be especially recognized as addressing approximate or limited knowledge because, historically, knowledge generated in traditional science through quantitative or technical means has been represented as "truth."

The model illuminates nursing and consciousness as mutually embedded in the same way that human beings and the world are mutually embedded—the integrality of the human–environment relationship. The model is holistic and organized to facilitate research of the complex patterns of nursing's relational caring. Although the model is presented linearly, the arrows indicate that it is dynamic. Each component of the model reflects the complex processes involved in researching life patterns and relational caring patterns in the human health experience. The model attends to the many research voices of nursing (Schultz, 1992).

LIFE PATTERN FORMS: PATTERN SEEING

Life pattern forms include all human–environment or human–field processes, both material and nonmaterial, that can be researched. The forms (structures of life processes) can be either ontological (ways of being), spiritual, religious, ethical, aesthetic, biophysical, or psycho-neuroimmunological and be concerned with nursing, health, healing, well-being, dying, or death. In addition, there are sociocultural forms: technical, historical, economic, legal, political, social, organizational, and educational. The forms obviously shift from abstractions to pattern manifestations (concrete life experiences) when they are considered for inquiry. They can be researched either separately or in conjunction with each other through the four ways of pattern seeing: technical, practical, critical, and creative. Each has its corresponding methodological framework and integration synthesis (see Figure 2.2).

Pattern seeing is a way of organizing or visualizing caring inquiry. Pattern seeing recognizes the movement of chaos (disorder and order) and change or transformation in nursing phenomena. It shows the critical bifurcation points at the edge of chaos and the movements to a new order in the caring relationship.

Within the four types of inquiry, the researcher seeks to generate data from different foci: through observation; interviewing; the use of instruments, questionnaires, or surveys; or contemplation. Data are recorded through handwritten notations, computerized digital devices, and audio/video tapes in preparation for data analysis. The technical, practical, critical, or creative caring dynamics are the lenses for generating data for pattern seeing.

1. **Technical caring dynamics**. The focus of the inquiry is primarily on the more mechanical, technical, or biophysiological aspects of life patterns. Quantitative analysis through technical reflective processes (including the role that intuition and interpretation play in the process of mechanical, mathematical, or technical research) is emphasized to identify and recognize patterns (findings). The patterns or findings in the research approach are designated as approximations rather than causal, predictable outcomes.
2. **Practical caring dynamics**. The focus of the inquiry is primarily on the meaning of patterns of actual, practical experience. Ethnographic, grounded theory, historic, phenomenological, and hermeneutic, or the caring inquiry approach (Ray, 1991) (singularly or in combination) are methods that can be used to describe and interpret diverse cultural life and self-patterns (Morse & Niehaus, 2009). Processes of intentional reflection (dwelling with data, intuition, and interpretation of past, present, and future possibilities) enhance understanding of choice patterns of life, self, and human–environment relationship in the movement of chaos (disorder and order), change, and transformation.
3. **Critical caring dynamics**. The focus of inquiry is primarily on the praxis of ethical and political life in the mutual human–environment process (Bernstein, 1971; Sumner, 2008). The meaning of "how ought we to live," or "how ought we to care," as we socially interact with culturally diverse persons or groups, is explored. The purpose is to ensure sociocultural change, emancipation, or liberation from oppressive conditions. Action, evaluative, or focus group research approaches are used with critical reflection (reflection focusing on issues that elicit ethical, transcultural, or feminist responses). In this way, new choices for change in human–environment relational patterning can be initiated.
4. **Creative caring dynamics**. The focus of inquiry is primarily on ideas, imagination, and experiential sociocultural life patterns. These can be of a philosophic and aesthetic nature (creative nursing, music, dance, visual art, theater), and spiritual, religious, or ethical nature or transcultural communicative spiritual–ethical caring (Ray, 1991, 2010). Creative caring dynamics, as a process of inquiry, facilitates wisdom. It encourages the use of traditional or creative research methods to integrate all interconnected

pattern-recognizing goals. The creative caring component of inquiry uses transformative reflecting and affirms the view that knowledge is always in process and continually being revealed. The mystery of the invisible, unfolding meaning of the universe and human experience is "sustained by Divine Imagining"—cosmic powers of space, time, and life, which are open to creation in the process of becoming (Harman, 1991, p. 76; Ray, 1997; Watson, 2005). Furthermore, human beings are anticipating, always in a state of hoping in relation to future possibilities for health, healing, and well-being (Rosen, 1988; Watson, 2005, 2008).

METHODOLOGICAL FRAMEWORKS: PATTERN MAPPING

Each component—technical, practical, critical, and creative caring dynamics—has a corresponding methodological framework for analytical pattern mapping. The methods corresponding to each dynamic view of the caring relationship are ways of capturing data by reflection (see Figure 2.2). Pattern mapping details plan out the strategies that a researcher uses to reveal the meaning of the caring relationship. Pattern mapping makes known the choice-making patterns used in the complex movement from chaos (disorder and order) to change and transformation.

Reflection is a process of thinking and apprehending human–environment integral patterning and responds to the question, "How do these data fit into the whole?" or "How is the whole intuited and interpreted by the data?" All methodological frameworks are subject to the reflective process, albeit differently with each particular caring dynamic. Reflection reinforces the active role that consciousness now plays in the process of inquiry.

PATTERN MAPPING PROCESSES

1. **Technical caring dynamics,** with its corresponding methodological framework of quantitative analysis and technical reflecting, requires special attention to the meaning of reflection and interpretation. Even with the use of quantitative tools, the observer cannot be separated from the observed, and interpretation enters into this form of inquiry. With this perspective, the researcher uses a qualitative understanding of the whole to obtain a richer view of a phenomenon with the use of different or new quantitative methods. Examples are the mixed quantitative and qualitative approaches to the study of complexity sciences, caring, and the science of unitary human beings in a technological human–environment relationship (Barrett, 1990a, 1990b; Davidson & Ray, 1991; the journals *Advances in Nursing Science, Visions, and Nursing Science*

Quarterly, which identify major research in the SUHB); chaos theory and nursing system research with multivariate time series (ARIMA) and Box-Jenkins models (Mark, 1994); the application of nonlinear dynamics in nursing research (Henderson, Hamilton, & Vicenzi, 1995), the role of environmental complexity in well-being using Shannon's information theory (Davidson, Teicher, & Bar-Yam, 1997), fractal thinking, and analysis (Briggs, 1992; Brown & Liebovitch, 2010; West & Deering, 1995); patterns, information, and chaos in life patterns and medicine (West, 2006); and spectral or cosigner analysis (Davidson, personal communication, December 1993).

2. **Practical caring dynamics** component has a corresponding methodological framework of qualitative analysis and intentional reflecting. This includes the use of qualitative methods of any type for pattern mapping (Morse & Niehaus, 2009). Examples are phenomenology, hermeneutics, or their combination; ethnography, grounded theory; story theory, the work of Smith and Liehr (2008); or other methods such as caring inquiry (Ray, 1991).
3. **Critical caring dynamics** component has a corresponding methodology of dialectical synthesis. It uses qualitative and quantitative analysis to comprise critical reflecting (Morse & Niehaus, 2009).
4. **Creative caring dynamics** component with its corresponding framework of interconnected holism uses transformative reflecting to map patterns (Ray, 1991).

INTEGRATIVE SYNTHESIS: PATTERN RECOGNIZING

Integrative synthesis or pattern recognizing relates to the goal of inquiry. Goals differ depending on which method is selected.

GOALS OF PATTERN RECOGNIZING

1. **Technical caring dynamics'** goal is patterning studied through different or select quantitative mappings. It seeks technical knowledge of patterns that is considered limited or approximate rather than seeking knowledge that is considered absolute or final.
2. **Practical caring dynamics'** goal is understanding. Understanding is achieved through recognizing caring in the human health experience and the descriptive and interpretative analysis of patterns of meaning. There is, however, no complete understanding of reality.
3. **Critical caring dynamics'** goal is change. Focusing research on praxis, the ethical–political life of interacting cultural groups in different contexts,

brings forth patterned knowledge of meaning and a commitment to facilitate liberation from situations that bind or destroy.

4. **Creative caring dynamics'** goal is wisdom. Creative and imaginative questions are addressed through this approach to research. Wisdom is the result of awareness or insight, which—through the use of any, all, or new research approaches—recognizes holistic and transformative choice patterning in the human–environment field processes. It is the fusion of complexity sciences, caring, aesthetics, and spirituality.

Transformative reflecting allows for the unfolding of a deep sense of personal, interpersonal, and spiritual knowing, a knowing wherein the spirit of the researcher and the research interest are nourished so that there is a true and profound attention to the "heart." By opening up to the totality of our being and to creative possibilities, the presence of the mystical can be felt in and through creative action. Faith, hope, and love are apprehended, and greater order or harmony of body, mind, and spirit and self-transcendence are anticipated (Ray, 1997, 2010; Reed, 2008; Watson, 2005, 2008).

COMPLEX CARING DYNAMICAL MODEL OF INQUIRY AND ITS IMPLICATIONS FOR NURSING

Key considerations of a Complex Caring Dynamical Model for Nursing Inquiry are summarized as follows:

1. The complex caring dynamical model of nursing inquiry was constructed as a unitive foundation for nursing because of nursing's complexity. The discipline of nursing is neither purely science nor purely art but is greater than a combination of the two. "David Bohm has proposed that the science and art in the future should move closer to art" (Briggs & Peat, 1989, p. 200). Complex caring dynamics points the way.
2. Complex caring dynamics illuminates the role of nurses/nursing in the human health experience by emphasizing the focus of nursing inquiry as caring in the human health experience within the unitive-transformative paradigm (Newman et al., 1991, 2008).
3. Complex caring dynamics identifies the caring relationship in the mutual human–environment process as the complexity in nursing science, education, research, and practice. This view offers a challenge to the concept of self-organization, which is a central thesis of complexity sciences. Scientists should entertain the view that all living systems in an interconnected universe do not self-organize at all but organize through the loving–caring relationship, both visible and invisible (Ray, 2010).

4. Complex caring dynamics calls for a reevaluation of and dialogue on nursing's fragmented philosophy (ontology, metaphysics, epistemology, ethics, and teleology).
5. Complex caring dynamics combines approaches to inquiry into a unified whole that responds to the many voices of nursing (Morse & Niehaus, 2009; Schultz, 1992).
6. Complex caring dynamics incorporates the different extant paradigms in nursing to facilitate an approach to inquiry that is in tandem with the holistic and complex nature of nursing (Ray, 2010; Watson, 2005, 2008).
7. Complex caring dynamics highlights the primordial nature of caring as the action of love, the magnetic appeal or pull of the universe (Ray, 1981, 1997; Watson, 2005, 2008).
8. Complex caring dynamics shows that the nature of complexity is always anticipating (Rosen, 1988), always hoping and is, therefore, the foundational process of creative evolution (Bergson, cited in Harman, 1991).
9. Complex caring dynamics shows that in accepting knowledge-in-process or emergence as the epistemological basis of reality, faith must be placed in both the esoteric phenomena of human consciousness, the deep intuitive inner knowing (interiority, Lonergan, cited in Webb, 1988), as well as the esoteric encountering of physical sense data in the mutual human–environment process.

CONCLUSION

In his book *Beyond the Quantum*, Talbot (1988) characterized the nature of truth as growing, changing, and transforming by stating, "The human race has reached a threshold of wisdom in which it can at long last abandon the lure of completeness and recognize that whatever form they take, there will always be new vistas to be discovered in science, and new worlds awaiting us beyond the quantum" (p. 224). In nursing, we hold the key to these new vistas because through caring for others, we hold "in the center of our hearts" the consciousness of the universe, love, and the magnetic appeal and pull of the universe (Ray, 1989, 1997, 2010; Watson, 2005, 2008).

REFERENCES

Arntz, W., Chasse, B., & Vicente, M. (2005). *What the bleep do we know*!? Deerfield Beach, FL: Health Communications.

Barrett, E. (1990a). *Visions of Rogers' science-based nursing* (pp. 31–44). New York: National League for Nursing (Publication No. 15-2285).

Barrett, E. (1990b). Rogers' science-based nursing practice. In E. Barrett (Ed.), *Visions of Rogers' science-based nursing* (pp. 31–44). New York: National League for Nursing (Publication No. 15-2285).

Bar-Yam, Y. (2004). *Making things work: Solving complex problems in a complex world*. Boston: Knowledge Press.

Benner, P., & Wrubel, J. (1989). *The primacy of caring*. Menlo Park, CA: Addison-Wesley.

Bernstein, R. (1971). *Praxis and action*. Philadelphia: University of Pennsylvania Press.

Bernstein, R. (1983). *Beyond objectivism and relativism: Science, hermeneutics and praxis*. Philadelphia: University of Pennsylvania Press.

Boykin, A., & Schoenhofer, S. (2001). *Nurse as caring: A model for transforming practice*. Sudbury, MA: Jones & Bartlett.

Briggs, J., & Peat, F. (1984). *Looking glass universe: The emerging science of wholeness*. New York: Simon & Schuster.

Briggs, J., & Peat, F. (1989). *Turbulent mirror*. New York: Harper & Row.

Briggs, J., & Peat, F. (1999). *Seven life lessons of chaos*. New York: HarperCollins.

Brown, C., & Liebovitch, L. (2010). *Fractal analysis*. Los Angeles: Sage.

Butcher, H. (2002). Living in the heart of helicy: An inquiry into the meaning of compassion and unpredictability within Rogers' nursing science. *Vision, 10*(1), 6–23.

Cannato, J. (2010). *Fields of compassion: How the new cosmology is transforming spiritual life*. Notre Dame, IN: Sorin Books.

Capra, F., Steindl-Rast, D., & Matus, L. (1991). *Belonging to the universe*. San Francisco: Harper.

Carper, B. (1978). Fundamental patterns of knowing in nursing. *Advances in Nursing Science, 1*(1), 13–23.

Davidson, A. (1988). Choice patterns: A theory of the human-environment relationship. (Doctoral dissertation, University of Colorado, 1988). *Dissertation Abstracts International*.

Davidson, A., & Ray, M. (1991). Studying the human-environment phenomenon using the science of complexity. *Advances in Nursing Science, 14*(2), 73–87.

Davidson, A., Teicher, M., & Bar-Yam, Y. (1997). The role of environmental complexity in the well-being of the elderly. *Complexity and chaos in nursing, 3*(1), 5–12.

Fawcett, J. (1993). From a plethora of paradigms to parsimony in world views. *Nursing Science Quarterly, 6*(2), 56–58.

Harman, W. (1991). *A re-examination of the metaphysical foundation of modern science*. San Francisco: The Institute of Noetic Sciences.

Henderson, J., Hamilton, P., & Viccnzi, A. (1995). Chaos theory in nursing publications: Retrospective and prospective views. *Complexity and chaos in nursing, 2*(1), 36–40. (Journal out of print and out of circulation).

Leininger, M. (Ed.). (1981). *Caring: An essential human need*. Thorofare, NJ: Charles B. Slack.

Leininger, M. (Ed.). (1984). *Care: The essence of nursing and health*. Thorofare, NJ: Charles B. Slack.

Leininger, M. (Ed). (1991). *Culture care diversity and universality: A theory of nursing*. New York: National League for Nursing Press.

Leininger, M., & McFarland, M. (Eds.). (2006). *Culture care diversity and universality: A worldwide nursing theory* (2nd ed.). Sudbury, MA: Jones & Bartlett.

Levinas, A. (1981). *Otherwise than being or beyond essence* (A. Lingis, Trans.). Boston: Martinus Nijhoff.

Lewin, R. (1992). *Life at the edge of chaos*. New York: MacMillan.

Lindberg, C., Nash, S., & Lindberg, C. (2008). *On the edge: Nursing in the age of complexity*. Bordentown, NJ: Plexus Press.

Marcel, G. (1949). *The philosophy of existence*. New York: Philosophical Library.

Mark, B. (1994). Chaos theory and nursing systems research. *Theoretic and Applied Chaos in Nursing, 1*(1), 7–14. (Journal out of print and out of circulation).

Mitchell, M. (2009). *Complexity: A guided tour*. New York: Oxford University Press.

Morse, J., Bottorff, J., Neander, W., & Solberg, S. (1991). Comparative analysis of conceptualizations and theories of caring. *Image: Journal of Nursing Scholarship,* 23(2), 119–126.

Morse, J., & Niehaus, L. (2009). *Mixed method design: Principles and procedures*. Walnut Creek, CA: Left Coast Press.

Newman, M. (1986). *Health as expanding consciousness*. St. Louis: MO.

Newman, M., Sime, M., & Corcoran-Perry, S. (1991). The focus of the discipline of nursing. *Advances in Nursing Science, 14*(1), 1–6.

Newman, M., Smith, M., Dexheimer-Pharris M., & Jones, D. (2008). The focus of the discipline revisited. *Advances in Nursing Science, 31*(1), E16–E27.

Nicolis, G., & Prigogine, I. (1989). *Exploring complexity*. New York: Freeman.

Peat, F. (2002). *From certainty to uncertainty: The story of science and ideas in the twentieth century*. Washington, DC: Joseph Henry Press.

Perkins, J. (2010). Convergence of caring and the Science of Unitary Human Beings (Manuscript in process) (Personal communication via email, May 3, 2010).

Polanyi, M. (1958). *Personal knowledge*. Chicago: The University of Chicago Press.

Prigogine, I., & Stenger, I. (1984). *Order out of chaos*. Toronto, ON: Bantam Books.

Ray, M. (1981). A philosophical analysis of caring within nursing. In M. Leininger (Ed.), *Caring: An essential human need* (pp. 25–36). Thorofare, NJ: Charles B. Slack.

Ray, M. (1984). The development of a nursing classification system of institutional caring. In M. Leininger (Ed. & Author), *Care: The essence of nursing and health* (pp. 95–112). Thorofare, NJ: Charles B. Slack, Inc.

Ray, M. (1987). Technological caring: A new model in Critical Care. *Dimensions of Critical Care Nursing, 6*, 166–173.

Ray, M. (1991). Caring inquiry: The aesthetic process in the way of compassion. In D. Gaut, & M. Leininger (Eds.), *Caring: The compassionate healer* (pp. 181–189). New York: National League for Nursing Press.

Ray, M. (1994). Complex caring dynamics: A unifying model for nursing inquiry. *Theoretic and applied chaos in nursing, 1*(1), 23–31. (Journal out of print and out of circulation).

Ray, M. (1997). Illuminating the meaning of caring: Unfolding the sacred art of divine love. In M. Roach (Ed.), *Caring from the heart: The convergence of caring and spirituality*. New York: Paulist Press.

Ray, M. (1998). Complexity and nursing science. *Nursing Science Quarterly, 11*(3), 91–93.

Ray, M. (2006). The theory of bureaucratic caring. In M. Parker (Ed.), *Nursing theories & nursing practice* (2nd ed., pp. 360–368). Philadelphia: F. A. Davis.

Ray, M. (2007). Technological caring as a dynamic of complexity in nursing practice. In A. Barnard & R. Locsin (Eds.), *Technology and nursing: Practice, concepts and issues* (pp. 174–190). United Kingdom: Palgrave Macmillan.

Ray, M. (2010). *Transcultural caring dynamics in nursing and health care*. Philadelphia: F. A. Davis.

Reed, P. (2008). The theory of self-transcendence. In M. Smith & P. Liehr (Eds.), *Middle range theory for nursing* (2nd ed., pp. 163–200). New York: Springer.

Roach, M. (2002). *Caring, The human mode of caring: A blueprint for the health professions* (2nd Rev. ed.). Ottawa, ON: Canadian Hospital Press.

Rogers, M. (1970). *An introduction to the theoretical basis of nursing*. Philadelphia: F. A. Davis.

Roger, M. (1990). Nursing science of unitary, irreducible, human beings: An Update 1990. In E. Barrett (Ed.), *Visions of Rogers' science-based nursing* (pp. 5–11). New York: National League for Nursing Press (Publication No. 15-2285).

Rosen, R. (1988). The epistemology of complexity. In J. Kelso, A. Mandrell, & M. Shlesinger (Eds.), *Dynamic patterns in complexity*. Singapore: World Scientific.

Schultz, P. (1992). Attending to many views: Beyond the qualitative-quantitative debate. *Communicating Nursing Research*. Boulder, CO: Western Institute of Nursing.

Smith, M. (1999). Caring and the science of unitary human beings. *Advances in Nursing Science, 21*(4), 14–28.

Speigelberg, H. (1982). *The phenomenological movement* (3rd Rev. ed.). The Hague, Netherlands: Martinus Nijhoff.

Sumner, J. (2008). *The moral construct of caring in nursing as communicative action*. Saarbrücken, Germany: VDM Verlag Dr. Müller.

Swinderman, T. (2005). *The magnetic appeal of nurse informaticians: Caring attractor for emergence*. Doctor of Nursing Science Dissertation, Florida Atlantic University, College of Nursing, Boca Raton, Florida.

Talbot, M. (1988). *Beyond the quantum*. Toronto, ON: Bantam Books.

Turkel, M. (2006a). Applications of Marilyn Ray's theory of bureaucratic caring. In M. Parker (Ed.), *Nursing theories & nursing practice* (2nd ed., pp. 369–379). Philadelphia: F. A. Davis.

Turkel, M. (2006b). What is evidence-based practice? In S. Beyea & M. Slattery (Eds.), *Evidence-based practice in nursing: A guide to successful implementation*. Marblehead, MA: HcPro Publishing.

Turkel, M. (2007). Dr. Marilyn Ray's theory of bureaucratic caring. *International Journal for Human Caring, 11,* 57–74.

Turkel, M., & Ray, M. (2000). Relational caring complexity: A theory of the nurse-patient relationship within an economic context. *Nursing Science Quarterly,* 13, 307–313.

Turkel, M., & Ray, M. (2001). Relational complexity: From grounded theory to instrument development and theoretical testing. *Nursing Science Quarterly,* 14, 281–287.

Turkel, M., & Ray, M. (2009a). Caring for "not-so-picture-perfect patients." In R. Locsin & M. Purnell (Eds.), *A contemporary nursing process: The (un)bearable weight of knowing in nursing*. New York: Springer.

Turkel, M., & Ray, M. (2009b). Caring for "not-so-picture-perfect patients": Ethical caring in the moral community. In R. Locsin, & M. Purnell (Eds.), *A contemporary nursing process: The (un)bearable weight of knowing in nursing* (pp. 225–249). New York: Springer.

Vicenzi, A. (1994), Chaos theory and some nursing considerations. *Nursing Science Quarterly, 7,* 36–42.

Vicenzi, A., White, K., & Begun, J. (1997). Chaos in nursing: Make it work for you. *American Journal of Nursing, 10,* 26–31.

Waldrop, M. (1992). *The emerging science at the edge of order and chaos*. New York: Simon & Schuster.

Watson, J. (1985). *Nursing: Human science, human care*. Norwalk, CT: Appleton-Century-Crofts.

Watson, J. (2005). *Caring science as sacred science*. Philadelphia: F. A. Davis.

Watson, J. (2008). *Nursing: The philosophy and science of caring* (2nd Rev. ed.). Boulder, CO: University Press of Colorado.

Watson, J. (2009). *Assessing and measuring caring in nursing and health sciences*. New York: Springer.

Watson, J., & Smith, M. (2002). Caring science and the science of unitary human beings: A trans-theoretical discourse for nursing knowledge development. *Journal of Advanced Nursing, 37*(5), 452–461.

Webb, E. (1988). *Philosophers of consciousness*. Seattle, WA: University of Washington Press.

West, B. (2006). *Where medicine went wrong: Rediscovering the path to complexity*. New Jersey, NJ: World Scientific.

West, B., & Deering, B. (1995). *Fractal thinking: The lure of modern science*. Singapore: World Scientific.

Wheatley, M. (1992). *Leadership and the new science*. San Francisco: Berrett-Koehler.

Wheatley, M. (2006). *Leadership and the new science: Discovering order in a chaotic world* (3rd ed.). San Francisco: Berrett-Koehler.

Response to Chapter 2

REFLECTIONS FROM A MIXED-METHOD DESIGN PERSPECTIVE

Janice M. Morse

EXPANDING THE CONSCIOUSNESS OF CARING

Reflections on Ray's Complex Caring Dynamics

I am wrestling with the title of my reflection: Should it be "The Expanding Consciousness of Caring" or as it is now written: "Expanding the Consciousness of Caring"? Marilyn Ray first addressed these issues in 1994 and revisits them in this chapter. In this reflection, I first address "the expanding consciousness of caring" issue, and second, "expanding the consciousness of caring."

THE EXPANDING CONSCIOUSNESS OF CARING

Ray's (1994) philosophy of caring is that caring is "complex choice making" occurring on the edge of chaos. Using complexity theory, Ray argues that the shared concepts in complexity science and caring forge a linkage of relevancy between the two. Thus, complexity science provides a perspective that enhances our ability to understand holistic caring in complex nursing relationships.

In nursing, Ray argues that the caring relationship is "attractive" and "pulls the [caring] system away from disorder." The patient and the nurse *choose* to move from the chaos of disease, in communal commitment to the other. Caring is a "sense of spiritual–ethical knowing," of "love and co-presence," and of "felt realness" in the relationship. Through love and trust, by "anticipating, working through knowledgeable caring-presence, (and) using technical competency, the nurse evokes, detects or facilitates reasoned moral choice" (Ray, 1994). Thus, in Complex Caring Dynamics theory, caring "co-creates opportunities for choice making" moving the relationship toward healing, health, well-being, or death and dying (Ray, 1994).

Ray argues that chaos theory enables the identification of patterning of the caring relationship, thereby bringing order to a very complex interaction. The inseparable link between the nurse and the patient, the mutual entering of the caring relationship, and the mutual choice for outcomes make the theory extremely pertinent to the world of praxis. But we need more: Is it the complexity that provides us with a choice to enter the relationship? Or is the nurse the agent of the choice? Are the "patterns of caring" initiated by the client, nurse to client, or mutually by both?

Additional questions need to be considered: Is caring *always* a mutual relationship? Or can caring be one sided? Certainly, if the nurse cared and the patient rejected his or her caring, "complex dynamics" would be in place. However, would Ray call such a situation "caring"?

There is no doubt that the theory of complex caring dynamics expands our consciousness of caring, of the relationship, and of the outcomes, the ethics of our practice. Although the caring relationships are patterned, the three patterns have not been identified or described in this chapter. One exemplar describes a mutual caring relationship. Is the discussion of only a mutual caring relationship a restriction or limitation of the theory or an opportunity that gives nursing practice space to develop?

EXPANDING THE CONSCIOUSNESS OF CARING

If we are to expand the consciousness of caring, Ray's model provides a comprehensive approach to exploring the "life patterned forms" of complex caring dynamics. Ray's (1994) model of "pattern seeing" (technical, practical, critical, or creative caring dynamics, matched with methodological frameworks (quantitative, qualitative, dialectical synthesis, or transformative reflection), and pattern recognition (approximation, understanding, change, and wisdom) *as a set*, provides a wonderful holistic approach to understanding the essence of nursing. Will investigating the model *in parts* violate its basic tenet of wholeness? Do we need to investigate Ray's Complex Caring Dynamics further, especially if this investigation fractionates the holistic model?

New research techniques, such as mixed-method designs (Morse & Neihaus, 2009), a multiple-method research program, or a metasynthesis of the work of many scholars, in part overcome the problem of fractionation and would meet the nine implications that Ray identifies, illuminating complex caring processes and outcomes in nursing. Thus, Ray has guided us toward *learning about caring from the perspective of complexity science*. Of course, such investigation does not change the nature of caring itself, only our understanding of it. It helps us to appreciate the power of caring, to use it as a treatment as well as an instrument for the provision of our nursing processes, and to communicate the affective component of care. It expands our own perspective on caring, moving it from a delineated task or a short interaction to a significant relationship—one that is lifesaving and key to the health care system.

REFERENCES

Morse, J., & Niehaus, L. (2009). *Principles and procedures of mixed-method design.* Walnut Creek, CA: Left Coast Press.

Ray, M. (1994). Complex caring dynamics: A unifying model of nursing inquiry. *Theoretic and Applied Chaos in Nursing, 1*(1), 23–32.

Response to Chapter 2

Complexity Science and Advancing the Discipline of Nursing

REFLECTIONS ON THE ONTOLOGY AND EPISTEMOLOGY OF COMPLEXITY SCIENCE AND NURSING SCIENCE

Pamela Reed

Complexity science is about change. Nursing is about change—understanding and facilitating the potentials to facilitate healing and well-being in the person–environment system. Complexity theory is appealing to nursing because it is a science perspective that logically supports the idea of human potential to make order out of disorder, to make meaning out of what at first seems meaningless, and to find strength in weakness. Today, what is regarded as only random error variance and disorder within linear models may be found to have purpose or explanation by nonlinear dynamics analyses of complexity science.

Kauffman (1995) suggested that complexity science may uncover a principle even more basic than Darwin's theory of selection for explaining the source of order and change in living systems, referring to self-organization. *Self-organization* refers to the capacity of a system to produce pattern and organization by taking in higher-order forms of energy from the environment and dissipating lower-order forms of energy into the local environment (Prigogine & Stengers, 1984).

Complexity science, as a conceptual framework and set of statistical methods, can help advance nursing as a discipline and professional practice. Complexity science offers ways to conceptualize and study change in complex phenomena across many disciplines—from physics and aerospace engineering to physics and from biology to psychology. Complex phenomena span units of analysis from the micro to macro; they may be embodied or embedded in an environment and undergo dramatic or gradual change. Although its roots date back a century, complexity science is a paradigm for developing 21st century ideas.

As revealed in Rogers' revolutionary 1970 publication on the foundations of nursing, the roots of nursing are congruent with what is now called complexity science. The terminology that has been used to describe complexity has varied over time, beginning with von Bertalanffy's (1968) *open systems theory*, Prigogine's chemical theory of *dissipative structures,* followed by the popularized term *chaos theory* (Prigogine & Stengers, 1984), then *complex adaptive systems* (CAS) (Kauffman, 1995), to the recent terms *complexity science* and *nonlinear dynamical systems* (NDS) theory (Guastello, Koopmans, & Pincus, 2009).

Complexity has become a central feature in defining our disciplinary perspective of human potential and change. Past worldviews depicted human change from a closed system and mechanistic perspective with the person holding a reactive stance against change and increasing complexity. Martha Rogers and the ontologic frameworks that she inspired (e.g., Newman's unitary–transformative and Parse's simultaneity paradigms and Ray's complex caring dynamics) depict complexity and change from an open system and integrative perspective with the person holding a transformative stance that *engages* the ongoing complexity and change.

Complexity science assists nurses to see human beings as systems not symptoms, and to appreciate rather than underestimate or misinterpret complexity. Bachman (2010) identifies four challenges posed by four modes of complexity in building design, which can be related to the work of nurses. Designing buildings is not unlike designing nursing care; both require creatively constructing environments that meet pragmatic and aesthetic needs and support well-being. Bachman's four modes have implications for designing nursing care that builds on the strengths and resources inherent in and among complex human systems.

- Mode 1 is *wicked complexity*, which poses problems that have no easy, single, or final solution. If designers fail to account for the indeterminacy of complex systems and limitations in human knowing, they are likely to limit their effectiveness by either simplifying the complexity or interpreting it as an irresolvable holy mystery.
- Mode 2 is *messy complexity*, which requires the designer to acknowledge the "holistic, inclusive, and authentic" nature of a given problem in the system (Bachman, 2010, p. 27). If this is not done, the designer may be inclined to impose a solution that is too biased and limited and fails to engage the richness and spontaneity that reside in the system.
- Mode 3 is *ordered complexity* that requires setting goals that address the essence of the system and that are relevant to and benefit the system. Failure to account for this can lead to unproductive planning and goals that "miss the target" (Bachman, 2010, p. 27).
- Mode 4 is *natural complexity* whereby the designer must work with the processes of nature, the person's flows and rhythms of energy, and self-organizing capacity. Attention to this provides for processes that have sustainability and beauty.

Embracing complexity—rather than controlling or simplifying it—is a frequent challenge for nurses in practice and research. It involves what Bachman

(2010) described as a "responsive inquiry" that is holistic and inclusive, and that acknowledges the essence while attending to the pragmatics of a problem (p. 33). The nature of nursing is such that nurses frequently encounter people in situations on the edge of chaos. Ray's chapter on complex caring dynamics alerts us to two areas in nursing—ontology and epistemology—that may be expanded by applications from complexity science. An ontology and an epistemology that embraces complexity enable nurses to design approaches that help patients transform chaos into opportunities for growth and well-being.

NURSING ONTOLOGY AND COMPLEXITY

Complexity science calls for new thinking about the ontology (or substance) of our discipline. Over a decade ago, I proposed a definition of nursing based on a synthesis of Rogerian (Rogers, 1970) and lifespan developmentalist (Werner, 1957) perspectives of human complexity and change. That is, that nursing is most fundamentally a process of well-being generated by changes in complexity and organization of the person–environment system (Reed, 1997). *Complexity* in this case refers to the number of parts and quantitative change in a system. *Integration* refers to a level of organization of the parts and qualitative change in a system. Illness can increase complexity by increasing the number of events or losses in a person's life. Meaning making and pattern recognition and other indicators of physical or emotional healing are examples of integration. Nurses help people integrate complexity in ways that promote well-being.

Nursing processes are inherent in living systems but, for various reasons, may require additional assistance from nurses and others to facilitate integration given health experiences or life choices that alter complexity in some way. Manifestations of nursing processes include interactions at all levels of analysis relevant to human systems, including the cellular, individual, cultural, and organizational although most nurses deal with nursing phenomena on the individual level.

Caring is one manifestation of a nursing process of well-being where the interactions help people organize complexity—whether by bringing a sense of order to their environment, meaning to a distressing event, or otherwise a focused energy and intent that supports the person through an irreversible change process. This idea was expressed almost two decades ago when Davidson and Ray (1991) wrote "Caring is the coherent radiation of nursing and the energy of nursing research" (p 83).

A NEW EPISTEMOLOGY AND PHILOSOPHY OF NURSING SCIENCE

Complexity science also calls for a new epistemology in nursing. It is an epistemology where there is keen awareness of the connectedness of events, open-ended explanations about underlying influences, and the nonlinear patterns of how things and people are related. Complexity in nursing phenomena necessitates a new process of knowledge development where practice and research are no longer arranged linearly; that is, where research occurs outside of practice with the findings handed down to the practicing nurses to translate and apply (Reed, 2006).

It is a new epistemology based on a new philosophical perspective of realism that is neither purely social (postmodern) nor purely natural (modern) in its scientific view of reality. Instead, the perspective of realism is primarily critical (between modernisms) in its view of a reality that has some ultimate status but where there is also always a degree of the unknown, mystery, and unpredictability, because scientific theories and instruments lag behind in their constructions and detections of reality.

It seems most prudent then, given a context of complexity, to practice nursing science by employing many ways of knowing, a broad definition of empirical, and with a patient-centered approach to knowledge development. The complexity framework necessitates a practice turn in epistemology, that is, a connection between theory, research, and practice where there is little if any slippage in the knowledge developed, used, and evaluated in practice.

NURSING'S NEAR FUTURE AND COMPLEXITY

An important next step in building knowledge is to move beyond using metaphor and analogy in applying complexity science ideas to nursing. It is time to move from these conceptual innovations to methodological applications of obtaining and analyzing data. Vicenzi, White, and Begun (1997) noted that the evolution of new knowledge, such as that regarding complexity science applications, moves from analogy and metaphor to empirical study. Similarly, editors Gaustello, Koopmans, and Pincus (2009) explained that the initial thrill of the early days of nonlinear dynamical systems has passed, and scientists across many disciplines are now focusing on the less glamorous but nonetheless essential work of theory development and empirical research into complexity.

NDS theory offers mathematical bases on nonlinear mathematical functions for statistical methods and techniques and analytic strategies to gain better understanding of complexity in nursing phenomena. These techniques will enable us to more accurately account for what we have already richly described metaphorically and qualitatively as nonlinear events. Older statistical approaches have allowed us to study change from a linear perspective; the theory of nonlinear dynamical systems of complexity science can enlarge our capacity to understand and study change in a way that brings the thrill of complexity science into nursing.

REFERENCES

Bachman, L. R. (2010). Embracing complexity in building design. In K. Alexiou, J. Johnson, & T. Zamenopoulos (Eds.), *Embracing complexity in design* (pp. 19–36). New York: Routledge.

Davidson, A. W., & Ray, M. A. (1991). Studying the human-environment phenomenon using the science of complexity. *Advances in Nursing Science, 14*(2), 73–87.

Guastello, S. J., Koopmans, M., & Pincus, D. (Eds.). (2009). *Chaos and complexity in psychology: The theory of nonlinear dynamical systems*. New York: Cambridge University Press.

Kauffman, S. A. (1995). *At home in the universe: The search for laws of self-organization and complexity*. New York: Oxford University Press.

Prigogine, I., & Stengers, I. (1984). *Order out of chaos: Man's new dialogue with matter*. New York: Bantam Books.

Reed, P. G. (1997). Nursing: The ontology of the discipline. *Nursing Science Quarterly, 10*(2), 76–79.

Reed, P. G. (2006). The practice turn in nursing epistemology. *Nursing Science Quarterly, 19*(1), 51–60.

Rogers, M. E. (1970). *An introduction to the theoretical basis of nursing*. Philadelphia: F. A. Davis.

Vicenzi, A. E., White, K. R., & Begun, J. W. (1997). Chaos in nursing: Make it work for you. *American Journal of Nursing,* 97(10), 26–32.

von Bertalanffy, L. (1968). *General systems theory*. New York: Wiley.

Werner, H. (1957). The concept of development from a comparative and organismic point of view. In D. B. Harris (Ed.), *The concept of development* (pp. 125–148). Minneapolis, MN: University of Minnesota Press.

CHAPTER 2
COMPLEX CARING DYNAMICS: A UNIFYING MODEL OF NURSING INQUIRY

- *Why is caring inquiry so important to the study and practice of nursing? What concepts come to your mind that illuminate the nature of caring?*
- *What research question related to your area of practice could you examine using this approach to methodology?*
- *What is the value of a mixed method design to your area of practice?*
- *Are the current outcomes' measures being used in your work environment congruent with the ideas discussed by Dr. Ray?*
- *In the response, Dr. Morse writes about caring. Are all individuals capable of receiving expressions of caring?*
- *Dr. Reed addresses the ontology (ways of being) and epistemology (ways of knowing) of complexity science and nursing science. How do you see the integration of these ways of knowing in nursing education, administration or nursing practice?*

THREE

Theoretical Issues and Methods for Increasing Understanding of Complex Health Care Systems

Alice W. Davidson
Stefan Topolski

How do we define or understand complexity? How do you study the complex phenomena of interest to health care? What methods developed to study complex systems are useful for the study of the mutual human–environment process? Each human being manifests patterns that are unique, and these patterns are linked to patterns in the environment. Complexity science provides new ways to examine the human–environment process including using different perspectives (scale), examining relationships in processes (pushing on a part changes something else), and discovering how these relationships give rise to collective behaviors (parts acting together). Constructing computer models of available parameters (micro) when combined with other perspectives to see meaning (meso) may provide valuable insights for health care (macro). For example, we could begin with a patient's lab values (micro); combine that data with what we have learned about the patient's lifestyle from our interactions with the patient and his/her family, which provides us information about the patient (meso); the nurse adjusts the care plan in order to increase the likelihood that the patient and family will be able to follow the plan of care on discharge and move toward the desired health outcomes (macro). Dr. West's response highlights the unpredictability in human conditions (complex systems), and how what we expect to happen may not happen at all once the patient is discharged.

INTRODUCTION

How Do We Understand Complexity?

Complex phenomena contain both regular and random behavior, with the highest complexity existing between randomness and order. Complexity increases with the number and variety of elements in a system and as the links or interactions between them increase. Small changes may result in very different patterns, and there is limited predictability. The familiar notion of cause and effect is not applicable. Dynamically unstable microscopic elements respond to the environment, yielding macroscopic phenomena with chaotic behavior. Processes repeat in a circular manner that utilizes self-reference. This feedback can either amplify or suppress disturbances depending on the system's past behavior (West, 2006). Complex systems can be characterized by their structure in space, the time it takes to respond to its environment, and how it self-organizes. In self-organization simple interacting units spontaneously emerge as nonequilibrium structural reorganizations on a macroscopic level. The scale at which the system is examined (the resolution) can make it appear very different. Something may appear very complex at high resolution but from afar appear smooth and simple. Likewise, a common feature of complex systems is their scale-free quality, where similar patterns reappear as one changes from high resolution to low resolution and back again. A significant source of complexity in biology, social systems, and health care (particularly nursing care) comes from these relationships and interactions on several scales. Complexity science principles and tools can help to provide insight and explain puzzling events or unexpected outcomes through examination of how simple rules or core values at a small scale can lead to observed system patterns and desired improvements at the big picture, macro scale.

Health care workers need better methods to study the phenomena upon which they focus—human–environment well-being. Nurses are trained to use the scientific method in assessing situations, planning action, and evaluating outcomes. But trying to fit the complex phenomenon that nursing studies, the mutual human–environment process, into a simple linear model of cause and effect (determinism) does not work well. The true joy and magic of nursing comes in those brief moments of emotional connection between nurse and a human–environment process. These moments can have huge impacts for good but are nearly impossible to predict in either timing or effect—because they are almost always dominated by uncertainty. It is presently impossible to analyze and control innumerable interrelated variables and separate them from others in the time available for a healing interaction. Such variables become important because complex phenomena contain nonlinear interactions resulting in both regular and random behavior that constantly evolve and unfold over time (Arthur, 1999). As we collect more data in these complex systems, the mean keeps changing, forcing us to conclude that the population mean may not exist and more data may not improve one's knowledge of a phenomenon (West & Griffin, 2004). Simpler but less useful phenomena that are linear may well fit our older methods and lead to a bias to study these easier problems. Complex phenomena, with their heavy-tailed distributions in frequency and/or probability density, which we do need to study, have been ignored or squeezed into an older linear worldview of symmetrically tailed bell curves, even though the results are questionable. Disturbed by this mismatch of complex data to older available research methods, many nurse researchers advocated abandoning quantitative methods for the qualitative methods of ethnography and hermeneutics, which respect the uniqueness of the individual.

Kelso (2005) and others suggest, however, that we do not need to throw out determinism because we embrace uncertainty. Rather, a science that "values the extremes but also the in-between" is needed (Kelso, 2005, p. 81). Between the idealized states of full cooperation and total independence of the component parts is a metastable intermediate regime where the difference between parts enables them to do their own thing, while the strength of coupling leads to a tendency to cooperate and create information. A newer understanding of this macroscopic collective behavior benefits from older microscopic knowledge of the forces of classical Newtonian mechanics and the probabilistic behavior of particles and waves in quantum mechanics. Of course, there is no one way to solve all complex problems, but there are methods of thinking about complex systems that can be very useful. Complexity science, such as it is, gives a theoretical foundation to support the wisdom of clinicians and managers successfully working in complex qualitative fields such as sociology, behavioral health, human resources, medicine, and other fields. These methods are useful for studying the mutual human–environment process with the goal of avoiding poorer results and creating positive outcomes instead. Still there can be pitfalls in using metaphors originating in other disciplines. It is important to remember that disciplines often rely upon metaphors to convey understanding of something they do not literally denote in order to suggest a similarity. Metaphors become the way we talk and think. Their pitfall comes when they become so entrenched in a scientific culture as to lose a rigorous meaning that can be critically tested by experiment. If complexity metaphors cannot be experimentally tested and falsifiable, they remain useful but have drifted from science toward philosophy.

MICRO, MESO, MACRO (OR SMALLEST, BIGGER, BIGGEST) LEVEL DESCRIPTIONS OF COMPLEXITY

Many scientists studying complex systems find it fruitful to examine the phenomena they study at various scales within the broader categories of micro, meso, and macro. In so doing, these scientists find new information that applies across disciplines and offers insights and new methodologies to study phenomena. Biologists note that chemical processes at the microscopic level are unstable and small changes in suitable parameters lead to large differences at the macroscopic level. Macroscopically, a system may appear to be in equilibrium, but that may not be the case. Even when the macrosystem is in equilibrium, the microscopic system is undergoing random fluctuations whose magnitude depends on the thermodynamic temperature. At the microscopic level, new and amazing adaptive abilities become possible with nanotechnology. Technology using the very small, nano level may soon lead to amazing applications including visualization within the human body.

The ultimate objective is to gain new information and understanding by linking the enabling technologies across all scales—nano, micro, meso, and macro. In the physical science, scales may be defined as a linear progression from the smallest nano (nm) to micro (1 mm), meso, macro (1 m), and larger. In nursing, medicine, sociology, and most fields of study, there is also a progression in size of observed events and relationships (see Table 3.1 for examples).

Conflict resolution at the smallest level involves teaching skills (conflict resolution training and nonviolent communication). Meso, at the intermediate scale, is community-wide intervention to support family structures, churches, and religious institutions, schools, media, or political systems. Macro, at the largest scale, is international and intercultural organization that makes decisions regarding war, environmental degradation, and global racism. Hegyvary (1991) applied these scales to nursing, viewing micro as individual responses to particular interventions (e.g., drug therapies), meso as the impact of interventions (clinical and organizational variables) directed to

TABLE 3.1 Examples of Micro, Meso, Macro Scaling in Health Care

SCALE	INFORMATION	HEARING LOSS	COMPUTATION	EVOLUTIONARY PROCESS	TIME	CATASTROPHE
Micro	Lab values	Left ear 50 dB hearing loss at 2000 Hz	Relation to norms	Change in values over time	Action	Catastrophic change in cellular protein synthesis
Meso	Relation to body signs	"You're mumbling" misunderstanding	Significance and relevance	Meaning	Learning	Mass extinction of species
Macro	Self perception Resilience Humor	Marital strife	Models Real-world phenomenon	Changes seen in outcomes	Evolution	Breakdown of ecological system and human society

a specific client group, and macro as outcomes focused on health of populations as related to the health care system and the environment (unemployment and poverty). There is again the size distinction, but the meso includes more. It involves meaning in what the impact becomes. Reeder (Chapter 1 responder and personal communication) says meso involves the experience of the human being in the environment (being, knowing, and caring), the cultural meaning of being in the world, and is central to spirituality, anthropology, ethnology, and phenomenology. A different lens, the mesoscopic one, is the realm of making meaning. The process is *dynamic and information moves among scales*, with recurrence, strengthening or weakening, and emergence. For instance, choice and emergent phenomena are seen at both the mesoscale "A Different View" and macro scale "Evolutionary Process" in Figure 3.1 that categorizes methods developed to study complex systems like the mutual, human–environment process.

Teague, Ferris, King, Mahoney, and Jeffs (2007) organized a program for the treatment of pressure ulcers within the Canadian National Healthcare system into micro, meso, and macro efforts, which also viewed meaning as involved at the mesolevel. In their vision, effective pressure ulcer treatment should involve all three scales.

Micro: Best practices for prevention and treatment including interprofessional teams, pressure mattresses, evidence-based care, wound care treatment clinics, and caregiver education.

Meso: Make it an organization priority (meaning) communicated through regional health networks, interdisciplinary teams, advanced practice

FIGURE 3.1 Complex Systems Methodologies

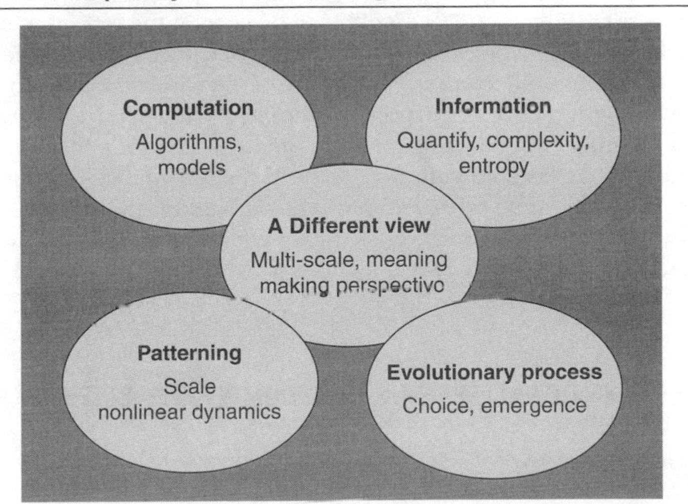

nurse champions, staff education, discharge planning, annual surveillance, content in undergraduate and continuing education courses, and faculty adjunct appointments to increase knowledge.

Macro: Meet benchmarks as shown through outcome databases, provide facility incentives, build interprofessional teams to manage across sectors, report individual hospital progress to the public, and transfer knowledge through research.

There is no mathematical theory of how the three scales interact; the best we do now is two scales that lead to stochastic equations of motion. Understanding multiple-scale phenomena is at the forefront of research.

How Do You Study the Complex Phenomena of Interest to Health Care?

- *Describe what we see (using different perspectives [scale])*
- *See relationships in processes (pushing on a part changes something else)*
- *See how these relationships give rise to collective behaviors (parts acting together)*
- *Understand the human–environment process*

How do you study the complex phenomena of interest to health care? Patterns can be observed in the mutual human–environment process across scales. In nursing, pattern is considered the distinguishing characteristic of a unitary field (Rogers, *The Science of Unitary Human Beings*, 1970). Rogers described the human life process as the human energy field embedded in the environmental energy field, continuous with the universe extending through space, toward greater diversity (Rogers, 1970). Health can be seen in the reflecting symmetry and harmony of physiological parameter patterns, choices made in the environment, and outcomes reflecting the integrality of the whole human–environment unity. Emerging through a "dynamic relatedness" (Newman, 1986) with one's environment, mutually affecting variables tend to co-effect themselves (as a whole) into stable, ordered patterns. Methods for seeing patterns, similarity or dissimilarity at different scales, and how the amount of information contained in the patterns changes in relation to things within or outside the system will be discussed in the following sections.

EXAMPLES OF PATTERNS USING A COMPLEXITY SCIENCE LENS

. . . a set of relationships that are satisfied by observations of a system, or a collection of systems

In an example at the olfactory sensory observation scale of a given patient, a strong smell could indicate to a knowledgeable practitioner that a certain ailment fits a diagnostic pattern (e.g., diabetic ketoacidosis). If associated with more information such as dry skin, and short and labored breathing.

. . . a simple kind of emergent property of a system, where a pattern is a property of the system as a whole but is not an obvious property of small parts of the system.

A forest is a pattern of trees—the forest pattern that we observe in a system of trees is really a property of the relationship between a tree and its environment.

. . . a property of a system that allows its description to be shortened (due to being self-similar in some degree) as compared to a list of the descriptions of its parts. (New England Complex Systems Institute, NECSI.org)

A simple pattern such as spots on a cheetah skin is hard to describe with complete precision but can be understood through computer models that simplify and observe a pattern by accepting that some inherent uncertainty always remains.

What Methods Developed to Study Complex Systems Are Useful to Study the Mutual Human–Environment Process?

Scale Collect and compare data on a system over space and time to see patterning.

Information Choose through observation and computation details that can be ignored to enable a simplified but still valid view of the whole.

Computation Use mathematics and computer science to examine relationships between and among elements.

Evolutionary Process Observe changes in the patterns, from stability to radical change, to evaluate how interdependent nonlinear dynamical processes are related (Sayama, 2008).

SCALE

Fundamental ideas for understanding complex systems include techniques for seeing patterning (relationships and change) at different scales, the amount of information in a system, and seeing what happens when parts are changed. Patterns may be observed at different scales of measurement, information, computation, or change over time. Self-similarity is a common quality of such patterns in complex systems.

Self-Similarity

Patterns that repeat and resemble themselves at different scales of observation are self-similar, and this is a common property of complex systems. Their repeating variability can be seen at different scales. Our observations depend on the resolution used to make the measurement and multiple measures at different resolutions or scales increase confidence in our description of the pattern. In looking at the lung, for example, similar branching structures can be seen in the bronchi, smaller in the bronchioles, smaller still in the alveoli. Looking at a few observations made at different resolutions provides more information than many observations made at a single resolution since the multiple scales may suggest how the elements are dynamically coupled. In nursing, an example could be measuring a person's neurologic function using electroencephalogram (EEG) (electrical activity), Mini Mental Status Exam (cognitive), tests of fine and gross motor skills (neuromuscular), and Activities of Daily Living (skills needed for daily life). Different data collection tools provide different kinds of information, but their pattern together often reveals an individual's strengths and weaknesses previously unseen.

Fractal Dimension

The fractal dimension measures self-similarity and can give more information than a measure of the mean used inappropriately. Fractal dimension measures how a complicated object fills space or a complex process fills time interval. The fractal dimension can be determined from the correlation of one part of a process to another (a scaling relationship). Mandelbrot (1977) discovered scale invariance (scaling), where magnification of a part reproduces the whole. The precision of a scaling relationship is determined by the resolution used to measure properties (e.g., finer resolution reveals smaller pieces). A common example of a fractal curve that resembles itself would be the coastline of England. Viewed from a satellite, the coastline is fairly smooth. As an observer descends in altitude, the English coastline becomes more and more twisted and tangled and longer as new indentations become visible and add to the measured length. While the actual outline of the coastline may change and become more detailed, the pattern of the wildly varying coastline appears similar at most scales of observation. Its fractal dimension is more than a one-dimensional line and less than a two-dimensional surface. Its dimension is a fraction between 1 and 2, hence it is fractal. Log-normal probability curves resemble some qualities of fractal curves and power laws.

Self-Organization

In self-organization, a large collection of simple interacting units spontaneously emerge as a different, not self-evident pattern on a macroscopic level. Some open systems' internal organization may increase in complexity without being guided or managed by an outside source. They usually display emergent properties when they are nonlinear and pushed far from (thermal) equilibrium (far from maximum entropy—as defined by the second law of thermodynamics). In the physical sciences, this change can be seen as a solid, like water ice changes to a gas. Hard ice is essentially a static system at equilibrium. There is a lot of entropy where atoms vibrate randomly and the large-scale structure sits still with fixed boundaries. As it warms, absorbs heat energy, and melts to a liquid, the weaker molecular attractions allow for some motion. Moving atoms drag neighboring atoms along and the liquid flows. There is small-scale order (local attractions) and large-scale disorder (uncorrelated over distance); as more energy enters the system, the patterns that result (e.g., whirlpools having large-scale order) are emergent.

In the earlier example of trees organizing into a forest, it is not self-evident from one original tree in a field that a collection of trees will lead to new ground flora and fauna, changed water cycles and temperature, climate, the smells and sounds that a forest make, nor the peace of mind it can give to human beings. Such collective interactions are the foundation of dynamic systems that are constantly changing within larger scale patterns that remain robust and resistant to change. In other words, trees may be cut down but new ones grow back and the forest remains. In clinical practice, it only takes one dysfunctional family to understand the constant flux within unchanging structures, which may often self-propagate from one generation to the next and be dysfunctional. An interesting example of the earlier cited risk of a metaphor misapplied would be the common practice of clinicians taking the well-defined term "complex" and using it as a mistaken metaphor when they mean complicated, difficult, dysfunctional families. Some clinicians mistakenly describe these dysfunctional families to be more complex when in fact their rapid and unpredictable changes combined with their lack of robustness to environmental challenges mark them as classically chaotic systems, very sensitive to initial conditions, and lacking in sufficiently redundant patterned complexity to be healthy.

INFORMATION

The complexity of something may be the length of the description necessary for us to describe it. To characterize something's complexity, one may create a sequence of numbers considered the computational resource needed to describe the object (algorithmic entropy). It helps to think of how many characters it would take to describe something in words. The number of symbols required to describe the complexity of the system may reflect the information content of a complex system.

Entropy

Entropy is the idea from the second law of thermodynamics that over time systems go from a state of order (low entropy) to a state of greater disorder (high entropy). Scholars may study different types of entropy: (1) Thermodynamic entropy (*S*) is increased disorder as a system does work (second law of thermodynamics) and (2) Shannon's entropy (*H*) attempts to describe a system's information content. According to Boltzmann's definition of entropy, if the system can be in one of many possible states (Ω), the number of bits needed to describe the one it is actually in is $I = \log_2 \Omega$. To describe the microstate of simple physical systems that are indistinguishable unless they differ by a discrete amount, the number of possible descriptions needs to equal the number of possible objects. The longer the description, the greater the number of possibilities that exist (Bar-Yam, 1997). Because this can become an overwhelming amount of information, coarse graining is a technique to smooth out the details of a distribution for all scales finer than some fixed size. This size should be determined by our limited powers of observation and computation to enable us to keep track of the details. In our everyday experience, we use only a small fraction of the variables necessary for a perfectly fine-grained description.

Entropy measures our lack of knowledge. In Shannon's information theory (Shannon & Weaver, 1948/1963), the quantity entropy (***H***) represents the expected number of bits required to specify the state of a physical object given a distribution of probabilities; that is, it measures how much information can potentially be stored in it. It is a quantitative measure of disorder. The disorder is in our head, in our incomplete knowledge of the situation, or the disarray of the system. The vastness of our lack of knowledge of the exact nature of the thing can seem daunting. Visualize a large mound of saltwater taffy sprinkled with dark and white chocolate bits being kneaded and stretched by a machine. The volume remains strictly constant, but the shape of the mound changes drastically. It stretches in some directions and contracts in others, it folds over again and again and the exact location of the chocolate bits becomes more uncertain. The macroscopic volume remains the same but the micro scale becomes smaller and smaller, and it is more and more difficult to describe how many dark chocolate bits are near a white one or clumped together in pockets of similarity. At the micro level, a small section could be occupied by a dark chocolate bit, a white bit, or no bit.

It is important to explain open and closed systems. The second law of thermodynamics (entropy or disorder never decrease) is only true in a closed system—the growth of complex life out of randomness would otherwise be impossible. If a system is at equilibrium, no spontaneous processes occur, and therefore, the system is at maximum entropy. This theoretical state of equilibrium is incompatible with biological life. Prigogine (see Prigogine & Stengers, 1984) showed that dissipative systems existed where entropy is "dumped" out of the system into the environment. These open systems have energy and/or matter continuously flowing through them. Continuously generating entropy, the system actively dissipates it into the environment, thus maintaining its organization at the expense of increased disorder in the surrounding environment. A small local system may briefly circumvent the second law of thermodynamics by exporting excess entropy to the surrounding environment. Overall, the second law of thermodynamics' entropy is still increasing in the universe. How can this concept be utilized in health care? Beautiful, clean, quiet, soothing environments are low in entropy and can easily absorb more physical and emotional entropy from a disordered ill individual.

Negentropy

Bertalanffy (1968) claimed that the life process of creatures is not running down, as indicated in the second law of thermodynamics, but is moving toward greater complexity and diversity. Erwin Schrödinger (1948) called this negative entropy "negentropy." This deviation from the usual situation in the universe can only be achieved in a "living" organism. Living things must seek negative entropy or negentropy to stay alive. To compensate for the entropy increase it produces and to maintain a fairly high level of orderliness, living things must suck order or potential order from the environment. This absorption of negentropy is similar to the removal of entropy discussed earlier. This view of the life process is espoused by nursing theorist Martha E. Rogers (1970) in *The Science of Unitary Human Beings* (SUHB). Organic compounds found in food are returned in a degraded form that plants can still use with the help of the sun. Plants derive their most powerful supply of negative entropy from sunlight (Schrodinger, 1948/1992). The earth and life on earth remain open systems far from equilibrium with hope that life will continue to thrive and grow.

Applying this information to well-being, a person on vacation gains information and decreases his/her entropy from being in a pleasant environment (a vacation setting). The person returns home and tells his/her friend about the environment, describing it in detail: auditory (rushing waves, breezes blowing through the palm trees), visual (purple and orange sunset, colorful parrots), olfactory (smells of salty air, heavy perfume of blooming flowers), tactile (gentle moist breezes on the face, warm sand between the toes), and gustatory (sweet guava, pineapple, spicy chilies, and succulent shrimp). The information is transmitted in sound waves and photographic visual patterns. Some information is lost in the telling as details of the experience are omitted. Still the friend's ability to gain information about the vacation setting creates brain patterns decreasing entropy within the person. After a while, she loses the information. If the original vacationer painted a picture of the place, wrote a piece of music from her inspiration, or created a charity to help the poor of the region, more information could be shared with another and conserved, yet the tendency of this natural process still remains a loss of information and increase in entropy over time.

Chaos

Chaos is behavior that appears to be random because we do not fully understand the rules driving the system being studied, but chaos is not random. Very sensitive to initial conditions and topological mixing, the system will evolve over time so that any given region or open set of its phase space will eventually overlap with any other given region. The deterministic rules of a chaotic system can create subtle patterns in the apparent randomness. If these can be identified, the system remains unpredictable (because we never fully know the initial starting conditions precisely), but to some degree it does become more comprehensible.

Chaos Theory is the field of study of dynamical, nonlinear systems characterized by sensitivity to initial conditions so that although the behavior is constrained within a particular range, the future behavior of the system is largely unpredictable. There will always be small, immeasurable differences in the system's initial states, small differences that may intensify and cease to be minimal. Small changes can have a huge impact. Chaotic behavior can arise in the simplest of systems. It is the norm rather than the exception, and studying chaos helps us understand complex behavior. We can see chaotic behavior in most systems, including electrical circuits, nerve cells, heated fluids, and social and financial systems, including health care. Chaos was first imagined by the French mathematician Henri Poincare around 1915. Later, the meteorologist Edward Lorenz "discovered" chaos in data runs on a computer program he was using to model weather dynamics. The word "chaos" came from mathematicians Li and Yorke (1975) describing a kind of aperiodic (random appearing) but bounded (not truly random) behavior in mathematical systems.

Most biological, technological, and social networks are complex—which means that they lie somewhere between completely regular and completely random. West (2006) used a standard inverted U-curve of complexity versus entropy to show that the highest complexity is between complete regularity (expected) and random (don't know what to expect).

Networks

Biological or social systems have decreased information when highly clustered, with small characteristic path lengths. Watts and Strogatz (1998) called them "small-world" networks. Through shortcuts, vertices are connected that would otherwise be much farther apart. Each shortcut has a highly nonlinear effect, contracting the distance between the pair of vertices and between their immediate neighborhoods (Watts & Strogatz, 1998) (see Figure 3.2).

Models of dynamical systems with small-world coupling display enhanced signal-propagation speed, computational power, and synchronizability due to shortened time and space constraints. In particular, infectious diseases spread more easily in small-world networks than in regular lattices. This model suggests the key role of shortcuts. To test this idea, Watts and Strogatz (1998) computed the collaboration graph for the electrical power grid of the Western United States and the neural network of the nematode worm *Caenorhabditis elegans*. Both graphs are small-world networks. They

FIGURE 3.2 Small-World Network

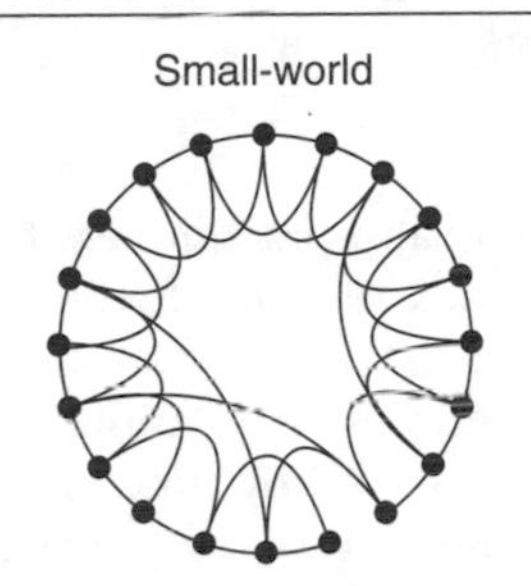

concluded that the small-world phenomenon is probably generic for many large, sparse networks found in nature with many vertices where even a tiny fraction of shortcuts would significantly contract the distance between neighborhoods. Network analyses, and particularly variants of a small-world model, prove very useful in explaining the success of biological systems and those agents such as disease, which succeed in damaging them.

Organizations that are not so robustly networked but are centrally controlled (hierarchical) can rapidly mobilize large resources but prove much less capable of highly complex tasks. When parts of such systems are more independent, those parts are freer to respond to independent demands of the environment but may not be able to call on sufficient resources of their larger system or environment like a hierarchical system can. Thus, complex systems that are most successful show a balance between these two characteristics of autonomy and scaled hierarchy. Most parts of complex systems function best when well connected but not dependent on other parts. Only when collective behavior is essential should parts be tightly connected to each other, as historically evidenced by group actions in times of socioeconomic crisis (Krugman, 2009).

COMPUTATION

Dynamical Systems

A large challenge to practitioners involved in intervening for the well-being of complex systems is their dynamic nature—how their patterns change over time. When parts of complex systems cooperate, however, a single describable behavior can sometimes emerge and continue to change with the environment. New patterns of behavior emerge from further instability and fluctuations, and internal and external competitive processes lead to pattern selection, which can be observed. Much of the system's potential information complexities (degrees of freedom) are suppressed, and only a few contribute to the behavior (Kelso, 1995).

Differential Equations

Although difficult for many health practitioners to use, differential equations are another method used to describe simple models for complex systems and their variables. In many areas of science and technology, differential equations are used when a deterministic relationship involving some continuously changing quantities (modeled by functions) and their rates of change (expressed as derivatives) are known or postulated. This is the controlled experiment where you pulse the system and note the response. It works very well in linear systems where superposition (the net response at a given place and time caused by two or more stimuli, which would have been caused by each stimulus individually) is appropriate and the results of many experiments can be laid over one another. It is not useful in nonlinear systems. To see the process evolving in time, we need (1) an approximation of initial conditions, (2) values of some parameters (e.g., x, y, z) describing the system, and (3) the time involved. The interplay of simple differential equations can create complex and chaotic system models. To partly show how such complex systems change over time, the process of the changing system is simplified (i.e., only a portion of its complex behavior is described) so that it can be shown on two- or three-dimensional space. A set of coordinates (a vector) defines the position of a system variable in space. A sequence of values in time can be drawn as an object in space—"phase space." Three values, such as temperature $x(t)$, potassium concentration $y(t)$, and electrical voltage measured from a cell $z(t)$ are plotted on the coordinates as they change over time. Patterns present in the positions (or other qualities) of these variables can be used to simplify the description of what we see. Each crude measure of the complex changes in the state of the subject system at a particular time t is considered a particular phase. Materials can change their properties suddenly when a control parameter, such as temperature, is changed even a small amount (e.g., water changing to water vapor upon reaching boiling temperature). These changes are called phase changes. More than one phase can coexist during phase transitions. Small changes may build before a sudden phase change can be seen (e.g., a tsunami, glacier calving, smoking intolerance, avalanche). Parameters for modeling near nonequilibrium phase transitions, where stability loss gives rise to new or different patterns parameters, are called order parameters. Parameters that lead the system through different patterns, but are not dependent on the patterns themselves, are called control parameters. Control parameters, like the choices people make, influence the order of the evolving system parameters (Haken, 1993). A result of the potential for change in self-organizing processes patterns can appear, change, and altogether disappear as a result of the ambient "noise" or perhaps by "genetic intervention" (Nicolis, 2005) in control parameters.

The values measured for $x(t)$, $y(t)$, and $z(t)$ may repeat over time, and this combination of values that correspond to a certain region of phase space is called an *attractor* (Liebovitch, 1998). The values change, the points denoting the system move about in space and leave a trail in time, a trajectory. Often, no matter where the orbits began, they end up on the same geometrical structure in phase space—an attractor. After an initial, unique course, all trajectories in this finite volume of phase space end up on the same attractor. A strange attractor is a picture of a system's behavior pattern that often repeats in a self-similar pattern until a core variable or value changes. The recurring patterns of alcohol abuse, domestic violence, teenage antisocial behavior, and innumerable other social behavior patterns are some metaphorical examples of a self-similar human behavior, which represents an underlying strange attractor.

Obtaining the variables in such a complex system as a human being, however, is the great challenge. There are times when approximations of the underlying variables or driving factors can be observed through long counseling with the individual patient or other members of their larger family system. When applied to living organisms, the order, structure, and directionality in a behavior pattern can be envisioned as the organism searching the environment for information that enables it to do what it can and wants to do (Thelen & Smith, 1994). Growth and development may be linear with incremental growth while being nonlinear and complex with new forms and abilities (Thelen & Smith, 1994).

An example in health care of a control parameter affecting order parameters on the macroscopic level is the nurse case manager. Newman, Lamm, and Mitchell (1991), using qualitative research methods, identified pattern recognition and successful resolution of client choices as themes evolving from nurse case managers' description of their client care. Nurses may guide their clients and then let go, allowing their clients to flow to critical choice points, facilitating expanded consciousness of the available choices while being sensitive to the rhythms within the relationship. A gifted healer will time their interventions to facilitate their client's becoming or development of greater insight into the self by relating to others (Buber, 1958; Fine & Peters, 2007). Using the art of knowing when to introduce more information for choice-making, when to encourage movement, and when to relax and let the person flow easily and timelessly with the environment requires a deep understanding of the mutual human–environment process.

Computer Modeling

We cannot currently model such complex qualitative phenomena in computer simulations, but much simpler dynamic systems can be modeled to aid pattern identification. They can suggest outcomes for certain interventions, portray radical changes in a system, and aid in decision making. Some models try to reflect a system's actual physical atomistic structure and dynamics while others try to abstract essential processes. The type and number of parameters used depends on a system's complexity as well as the scale at which one wants to observe and describe it. To examine possible outcomes of a particular system's behavior based upon existing theory and observations of real life, a computer model allows manipulation of the spatiotemporal parameters described earlier. Instead of a static model of variables in a system's phase space calculated over time, we observe the same variables changing in time in an active computer model. Computers multiply the number of possible data points to see patterns of order evolving within the behavior. The model, data, and computer program enter into the usefulness and validity of the conceptual modeling of a complex system. While no model is valid in all domains, this is neither the intent nor requirement for modeling to be useful in decision making.

Bar-Yam (1997) suggested the following key considerations for modeling complex systems:

- Do not take it apart. Interactions and context or environment are very important in complex systems.
- Do not assume it is a linear, smooth pattern. Complex, dynamic patterns that characterize life are irregular and punctuated by critical, seemingly small things that change the entire system.
- Many parameters may be important.

Visualization

Traditional diagrams, which are simple and linear, cannot convey function as directly as they convey structure because depicting function requires showing changes in force, movement, and state, which is difficult in static, two-dimensional diagrams. Depictions of complex systems use extensive graphical aids to more accurately and usefully demonstrate their dynamism. Arrows and shading can depict gradations, direction, movement, and order of events. Computer animation of a diagram can more clearly show these structural transitions over time. There are other computer methods that add rigor to such an approach. Computer-assisted reasoning uses modern computing to generate powerful interactive visualizations, even an ensemble of plausible models, rather than one single model, to represent the available information about the future system. The following figure (Figure 3.3) would be an example of a traditional linear approach to modeling the physician educational process.

Visualizations, often generated by algorithm, help users to generate hypotheses for desirable strategies. An algorithm is a clear rule for solving a class of problems. Good models have a better understanding of the underlying physics, mechanisms, and interactions of the phenomenon, but all models require some approximation as they reach toward unfamiliar territory. Complex problems give no guarantee that any particular conclusion is

FIGURE 3.3 Linear Model of Medical Education*

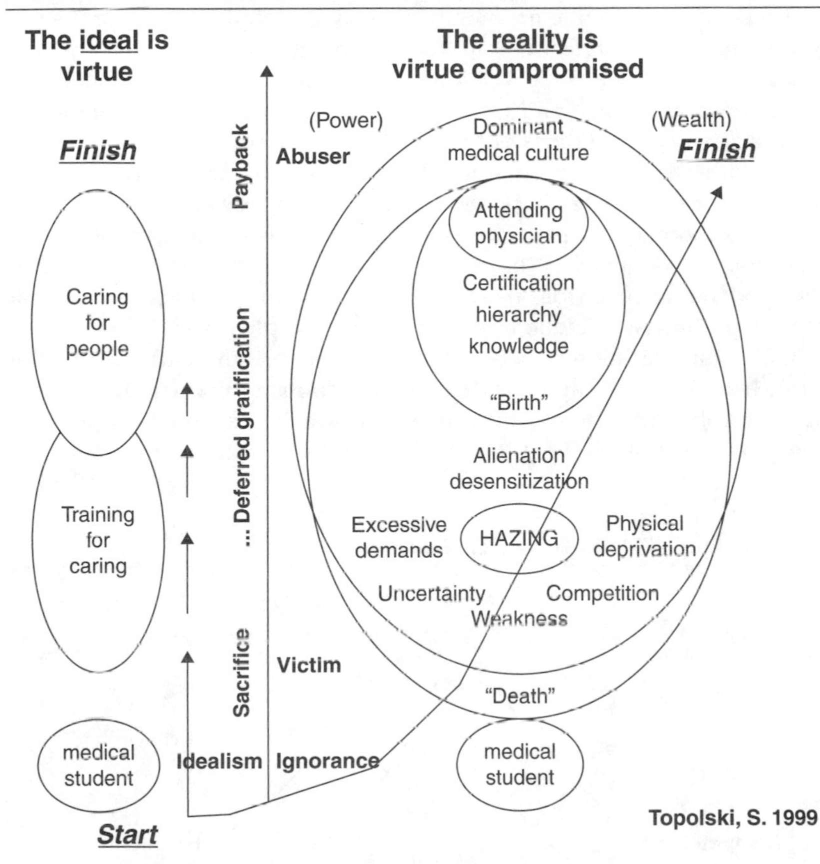

* This is a more traditional linear process model than a chaotic attractive approach. We also borrow from set theory so show social transitions.

correct, yet choices are explored and solutions are sought, which do improve our understanding of pressing issues today (e.g., Kobti, Reynolds, & Kohler's [2004, Santa Fe Institute] use of agent-based modeling of cultural change in swarm, using cultural algorithms to study drought during Anasazi times).

Agent-Based Models

Cellular Automata, such as Netlogo, Repast, and others, successfully model complex systems due to the following principles: (1) all complex systems share some profound similarities given by many quasi-independent agents interacting at once a massively parallel arrangement, (2) agents are adaptive (constantly responding to each other and the environment), (3) no one agent is completely in charge (decentralized), and (4) overall behavior of the whole emerges spontaneously from many low-level interactions (emergence). A grid in two or three dimensions of many "automata" whose states are finite and discrete are simultaneously updated by a uniform state-transition function that refers to states of their neighbors (Sayama, 2008). The computer game of "Life" is a well-known example of this simple concept. Each cell looks at its own and nearest neighbors' states before changing to the local majority state. Effects of boundaries can be seen.

Both differential equations and cellular automata can be broadened to moving (kinetic) models and applied to large systems where individuals are visualized as the microscopic actors interacting to yield macroscopic observations. Boltzman equations (basis of modern kinetic theory) can describe the thermodynamic behavior of a macroscopic system when the microscopic structure is imprecise. Kinetic models are being applied to traffic flows on roads and communication networks (see Bellomo & Pulvirenti, 2000). They resemble health delivery systems that are collections of many individuals.

Respectably large simulation studies can now be executed on inexpensive personal computers. What follows (Figure 3.4) would be an example of

FIGURE 3.4 Fractal Model of Medical Education*

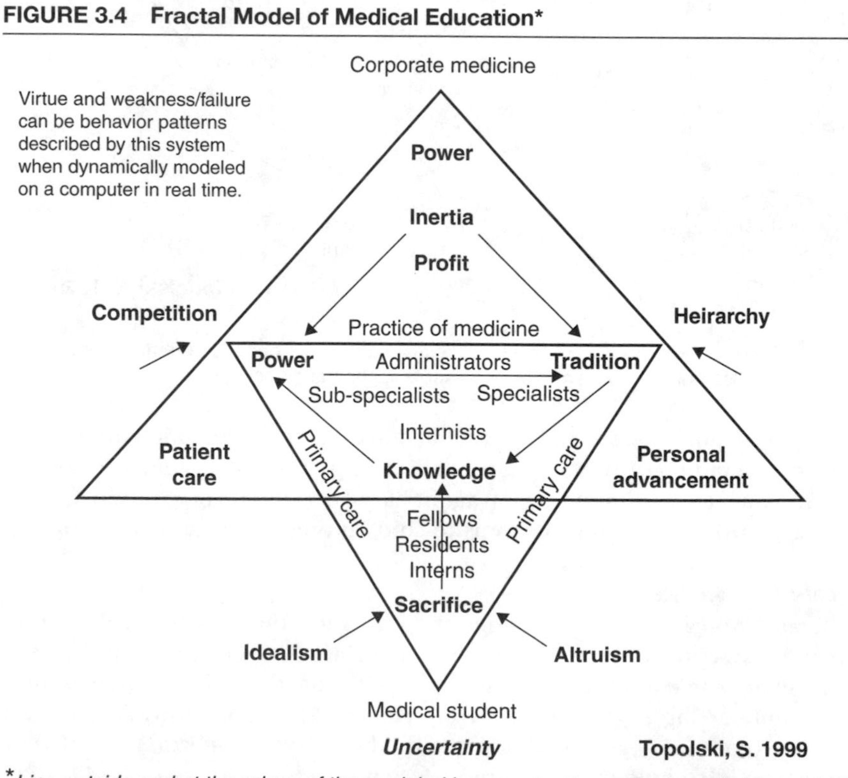

Lies outside and at the edges of the modeled human structure—the boundaries are fractal.

a complex systems approach with computers to model the same process of physician education from the prior figure.

Nonlinear Estimation

Finding the best-fitting relationship between the values of a dependent variable and the values of a set of one or more independent variables can be difficult in nonlinear relationships. Computing the relationship between the dose of a drug and its effectiveness are commonly addressed by such techniques as multiple regression or analysis of variance (ANOVA/MANOVA) that assume that the relationship between the independent variable(s) and the dependent variable is linear in nature. Nonlinear estimation leaves it up to you to specify the nature of the relationship. See http://www.statsoft.com/textbook for an excellent index and explanation of statistical techniques.

Data Mining

Data mining is an analytic process designed to explore data to search for patterns and then to validate findings by applying the detected patterns to new subsets of data. The ultimate goal of data mining is generating useful predictions. Data mining consists of three stages: (1) the initial exploration, (2) model building or pattern identification with validation and (3) deployment. Exploratory data analysis is used to identify the most relevant variables and determine the complexity of models that can be taken into account in the next stage. Model building next involves applying different models to the same data set and then comparing their performance to choose the best. Deployment involves using the model selected as best in the previous stage and applying it to new data to estimate or predict expected outcomes.

Modeling Tools

Mathematicians and biologists use a variety of tools to write their models, including Mathematica, MATLAB, and SBML. The Bayesian approach to statistical analysis models both observed data and any unknowns as random variables with a framework for combining complex data models and external knowledge or expert opinion (Clark & Gelfand, 2006). Permitting existing modeling environments to talk to one another, through standard run-time interfaces, enables each modeler to work in the environment they are used to. This is made possible by generic middleware systems such as web services like SOAP (Box, Ehnebuske, Kakivaya, & Layman, 2000) and XML-RPC (Winer, 1998–2003). XML is the gold standard here as it is the most widely supported. SOAP uses XML meta objects and a communication system abstraction for easily sharing objects through XML.

An open source tool that runs on Mac Mathomatic is a small, portable Computer Algebra System (CAS) written entirely in the C programming language. It is free software, published under the GNU Lesser General Public

License (LGPL version 2.1). This is a console mode application that does symbolic math and quick calculations in a standard and generalized way.

Metamodels and Modules

Model integration connectors allow two separate models to be executed together. Some (Hetherington et al., 2007) advocate the use of modules to represent the same system at different scales to provide insight into the behavior of the whole system. They provide tools for constructing metamodels as modules and allowing them to be used together. One of the factors limiting the development of models of complex systems, like ecosystems, has been the inability of any single team of researchers to deal with formulating, building, calibrating, and debugging complex models. Realistic models are too complex for any single group of researchers and require collaboration between specialists. Structuring the model instead as a set of distinct modules with well-defined interfaces enables teams of specialists to work independently on different modules with minimal risk of confusion. A site at the University of Maryland, http://www.uvm.edu/giee/SME3/MetaModelsF.html, describes an integrated environment for high performance spatial modeling, called the Spatial Modeling Environment (SME). This environment links icon-based modeling environments with advanced computing resources to allow modelers to develop simulations in a user-friendly, graphical environment, requiring no knowledge of computer programming.

There is a growing need to develop systematic modeling and simulation approaches for multiscale problems (see the journal *Multiscale Modeling and Simulation*). Traditional monoscale approaches have proven to be inadequate, even with the largest supercomputers because of the range of scales and the prohibitively large number of variables involved. Microscopic models are often too complex to be dealt with, and the answers contain too much information that is of little interest, complicating the task of extracting useful data. In many fields (engineering, mathematics, and biology), it is found that boundary conditions are very important to define and include, but they are also very difficult due to the very important and unpredictable turbulence occurring at boundaries. Adaptive and heterogeneous multiscale methods (HMM) (Engquist, Li, Ren, Vanden-Eijnden, 2003) allow better use of the knowledge available at all scales, macro and micro, as well as the special features that the problem might have, such as scale separation or self-similarity.

In the medical field, the National Institutes of Health has created a Resource Center for the Development of Multiscale Modeling Tools for Structural Biology, http://mmtsb.org/. Cancer is one of many biological processes in which coupled mechanisms interact across multiple spatial and temporal scales: from the gene to the cell to the whole organism, from microseconds to years. Mathematicians are now working on the difficult task of building practical multiscale models that capture these complex dynamics. For example, one new cancer model attempts to use information about the cell cycle, genes, cellular kinetics, and tissue dynamics to test predictions of how the timing of radiation therapy might influence its effectiveness.

A combined National Institutes of Health (NIH), National Science foundation (NSF), and National Aeronautics and Space Administration (NASA) solicitation in 2004 provided an opportunity to develop tools to enhance computational modeling of biological, biomedical, and behavioral sciences at multiple scales ranging from the molecular level to an entire human population. The use of multiscale algorithms, while encouraged, presents unique problems for validation against data and observations. The potential for rapid propagation of error presents significant challenges for the nonlinear modeler and requires a robust integration of theory with model design. The granting agencies encouraged the formation of a Consortium of Investigators for information exchange, model intraoperability and evaluation, and discussion of critical issues pertaining to multiscale modeling.

The Physiome Projects, http://bioeng.washington.edu, are diverse research projects addressing integrative systems physiology and biology conducted by international teams and individual investigators. Large databases are needed for this work and efforts to organize suitable databases have been made through the new Systems Biology Markup Language (SBML), CellML, and JSimFor multiple spatial scales and wavelet transforms and pyramidal approaches aiming to discriminate between local and global characteristics. Their challenges are to discriminate long-term dependencies from immediate responses to local events; to identify responses to both short impulse (sub-millisecond) and long-range time horizon effects over seconds, minutes, and much longer periods. Physiome efforts use top-down, quantitative approaches to describe molecular, cellular, organ, and overall system behavior from the bottom-up. The goal is to elucidate the paths from genes to organisms, from intra- and intercellular dynamics of genotype to phenotype. It remains to be seen if they will succeed with an emphasis on signaling, regulation, information, and mathematical modeling to predict behavior.

Embedding

One can ask how many dimensions are necessary to describe a system. This act of transforming a sequence of values in different directions into an object in space is called an embedding. Embedding describes the space that contains the object (line, area, volume). In mathematics, an embedding is one instance of some mathematical object contained within another instance, such as a group that is also a subgroup. Mathematical, graphical, and software techniques are often used to make patterns more apparent. Superposition refers to overlapping separate properties of parts or aspects of a system without considering how they affect each. When this happens, the sum of the properties of the parts is a property of the system (NECSI.org). What remains is to describe a dynamic expression of the relationships between these parts.

Stochastic Process

Patterns in computer modeling become more realistic when some stochastic or random values are added to each cycle of a model run. Thus, the next state of a stochastic process becomes determined both by the process's predictable actions (deterministic) and by random elements. In manufacturing, many processes are stochastic and are monitored to observe control parameters that can drastically affect the process over time. Many parameters are tracked simultaneously and statistical models are used to define limits and identify when corrective action must be taken to bring the process back within acceptable parameters. Experience in both nursing and medicine reveals that the human–environment process is complex and expert practitioners acknowledge that outcomes often appear stochastic, resulting from both known and unknown causes. It is common for the same intervention in different patients with similar symptoms to produce very different outcomes.

In the physical sciences, a process view regards system qualities as dimensions and includes length, width, height (size/shape), momentum, frequency, or electrical charge. In the human–environment process, an example of such dimensions of freedom could include sensory variables (e.g., hearing, visual acuity), resources (e.g., friends, finances, accessibility), and conflicts (stressors, challenges, and degree of self-transcendence) among many possibilities. Reed (1991, 2002) proposed that self-transcendence may occur during life experiences that are significant and increase awareness of personal mortality. Her work indicated that self-transcendence is a correlate of both increased complexity and well-being.

Catastrophe Theory

Catastrophe theory (Bak, 1996) includes a set of methods used to study how systems may undergo sudden large changes in behavior as variables that control them are changed only steadily and very little. Despite a gradual, smooth change in parameters affecting the system gently and smoothly, a point is reached when the system suffers a sudden and completely contrasting reaction. Katerndahl (2008) compared the fit of linear and Cusp Catastrophe Modeling in explaining changes in health care system utilization for panic symptoms in emergency, general medicine, mental health clinics, and self-treatment. Cusp catastrophe modeling accounted for more variance in human behavior than all linear models when describing use of mental health settings and self-treatments. Cusp catastrophe may also explain bimodal distributions in behavior, delays in behavior change, and sudden shifts in behavior in stressful situations (http://www.ncbi.nlm.nih.gov/pubmed/18765074).

These theories have been used in many disciplines. Wagner and Huber (2003) applied a cusp catastrophe nonlinear model with the goal of predicting turnover for managerial decision making in nursing care delivery systems. In several studies completed in the late 1970s and early 1980s, Sheridan (1985) and colleagues used the concept of catastrophe theory to postulate predictive relationships in the dynamic changes in work behaviors of nurses including declining job performance, frequent absenteeism, and turnover. The cusp catastrophe model suggests that turnover represents a discontinuous change occurring slowly and unnoticeably until the employee reaches a threshold that results in qualitatively different behaviors. This is an important advance in validity over older linear models that assume equal increments of behavior in a linear direction. The implication of the catastrophe model is that an individual's inclination to turnover is dynamic; that is, fluid and variable over time. At the same time, a small intervention at critical points in time can make a large difference in human behavior. This shows again that scale is an exciting new way to look at phenomena. At a different spatial or time scale, a catastrophic event at one level may be inconsequential (e.g., the death of a tree in a forest) from another broader viewpoint.

How Does Complexity Science Not Only Help Us to Make Sense of Our World But Also Guide Action?

Goal: To create positive outcomes or to lessen poor outcomes.

Variability and Complexity Are Related to Health Care Outcomes

The consequence of diminished complexity and variability is less ability to adapt to the environment.

To study and apply this concept, it is necessary to measure the complexity of your subject and examine how complexity relates to various measures of well-being. Research studies in increasing numbers show that complexity is healthy in both the environment and human beings (see, e.g., West, 2006). A complex environment exercises the human ability to respond to challenge. Gifted children, even as infants, demand and create complex environments around themselves (Gottfried, 1994). From his studies of human brains, correlating the development of dementia with intellectual ability and lifetime habits, Snowdon, Greiner, Kemper, Nanayakkara, and Mortimer (1998) concluded that mental exercise maintains mental functioning, similar to the effect of physical exercise on muscular health. Davidson, Teicher, and Bar-Yam (1997) explored the use of quantitative complexity to evaluate the effects of environmental stimulation and deprivation on the well-being of the elderly. Photographs of elderly persons' homes were analyzed for quantitative complexity. The quantitative measure was based on the number of visually distinct environments characteristic of the objects in the room. This measure was correlated with established quantitative measures of their well-being. It was found that more complex environments correlated with higher cognitive function and greater locomotor activity. An environment sufficiently complex to offer choices may prolong a high quality of life. In contrast, an empty environment arranged to provide order, comfort, and safety removes much of the challenge of disorder, and becomes unhealthy and unsafe without the stimuli essential to the generation of diverse neuronal impulses which maintain cognitive function and well-being.

Evolultionary Process

Emergence—Radical Change May Arise from Intense, Complex Situations (e.g., Health Crises) and Provide Opportunity

A seemingly small change can have a very large effect in a nonlinear system. Lorenz developed a meteorological model that implied that uncertainty equal to a butterfly flapping its wings in one hemisphere's remote jungle can lead to a hurricane months later in a different hemisphere. Chaotic processes are inherently that unpredictable. Complex systems are less so, but processes still can accumulate to the point that a radically different behavior occurs. At a critical "bifurcation point," the system chooses a new path, leading to very different outcomes. Emergence is what parts of a system do together that they would not do by themselves (collective behavior) and what a system does by virtue of its relationship to its environment that it would not do by itself (its function). Similarly, the action a nurse takes may emerge suddenly from many small pieces of information on many scales—from laboratory data on health at the cellular and organ level to the "story" of how the current health event fits into the person's life. Walking into an intensive care room, an experienced nurse may have an immediate intuition that something is very wrong and quick action is needed. This intuition can seldom be fully explained by the available information; somehow it emerges from the gestalt of the event. In the language of complexity science, it is termed "strong emergence." A strong emergent property can only be found in observations of the system as a whole. The massively parallel processing power of the human mind makes this possible at the qualitative human level. When the patient interacts with the environment to select the allowable states, oscillations of multiscale variety with even negative values may occur. Such a strong emergent property cannot be "found in the properties of the system's parts or in the interactions between the parts" but in the properties of the ensemble (Bar-Yam, 2004a, 2004b, p. 15). Then the behavior of the whole may further determine the behavior of the parts. Opposite of the concept of the universe built on law, like a machine, is the vision of a "world self-synthesized" (Wheeler, 1999, p. 314), where the observer-participants create bit by bit the world of space, time, and things (Barrow, Davies, Harper, 2004).

Using the Environment to Promote Human–Environment Well-Being

When a complex system is observed, we wonder how much the information (often a time series) we collect contributes to the behavior observed and how mutual dependencies with other systems are reflected in the observations. Time series in health care are collected on many body functions using monitoring technologies such as electrocardiograms and electroencephalograms and other intensive care unit equipment. Continuity of care in primary care is the most effective way to collect the most complex qualitative time series data. Complex variability (a consequence of multiple waves interacting or multiple oscillators coupled together) can be seen in waves (e.g., neurological waves, respiratory waves, cardiovascular waves, and sound waves). Waves are key to the self-organization of energy and information within living organisms. Repeating wave patterns are called rhythms. There is extensive biomedical research on wave patterns and time series supporting the presence of chaos and complexity in human health and disease. Davidson, Kane, and Tierney (Davidson, 2001) compared the rhythms manifested in 72 hours of blood pressure and skin temperature data collected on two groups hospitalized in intensive care with suspected myocardial infarction. They altered the intensive care unit environment, decreasing noise, bright lights, and unpleasant odors for only one group of patients. Their results showed a statistically significant difference in the circadian rhythms (repeating 24-hour patterns) between the two groups. The group that experienced the environment with lower sensory interference manifesting rhythms were more robust and more closely correlated with those of nonhospitalized, healthy controls. They theorized that decreased interference from an environment disruptive of the individual's unique rhythms might enable persons to heal faster.

Intuition—An Example of Strong Emergence

How can we learn more about nursing intuition so that we might better teach it to novice nurses and properly weight it in making health care funding decisions? Benner's original work entitled *Novice to Expert* (1984) and further developed in Benner, Tanner, and Chesla (1996) uses phenomenology to examine the expert nurse's intuition. A nursing intuition mathematical model developed by Davidson, Ray, Cortes, Conboy, and Norman (2006) uses methods from complexity science to model the phenomenon. It is built on the belief that although it is very difficult to "extract the large-scale view from the highly detailed fine scale information, the conventional and natural assumption is that it is possible in principle" (Bar-Yam, 2004a, 2004b, p. 18). To describe the system at a particular resolution (e.g., the macro level of the nurse/patient relationship), the system is reduced to one less bit of information than the number of variables. That one is the collective behavior of all the variables. All the variables are independent, but together are mutually dependent. The nurse intuits the whole from information on the subsystem (five or six variables that were very important in determining the action a nurse would take). A nurse walking into a client's room in the intensive care unit (ICU) would consider different action based on age, medical diagnosis, and information that could be gathered quickly with four senses as abnormal or normal. Examples of such data include the following:

- Visual (skin color; restlessness; relationship to the environment, for example, in bed, tubes connected, temperature)
- Olfactory (odors suggesting ketoacidosis, incontinence, alcohol)
- Auditory (moaning in pain, incoherent, rattling respirations)
- Tactile (moistness, temperature, and turgor of skin) (may also be important but to a lesser extent, and is not included in this beginning model)
- Gustatory (absence of or distorted, unusual taste such as metallic or salty taste)

From nursing reports, she would already know what others had observed and what actions had already been taken. From the medical records (primarily collected in the chart), she could learn lab values and other doctor discussions. From the family, she could learn the person's history and the event that brought them to the ICU, their concerns and the members' ability to be involved, links they had to other resources, and how they were being involved. In daily practice, quality nurses routinely use immediate pattern recognition to synthesize vastly complex qualitative and quantitative data into a clinical intuition that good doctors and nurses can rely on and patients benefit greatly from.

Tanner (2006) concluded that nursing clinical judgment is influenced by (1) what nurses bring to the situation, (2) the patient's unique patterns and engaging with their concerns, (3) the context of the situation and culture of the unit, (4) a combination of reasoning patterns, and (5) a breakdown in clinical judgment that can trigger reflection on practice and possibly result in a change in practice. These findings are similar to Bar-Yam's (2004a, 2004b) list of factors involved in emergence (a creative leap that is not logically derived from the parts or facts). These include the following:

- the relational interactive process of synthesizing data from many dimensions
- the recognition of patterns
- the combinatorics or combining for the emergence of something new

Another factor that is important is time, free time, to reflect and be open to new ways to synthesize data in previously unforeseen ways. The lack of time in rushed clinical settings is the greatest inhibitor of the creative leap, and nurses are being denied this time by institutional structures in health care today. Understanding how nurses take their knowledge of patterns in the human being and the environment and combine it with unseen information from other dimensions to allow new understanding and action to emerge is a quest for the future.

SUMMARY

We discussed some of the methods being used to study complex phenomena. Some have been applied in research on the mutual human–environment process. Information from these studies incorporated into models may be tested and improved to discover a greater understanding for nursing human beings to health. Complexity methods can improve our vision of health in the connections between individual, their family, and the greater society. Complexity has made strong inroads into more effective scientific team and organization management. Its application to medical care systems and health care in the United States can guide action to improve mutual human–environment well-being.

REFERENCES

Arthur, W. B. (1999). Complexity and the economy. *Science, 2*(284), 107–109.

Bak, P. (1996). *How nature works: The science of self-organized criticality*. Berlin, CT: Springer.

Barrow, J. D., Davies, P. C. W., & Harper, C. L., Jr. (Eds.). (2004). *Science and ultimate reality: Quantum theory, cosmology and complexity*. Cambridge, MA: Cambridge University Press.

Bar-Yam, Y. (1997). *Dynamics of complex systems*. Reading, PA: Addison-Wesley.

Bar-Yam, Y. (2004a). Multiscale variety in complex systems. *Complexity, 9,* 37–45.

Bar-Yam, Y. (2004b). A mathematical theory of strong emergence using multiscale variety. *Complexity, 9,* 15–24.

Benner, P. (1984). *From novice to expert: Excellence and power in clinical nursing practice*. Menlo Park, CA: Addison-Wesley.

Benner, P., Tanner, C., & Chesla, C. A. (1996). *Expertise in nursing practice: Caring, clinical judgment and ethics*. New York: Springer.

Bellomo, N., & Pulvirenti, M. (Eds.). (2000). *Modeling in applied sciences: A kinetic theory approach*. Boston: Birkhauser.

Bertalanffy, L. (1968). *General systems theory foundations, developments, applications*. New York: George Brazziler.

Box, D., Ehnebuske, D., Kakivaya, G., & Layman, A. (2000). *Simple object access protocol (SOAP) 1(1)*. W3C Note.

Buber, M. (1958). *I and Thou* (R. Smith, Trans.) (2nd ed.). New York: Collier Books, Macmillan Publishing Company.

Clark, J. S., & Gelfand, A. E. (Eds.). (2006). *Hierachical modelling of the enviromental sciences: Statistical methods and applications*. Oxford, UK: Oxford University Press.

Davidson, A. W. (2001). Person-environment mutual process: Studying and facilitating healthy environments from a nursing science perspective. *Nursing Science Quarterly, 14,* 108–201.

Davidson, A. W., Ray, M., Cortes, S., Conboy, L., & Norman, M. (2006). Summer School Modeling Project. *Complexity for Human-environment Well-being*. Cambridge, MA: New England Complex Systems Institute.

Davidson, A. W., Teicher, M. H., & Bar-Yam, Y. (1997). The role of environmental complexity in the wellbeing of the elderly. *Complexity and Chaos in Nursing, 3,* 512.

Engquist, W. E. B., Li, X., Ren, W., & Vanden-Eijnden, E. (2003). Capturing the macroscopic behavior of a system with the help of micro-models: The heterogeneous multiscale method (HMM) and the "equation-free" approach to multiscale modeling. *Mathematics. Science, 1*(2), 385–391.

Fine, M., & Peters, J. W. (2007). *The nature of health: How America lost, and can regain, a basic human value.* Oxford: Radcliffe Medical.

Gottfried, A. W. (1994). *Gifted IQ.* New York: Plenum.

Haken, H. (1993). *Advanced synergetics: Instability hierarchies of self-organizing systems and devices.* New York: Springer-Verlag.

Hegyvary, S. (1991). Issues in outcomes research. *Journal of Nursing Quality Assurance, 5*(2), 1–6.

Hetherington, J., Bogle, I., Saffrey, P., Margoninski, O., Li, L., Varela Rey, M., et al. (2007). Addressing the challenges of multiscale model management in systems biology. *Computers & Chemical Engineering, 31,* 962–979.

Katerndahl, D. (2008). Explaining health care utilization for panic attacks using cusp catastrophe modeling. *Nonlinear Dynamics Psychology and Life Sciences, 12,* 409–424.

Kelso, J. A. (2005). The complementary nature of coordination dynamics: Toward a science of the in-between. In R. R. McDaniel Jr., & D. J. Driebe (Eds.), *Uncertainty and surprise in complex systems: Questions on working with the unexpected* (pp. 79–85). Berlin, CT: Springer.

Kelso, J. A. S. (1995). *Dynamic patterns: The self-organization of brain and behavior.* Cambridge, MA: MIT Press.

Kobti, Z., Reynolds, R. G., & Kohler, T. (2004). *Agent-based modeling of cultural change in swarm using cultural algorithms.* Santa Fe, NM: Santa Fe Institute.

Krugman, P. (2009). *The return of depression economics and the crisis of 2008.* New York: Norton.

Liebovitch, L. S. (1998). *Fractals and chaos: Simplified for the life sciences.* New York: Oxford University Press.

Li, T. Y., & Yorke, J. A. (1975). Period three implies chaos. *American. Mathematical. Monthly, 82,* 985.

Mandelbrot, B. B. (1977). *The fractal geometry of nature.* New York: W. H. Freeman.

New England Complex Systems Institute, NECSI.org. Retrieved September 29, 2010, from http://necsi.org/guide/concepts/patterns.html

Newman, M. (1986). *Health as expanding consciousness.* St Louis, MO: Mosby.

Newman, M., Lamm, G., & Michaels, C. (1991). Nursing case management: The coming together of theory and practice. *Nursing and Well-being Care, 12*(8), 404–408.

Nicolis, J. S. (2005). Super-selection rules modulating complexity: An overview. *Chaos, Solitons & Fractals, 24,* 1159–1163.

Prigogine, I., & Stengers, I. (1984). *Order out of chaos.* New York: Bantum Books.

Reed, P. G. (1991). Toward a nursing theory of self-transcendence: Deductive reformulation using developmental theories. *Advances in Nursing Science, 13,* 64–77.

Rogers, M. E. (1970). *An introduction to the theoretical basis of nursing.* Philadelphia: F. A. Davis.

Sayama, H. (2008). *New England complex systems modeling summer school.* Boston: Unpublished Presentation.

Schrödinger, E. (1948/1992). *What is life?* Cambridge, MA: Cambridge University Press Canto Edition.

Schrödinger, E. (1944). *What is life—The physical aspect of the living cell.* Cambridge, MA: Cambridge University Press.

Shannon, C. E., & Weaver, W. (1963). *The mathematical theory of communication.* Urbana, IL: University of Illinois Press (Originally published by CE Shannon [1948, July/Oct] *Bell Systems Technical Journal*).

Sheridan, J. E. (1985). A catastrophe model of employee withdrawal leading to low job performance, high absenteeism, and job turnover during the first year of employment. *Academy of Management Journal, 28,* 88–109.

Snowdon, D. A., Greiner, L. H., Kemper, S. J., Nanayakkara, N., & Mortimer, J. A. (1998). In J. M. Robine, B. Forrette, C. Franchesci, & M. Allard (Eds.), *The paradoxes on longevity.* New York: Springer.

Tanner, C. (2006). Thinking like a nurse: A research-based model of clinical judgment in nursing. *Journal of Nursing Education, 45,* 204–211.

Teague, Ferris, King, Mahoney, & Jeffs (2007). *Pressure Ulcers Micro, Meso and Macro Solutions.* Canadian National Health Care Conference. June 11, 2007.

Thelen, E., & Smith, L. B. (1994). *A dynamic systems approach to the development of cognition and action.* Cambridge, MA: MIT Bradford Books.

Wagner, C. M., & Huber, D. L. (2003). Catastrophe and nursing turnover: Nonlinear models. *Journal of of Nursing Administration, 33,* 486–494.

Watts, D. J., & Strogatz, S. H. (1998). Collective dynamics of 'small-world' networks. *Nature, 393,* 440–442.

West, B. J. (2006). *Where medicine went wrong: Rediscovering the path to complexity.* Hackensack, NJ: World Scientific.

West, B. J., & Griffin, L. A. (2004). *Biodynamics: Why the wirewalker doesn't fall.* New York: Wiley.

Wheeler, J. A. (1999). Information, physics, quantum: The search for links. In G. Hey (Ed.), *Feynman and computation: Exploring the limits of computers.* Cambridge, MA: Perseus Books.

Winer, D. (1998–2003). *XML-RPC specification.* Retrieved September 29, 2010, from http://www.xmlrpc.com/spec

Response to Chapter 3

Bruce J. West

Part of the difficulty with understanding the complexity patterns in social phenomena, particularly those patterns observed in health care networks, is the inappropriate nature of the models we apply to their description. The modeling strategies typically come out of the physical sciences where the phenomena are described by equations of motion that determine how the network changes over time. These equations in turn come from the energy transport in such systems, but in health care networks, it is the information that is flowing and forming patterns, not the energy. These are information-dominated networks, not energy-dominated networks, and consequently we know little about how to model them. The authors outline what we know about how to develop complexity models in a general way and then they apply certain aspects of that knowledge to understanding health care. Specifically, the complexities of health care networks are manifest in inverse power-law distributions. Attempts to replace these heavy-tailed distributions with narrowly focused bell-shaped ones are doomed to fail because the central averages are unsuitable for characterizing complex social phenomena. What is needed is a complexity science, but it is not clear that such a science does exist, or, in fact, can exist.

It is of value to explicitly lay out what we have learned about the properties of complex networks over the past two centuries. In the "normal" worldview, the response of a network to a stimulus is proportional to the size of the stimulation, which means that the network is linear. Consequently, a 50% change in input results in a similar change in the output. In this way, a network is seen to be stable and therefore its future behavior can be predicted, at least in the sense of a distribution of possible futures. The second property is that networks are additive, which results from the principle of reductionism, so that the weak interactions at the micro scale bubble up to the macro scale to form a bell-shaped distribution. This distribution is normal, with a single peak denoting the average and a rapid decline of probability on either side. Consequently, complex networks were thought to consist of linear additive processes that are stable when exposed to the environment. This world of the normal distribution is no longer thought to describe the world in which we live, specifically not social networks. Health care is much too complex to satisfy the simplifying assumptions necessary for normal statistics.

The point missed in the development of social interactions is that energy is the wrong variable; information is the quantity of interest and not the energy. As one proceeds up the modeling chain from data that are closest to reality, to information, the first level of abstraction, to knowledge, the second level of abstraction, it becomes clear that information consists of the patterns observed in the data, and knowledge is the interpretation of those patterns provided by theory. Most discussion relating to health care is at the first level of abstraction with little or no success in formalizing the theoretical relations.

Complexity science is envisioned to incorporate randomness and determinism in balance, so that complex phenomena are neither absolutely predictable nor are they completely unpredictable (random). Absolute predictions of dynamical event outcomes are replaced with distributions of possible futures. Unfortunately, such distributions are all too often interpreted as metaphors where the quantitative weights are interpreted as the qualitative future. This loss of predictability is a consequence of microscopic variability being coupled to the macroscopic world as random fluctuations. In physical networks, these fluctuations are produced by thermal (microscopic property) motion and are related to the thermodynamic concept of temperature (macroscopic property). However, in the social networks of interest here there is no temperature analog; there are fluctuations, which are not related to a "social temperature" but rather to the fact that the environment has a finite number of variables. The fluctuations in social networks are unlike those in physical networks where the environment has 10^{23} particles, essentially an infinite number, influencing the dynamics of the network of interest. However, the coupling across scales is probably more important in the social network than in the physical network; at least in most cases.

The micro, meso, and macro coupling pursued by the authors is one type of taxonomy for social activity. But this is a choice of convenience for describing multiple scales and the coupling across those scales. In physics, multiple scale analysis has been restricted to the two-scale case of micro and macro, fluctuations and averages. Much of physics concerns the macro parameters that determine the transport of various energy forms within the physical network since that is where most of the measurements have been made historically. In the social sciences, this focus on the energy would be replaced by the various forms of information transport. But as the authors point out, there is no mathematical theory of how even the three scales in social phenomena interact with one another.

So what does complexity science bring to the discussion? That is, assuming there is such a thing as complexity science and that it is more than just a sensitivity to the complexity in networks that traditional science does not have the mathematical infrastructure to explain, for example, such things as emergent properties, what can we learn from the complexity perspective? There are a few general principles that have been developed to characterize complex phenomena; one is self-similarity, such as that measured by the fractal dimension. Such scaling determines how spatial structures are coupled across different-sized heterogeneities and how time series are coupled across various sized bursts of events. These structures can be organized by the internal nonlinear network dynamics, but in ways that have no thermodynamic analogue.

Entropy has historically been the measure of information within complex phenomena both in and out of physics. The negentropy of a network is a measure of information, particularly in living systems where the degree of order is very high. The negentropy concept has to be generalized to describe information-dominated networks, which are very far from equilibrium. This is not a simple task, in part, because entropy is itself an equilibrium concept. Nonequilibrium statistical physics has developed a number of techniques for addressing this issue, but even these advanced concepts require further generalizations to be made applicable to social and life science networks.

However, the application of these recently formulated (in a historical sense) techniques must be made with caution. For example, the authors maintain that chaos is not random, but this is a difficult thing to establish. In Russia, chaotic dynamics is called stochastic dynamics, indicating that at least one school of thought does not see a formal difference between the dynamics of chaos and the stochastic processes described by stochastic differential equations. The sensitive dependence on initial conditions used to define chaos manifests micro fluctuations on the macro scale and, consequently, the micro nonlinearities amplify microscopic uncertainty to the level of macroscopic dynamics. In this way, small changes can have large effects or the differences in individual activity can have consequences on the level of the organization. The authors discussed this in terms of nonlinear differential equation sets and solutions known as attractors in phase space. Of course when the attractor's geometric structure is fractal, the solution is chaotic and the attractor is "strange." What is strange is that the attractor's geometric structure in phase space is a fractal, and as the trajectory describing the network's dynamics sweeps across the attractor, it traverses layers that are close to one another in a Euclidean sense, but exponentially far from other network trajectories that were arbitrarily close at earlier times.

The perspective presented in this chapter displays the wide gap between real-world health care applications and the abstract mathematical concepts of complex social phenomena. It is clear that there exist mathematical formalisms to describe complex structures and complex interactions and that health care networks are certainly complex. But it is not clear how to select the proper tool from the mathematical toolbox when faced with a particular health care problem. One possible strategy for making the correct choice is through appropriately processing the data. The authors discuss data processing techniques that acknowledge the possible nonlinear dynamics of the underlying process, but the brush strokes are very broad. Consequently, the discussion is not prescriptive as to the what, where, when, and why of using particular techniques to answer specific questions.

Like many other complex networks, those of health care are not restricted to a single discipline but cut across a number of disciplines. In this sense the multidisciplinary nature of complexity science is more than the operation of mechanisms at the interface between biology and physics to form biophysics or similarly for biochemistry. It is the deeply integrated concepts of disparate disciplines that are required to understand complex phenomena such as health care. Of singular importance is the collaboration across disciplines because complex phenomena do not fit neatly into our carefully crafted disciplines. This too feeds into the need to develop multiscale models of complex networks because, like our disciplinary categories, the resolution sizes of our disciplinary networks have been chosen for convenience. Complexity networks are not necessarily chaotic, even though they typically balance the routine of determinism and the unpredictability of randomness. The balance between regularity and randomness is most readily described by variability, and for complex networks that variability is manifest in an inverse power-law distribution. The health of an individual or an organization is quantified by this variability and pathology is shown to reduce the variability thereby reducing a person's ability to adapt or an organization's flexibility and responsiveness.

Of course, not all the phenomena of interest are continuous. The bursting of a bubble, laughing at a joke, and the collapse of a business are produced by an abrupt change in the qualitative character of the dynamics. This qualitative change might be due to bifurcations, phase changes, or cusp catastrophes, all nonlinear mathematical mechanisms that describe discontinuous change. The discontinuity describes loss of control, resulting in the emotional breakdown of an individual or the irrational tightening of organizational control; moreover, the experimental evidence for disease being the loss of variability is quite strong. On the positive side, individual variability resonates with the complexity of the environment to reinforce a high quality of life. As the authors point out, a lack of environmental complexity may lead to degradation in cognitive function and well-being. Similarly, on a macro level, an organization's variability enables it to respond to the vagaries of market forces and adapt to changing conditions. Moreover, the flexibility at the organizational level must be sufficient to accommodate the individual's variability while still carrying out its function.

CHAPTER 3
METHODS FOR INCREASING UNDERSTANDING OF COMPLEX SYSTEMS

- *Drs. Davidson and Topolski wrote about the moments of joy and magic at work being unpredictable in terms of timing and effect. Think about an experience at work where you experienced joy and magic. What was this moment like?*

- *Can you identify examples of observed events and relationships in your work environment at the Micro, Meso, Macro levels of scaling?*
- *What patterns of relationships do you observe in your work environment?*
- *Can you identify a situation at work where a small change had a large non-linear effect?*
- *In the response, Dr. West talks about bell-shaped distribution. What data in your organization is displayed this way? Are there other ways to display types of data?*

FOUR

Relational Caring Complexity: The Study of Caring and Complexity in Health Care Hospital Organizations

Marilyn A. Ray
Marian C. Turkel
Jeffrey Cohn

In the current organizational culture, health care nurses are challenged to understand their relationship to the system with both its humanistic and spiritual–ethical dimensions and its bureaucratic dimensions, such as the technological, economic, political, and legal as an integrated whole (Ray's Theory of Bureaucratic Caring). The meaning of caring and contemporary health care culture was explored through the research journey of Ray and Turkel in the complex organization of the hospital. System phenomena are powerful forces and are part of the human–environment mutual process. The cocreative process of nurse–patient–physician experiences often move among disorder and order (chaos), struggle, and transformation. The picture of transformation is illuminated in the theory of complex relational caring dynamics, which shows how access to the part (e.g., the economic dimension) has meaning for the whole (the spiritual–ethical caring dimension), and how knowledge of the whole has meaning for the part. Cohn, as a physician and Chief Quality Officer, shows how a patient's story self-organizes and transforms not only the life of a patient but also the life of the caregiver in the caring moment. Any organizational culture must be significant in the lives of nurses and other professionals who are guided by compassion and justice; therefore, new approaches to practice, such as, initiation of Watson's caritas processes and other relationally oriented program developments that improve the lives of people must be integrated into contemporary health care systems. Eggenberger's response addresses the nature of bureaucratic caring as complex relational caring dynamics in the hospital organizations and how organizational leadership and caring education in complex organizations can enhance care outcomes and improve systems themselves.

INTRODUCTION

Complexity sciences and theories tell the story of the quest for knowledge of reality. Twentieth-century science completely rewrote the story of our universe by illuminating the idea and science of interconnectedness of all creation (Cannato, 2006; Peat, 2003). The new universe story highlights the notion of belongingness, nature as organic and alive, and the human–environment integral relationship (Al-Khalil, 2003; Arntz, Chasse, & Vicente, 2005; Briggs & Peat, 1989, 1999; Capra, Steindl-Rast, & Matus, 1991; Rogers, 1970, 1994). Consciousness or "mind giving rise to or relating to matter" is the new metaphysics. Complex systems and their properties are not comprehended strictly from an objective viewpoint or isolated parts as in traditional science, or from just subjective experience as in postmodernism, but from understanding the meaning of the dynamics of the whole, the network of relationships that capture the history of choices toward relational self-organization emergent within chaotic systems. The entire web of the mutual human–environment relationship is inherently interconnected, dynamic, and emerging. Complexity sciences thus include belongingness, mutuality, uncertainty, unpredictability, irreversibility, choice, and emergence (Arntz et al., 2005; Lindberg, Nash, & Lindberg, 2008). From the time of Nightingale to the present, nursing has been advanced as a human–environment mutual process; it is a holistic and unitary discipline focused on caring for others (Davidson & Ray, 1991; Newman, Sime, & Corcoran-Perry, 1991; Newman, Smith, Dexheimer-Pharris, & Jones, 2008; Nightingale, 1992; Ray, 1994a, 1994b, 1998, 2010a, 2010b; Roach 2002; Watson, 2005, 2008). Relational caring was identified as the essence of nursing and central to the human health experience (Newman et al., 1991), and a perspective after review of knowledge development within the unitary-transformative perspective (Newman et al., 2008). In the complex organizational culture of the hospital, Ray's research showed that there is a dynamic and holographic interrelationship between nursing as a caring way of being with humanistic, ethical, and spiritual dimensions, and the environment or social structure of systems, the economic, political, legal, and technological dimensions. The research emerged as the theory of bureaucratic caring (Coffman, 2006, 2010; Ray, 1981a, 1984, 1989a, 2001, 2006, 2010a, 2010b; Turkel, 2007). Nurses and others as caring persons were connected with their organizational environment (social network). In today's organizational health care culture, nurses are challenged *to recognize, and understand the meaning of* not only their relationship to patients and other professionals but also to complex organizations, especially understanding how values and beliefs are shaped within the network of a moral community (Bell, 1974; Britan & Cohen, 1980; Ray 2010a, 2010b; Turkel & Ray, 2000, 2001, 2004, 2009).

This chapter explores the evolution of the journey of relational caring complexity (see Figure 4.1) beginning first with a presentation of the dynamic process of development of bureaucratic caring theory; second with the theory of relational [caring] complexity; and third with the continual evolution to the study of complex caring dynamics in organizations. An image of technology and economics as powerful energy fields within the whole of a hospital organization will be highlighted to show the cocreative processes nurses experience as they move between disorder and order (chaos), and the struggle and transformation in complex systems. Exemplars from practice and an interview with and story from Dr. Jeffrey Cohn, Chief Quality Officer, in a complex hospital system will provide an opportunity for the reader to understand the theory and research trajectory highlighted in this chapter.

FIGURE 4.1 Evolution of the Journey of Relational Caring Complexity

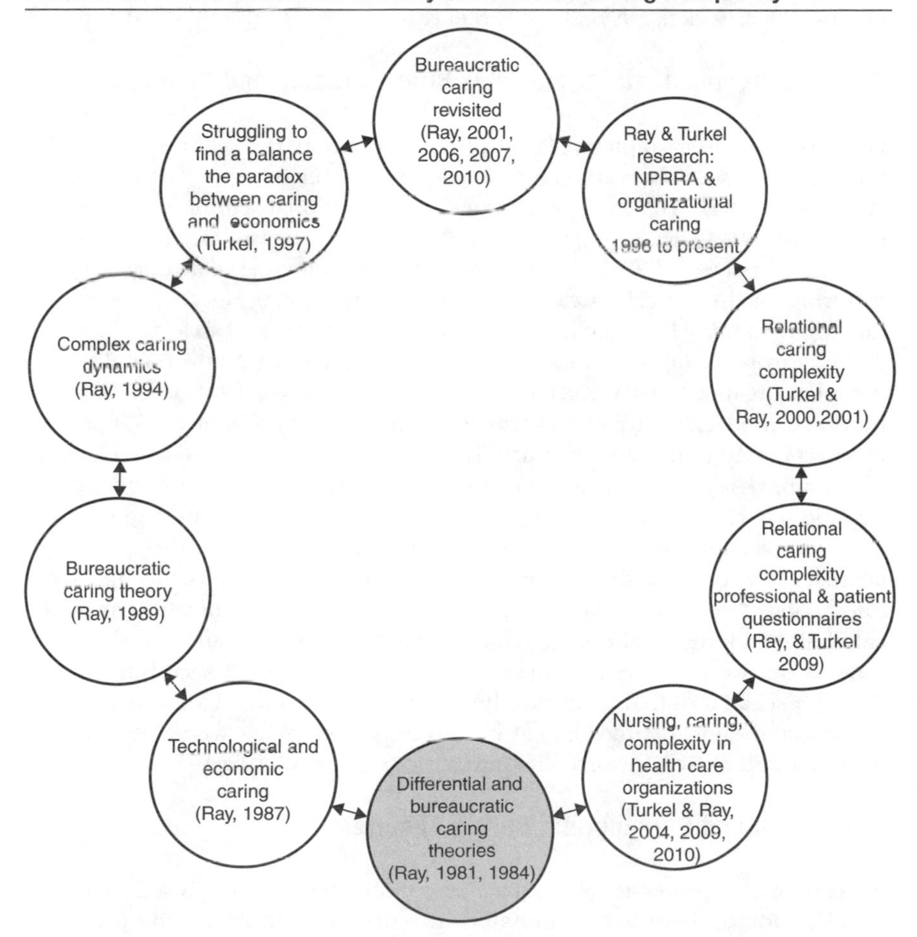

The Evolution of the Journey of Relational Caring Complexity

Like most important journeys, the authors began their journey in hospitals where they committed themselves first to caring for patients and then to the study of the relationship between caring and complex systems. As a new researcher in transcultural nursing, Ray (1981a, 1981b, 1984, 1987a, 1987b, 1989a, 1989b, 2010a, 2010b) was interested in the study of caring in the human cultures of the world, but was drawn to researching the meaning of caring in the complex culture of the hospital. Turkel (1997) was drawn to the study of hospitals as complex systems with a specific emphasis on economics and caring. Their shared interests in complex caring dynamics brought them together to continue the pursuit of knowledge in, and research of, professional and patient caring in complex organizational cultures (Ray & Turkel, 1996, 2009, 2010; Ray, Turkel, & Marino, 2002; Turkel, 1997, 2007; Turkel & Ray, 2000, 2001, 2004, 2010).

Complex Organizational Systems as Bureaucracies and Cultures

Organizations are complex cultural systems consciously developed by people for the purposes of coordinating activities for specific ends (Downs, 1993; Wheatley, 2006). Highly concentrated organizational cultures are woven into meaningful social dimensions of the social structure. Organizations are grounded in diverse beliefs and values about government, management, manufacturing, technological efficiency, and human relational employer–employee ideologies ranging from the military to hospital cultures and so forth (Bar-Yam, 2004; Perrow, 1986). Developed over many centuries of Western civilization, complex organizations were recognized as rational–legal forms of bureaucracy, and a bureaucratic model was identified by Max Weber (1947) and his followers early in the 20th century. The process acknowledged key elements of a rational–legal system that included the major areas of economic, political, social, technological, and legal dimensions that we recognize in complex systems today. Within a bureaucracy (organizational culture), diverse people cocreate and construct their complex and dynamic social reality through social interaction. Within a bureaucracy, the interplay of communication and ethical choice making within relationships forms a community and sets the moral tone of the work–life environment (Turkel & Ray, 2004, 2009). Bureaucracy thus is a social tool that identifies the dynamics of relational authority, power, control, and ethical behavior that has an impact on what we value and how we think and reason in work–life interactions (Perrow, 1986).

Differential and Bureaucratic Caring Theories

As part of her professional journey, Ray embarked upon a research career, and her doctoral studies culminated in a dissertation regarding the study of the meaning of caring in the complex culture of a hospital. Caring is considered a central focus of the discipline of nursing (Leininger, 1981; Newman et al., 2008; Ray 1981a, 1981b, 2010a, 2010b; Watson, 1988, 2005, 2009) and "unfolds within a participative process [caring relationship] formed by the informational patterns of the nurse and the patient" (Newman et al., 2008, p. 4). Thus, the sphere of knowledge in nursing captures the views of caring, copresence, wholeness (health and well-being—integration of body, mind, and spirit), choice, and emergence in the human–environment relationship. For her doctoral research, Ray (1981a) designed a qualitative research approach using three methods, ethno-nursing, phenomenology, and grounded theory to describe, intuit, and discover substantive and formal theories of the meaning of caring in a hospital (Elliston & McCormick, 1977; Geertz, 1973; Glaser & Strauss, 1967; Husserl, 1970; Leininger, 1978). The interplay between the microculture (the caring relationship) and the macroculture (the organizational cultural context) was identified, and a substantive theory called differential caring was discovered, and a formal theory called bureaucratic caring (1981a, 1984, 1989a, 2010b) was elucidated. Within differential caring theory, caring in the complex organization of the hospital differentiated itself in terms of meaning by its context—dominant caring dimensions related to areas of practice or units wherein professionals worked and patients resided. Differential caring theory showed that different units espoused different (dominant) caring models based on their organizational goals and values (Ray, 2006) and were organized into a hierarchical structure: psychological, practical, interactional, and philosophical (Ray, 1981a, 1984). The data showed that differential caring theory is an expression of beliefs and behaviors relating to the competing technological, ethico-religious, political, legal, economic, educational, humanistic, and social factors of the organization and dominant culture. The formal theory of bureaucratic caring was developed by means of using the qualitative data of interviews, participant observation of the social world or the organization, and the interpretation of meaning using an Hegelian (Baum, 1975) philosophy illuminating the thesis of caring (a positive element in the organization) in relation to the potential antithesis of caring (the organization as bureaucratic) to a synthesis, the theory of bureaucratic caring. By researching and understanding the meaning of the human–environment mutual process, nurses [and others] ". . . become subjects rather than objects of their own history" (Ray, 1981a, p. 251).

The Continuing Journey: Further Research of the Evolution of Complex Caring Dynamics in Nursing

Evolving perspectives on the nature of nursing within complex systems and the sciences of complexity incorporating the view of nursing as a human–environment mutual process (Rogers, 1970, 1994), the metaphysical reexamination of nursing as relational caring, and organizations as living systems promised to revolutionize nursing science and practice. Nursing paradigms both in the recent past and now include the concepts of unitary human beings, energy patterns, human field patterning, holism, relational caring, caring as sacred science, and transcultural caring as central phenomena of the discipline (Davidson & Ray, 1991; Leininger, 1981; Leininger & McFarland, 2006; Lindberg et al., 2008; Newman et al., 2008; Ray, 1994, 1998, 2010; Ray & Turkel, 2009; Rogers, 1970; Sumner, 2008; Watson, 2005, 2008).

Ray (1987a) continued with research in the complex organization of the hospital. She chose a phenomenological approach to study the meaning of caring in the intensive care unit (ICU) generating five themes and a moral–ethical conceptual framework. A model of the relationship between technology and caring within ICU nursing revealed the synthesis between experiential and principle-based ethics. Thus, the unity of meaning between the experience of critical care and technological caring has its foundation in ethics, in the coexistent relationship between experience [human caring and understanding of suffering] and principle [beneficence—to do good and to do no harm autonomy—to allow choice and justice—to be fair] (Ray, 1987a).

Economics was a significant dimension of Ray's (1981a, 2010b) initial research of the study of caring in complex organizations. Nurses are confronted with an ethical paradox: on the one hand, by the demand to be responsible to the organization and to society to define themselves in economic terms in health care organizations, or colloquially to respond to the "bottom line," and on the other hand, with the need to preserve the humanity of patients through human caring activities and maintain the economic well-being of a hospital (Ray, 1987b). "Nurses have the ethical choice to attend to human caring as a fiscally viable resource or allow it to become a non-specific entity in the healthcare system today" (Ray, 1987b, p. 36). The synthesis of economic caring must be an answer to the moral conflict. How do nurses measure intangible benefits, such as caring or the practice innovations that they use to comfort patients, relieve pain, provide comfort for distress, and cocreate ways to deal with suffering and grief? Ray identified that "Foa (1971) attempted to reconcile this difficulty by proposing an exchange model that groups can use to distinguish resources in a way that reflects similarities and differences in the exchange behaviors associated with them" (Ray, 1987b, p. 39). Foa grouped interpersonal and economic resources on two coordinates and six types; for example, *interpersonal resources* such as love ([caring] as an expression of affectionate regard, warmth, or comfort), status (evaluative judgments conveying high or low prestige, regard, or esteem), information (advice, opinion, and instruction), and *economic resources*, such as money (currency that has some standard of economic value), goods (tangible products, objects, or materials), and services (activities performed on the body constituting labor for another) (Ray, 1987b). Although nursing has made some headway in terms of studying economic nursing (see subsequent research of Ray et al., 2002; Ray & Turkel, 2009; Turkel, 1997; Turkel & Ray, 2000, 2001, 2004, 2009; and the journal *Economic Nursing*), there continues to be a dearth of data related to the study of human caring, economics, and nursing practice in complex organizations

(Eisler, 2007). A synthesis of noneconomic resources and traditional economic resources, that is, economic caring, can contribute to solving complex economic problems and contribute to understanding the human–environment mutual process in nursing and health care organizations.

Complexity Sciences and Human–Environment Well-Being

Researching caring in organizational cultures and working with Davidson (Davidson & Ray, 1991) on human–environment integrality and the holistic and unitary nature of nursing, Ray (1994a, 1994b; Ray et al., 1995, 1998) was continually intrigued by complexity sciences in terms of its focus on patterns of living and nonliving systems, how these patterns are interconnected in a network of relationships, and how they are linked to corresponding principles of order. The fundamental idea in the sciences of complexity is that all things in nature are interrelated and organize themselves into patterns. Over time, these patterns exhibit no rational predictable behavior but produce new, more complex structures, known as nonlinear, emergent properties. This philosophy thus changed the view of science. The universe is considered holistic, unpredictable, complex, and dynamic rather than mechanistic, controllable, objective, and predictable. Scientists of complexity discovered chaos theory (Briggs & Peat, 1989, 1999). In all evolving human and material systems, dynamic turbulent activity takes place at the boundary (the "edge of chaos") where forces of reciprocity between order and disorder or new forms of order emerge. The tension between order and disorder drives change, and a creative reordering (self-organization) in systems emerges. The "edge of chaos" is a communication and information process, which feeds back on itself and coordinates system behavior through continual mutual interaction. At the bifurcation point (the edge of chaos) between disintegration and transformation, the system encounters a future that is wide open. One fluctuation then becomes dominant and a new pattern forms with greater order (self-organization). The system then remains stable until at some point too much change again creates a chaotic system and a new arrangement for order is activated. In other words, at the self-organizing process point or at the point for potential entropy (disintegration or decay) or equilibrium, in what is known as a phase space, there is a magnetic appeal for a system, seemingly pulling the system toward it. (In nursing, the magnetic appeal is identified as caring [Ray, 1994, 2010; Ray et al., 1995, 2002; Swinderman, 2005; Turkel & Ray, 2004]). The system seems to hesitate in the phase space, seemingly being offered a choice among various possible directions of evolution. Through the interweaving of iterations or feedback loops, the chaotic state (disorder and order) containing the possibilities for self-organization—structure, patterns, and process—is free to seek out its own solution to the current environment. The unpredictable system "chooses" one possible future leaving the others behind. Chaos, rather than merely a mindless movement, has awareness; it seeks out and chooses its own subtle form of order. Thus, bifurcation points constitute a living history of the choices by which people evolved from primordial beginnings to the complex cellular and social forms of today (Ray, 1994).

In human systems, our personhood is defined through relationships. This new-paradigm thinking is embedded in the social paradigm, which is also a pattern or network of relationships. The new epistemology recognizes wholeness, that is, we individuals (the part) can only be understood from the dynamics of the whole (system or society). For example, a complex, global society with individual and different coexisting cultures illuminates our social interrelatedness. The only way to understand any one culture (the culturally diverse) is to understand its relationship to the whole (the universal) (Leininger, 1991). Systems self-organize through choices made within the context of the human–environment relationship. However, in nursing, the notion of self-organization can be understood as *relational self-organization* through the caring relationship (Ray, 1994). As nursing scholars, we see that the notions advanced in complexity sciences simultaneously were advanced in the Rogers' (1970, 1994) science of unitary human beings (the universe), Leininger's (1991) theory of culture care diversity and universality (the culture), and Ray's (1981a, 1989; Turkel & Ray, 2000, 2001) theory of bureaucratic caring (complex organizations).

Research of Ray and Turkel (Bureaucratic Caring Revisited)

Turkel (1997) further studied complex organizational cultures by her grounded theory methodological analysis and substantive theoretical development of caring and economics. The emergent substantive theory of struggling to find a balance refers to sustaining the caring ideal in a health care reality controlled by costs. Nurses are practicing caring in an environment where the economics and costs of health care permeate discussions and impact decisions. It is a paradox in practice that nurses have less time to provide patient education, while at the same time the need is greater because of higher patient acuity levels. As patients have less information to guide future health care decisions, the readmission rate is higher.

After Turkel's initial nursing research, and upon receiving federal research grant funding from 1996 to 2004, Ray and Turkel formally studied organizational caring with an emphasis on economic resources with nurses, patients, and administrators in military and civilian hospitals using qualitative and quantitative research methods. By means of a grounded theory qualitative method, a questionnaire was designed to measure patient and professional responses and test the theory of relational [caring] complexity (Turkel & Ray, 2000, 2001). Relevant data from the grounded theoretical categories of relationships, caring, and costs (subscales) and the related subcategories and properties served as the basis for items on the questionnaire. Psychometric evaluation revealed that refinements to the questionnaires should be made; thus, two forms, one for patients and one for professionals (nurses and administrators), subsequently were designed for continued research, and testing of the relational [caring] complexity theory (Ray & Turkel, 2009). The findings from the questionnaire data validated what was revealed in the initial grounded theory study. Human caring, not cost, was the predictor of the *value* of the nurse–patient relationship in both civilian and military organizations. The research pointed to the fact that humanistic relational caring in an economically driven health care system, although valued, must continually be studied as an economic resource. The theoretical testing of the relational caring complexity theory illuminated that nursing practice as caring in complex organizations is complex and nonlinear. It is a synthesis of both process and outcome, that is, nursing practice is relationally self-organizing and emerging through the dynamic ethical choices that are made in the human–environment mutual process (Ray & Turkel, 2009; Turkel & Ray, 2001, 2004, 2009; Turkel, Ray & Marino, 2002). The funded research and further reflection on the theory as a complex relationship of interconnectedness illuminated the holographic nature of the bureaucratic caring theory (BCT) (see the model of BCT in chapters in Coffman, 2006, 2010, and Ray, 2001, 2006, 2010b; Ray & Turkel, 2010; Turkel, 2007). Revisiting the theory and capturing it as a holographic theory revealed to Ray that not only is caring represented as parallel with the economic, political, technological, legal, physical, social–cultural, educational, ethical, and spiritual dimensions, but now it is *spiritual–ethical caring* that is represented as parallel with the social structural dimensions. Continued research illuminates caring as spiritual in the sense that it is holistic (integrative and relational and integrally connected to organizational life) and ethical, that is, caring is the conscience of organizational life. Nurses commit to their moral obligation to others and the organization, sometimes in conflict but always in relationship. Both the bureaucratic caring theory and the relational caring complexity theory are the synthesis of emergent properties, the order or unity that is cocreated through spiritual–ethical caring—patterns of the mutual human–environment integration that are continually enfolding and unfolding in complex health care systems in particular, and the life world in general.

Relational Caring Complexity: Professional and Patient Relational Caring Questionnaires

From the culmination of qualitative and quantitative research and theoretical testing, Ray and Turkel refined two relational caring questionnaires (2009; Watson Caring Science Institute [WCSI], International Caritas Consortium website) that can be used to study the value of caring among nurses, patients, and administrators within complex organizations. What value does relational caring have? As we know, the value of nursing is perceived by professionals as caring for others (an intangible resource) and is considered a tangible economic resource in an economically driven contemporary health care system, that is, money spent. For nurses, the value of caring as an interpersonal resource has both human and economic ramifications; it is necessary and sufficient for nurses, patients, and hospital administrators. The relational caring questionnaires, generated over the decade-long research, can be used to study not only the economics of caring, but also nursing and administrative caring practices as they continuously emerge in complex systems.

The Continual Emergence: Nursing, Caring, and Complexity in Health Care Organizations

Nursing, caring, and complexity in health care is a dynamical model of nursing practice and inquiry (Ray, 1998; Turkel & Ray, 2000, 2001, 2009). Conducted research recognizes the unitive foundation of nursing, the human–environment integral relationship, and self-organization in nursing practice as relational self-organization for health, healing, and a peaceful death. As complexity sciences are promoted as the "sciences of quality" and represent a vision of reality that is relational, complexity nursing can be understood in the context of the caring choices that are made in the network of relationships. Nursing is neither purely science nor purely art, but rather a combination of the two. Because the future state of the complexity sciences will be more closely aligned with art (Bohm in Briggs & Peat, 1989), what does that truly mean? Art or aesthetics embraces the *essence* of patterns of meaning (Langer, 1953). The meaning of complexity science as the science (and art) of quality is similar to Langer's (1953) notion that can lead to a new philosophy of living form, living nature, mind, and ethical issues in human society. Bureaucratic Caring Theory (2001, 2006, 2010a, 2010b), complex relational caring theory (Turkel & Ray, 2000, 2001, 2004), and the theory of existential authenticity advanced by Ray (1997a, 1997b) studying nursing administration addressed the integration of political, legal, technological, and humanistic, spiritual, and ethical dimensions of hospital systems as a living form. Knowledge and the essence of these patterns are critical to understanding nursing within complex organizations. As a follow-up to the initial research, Turkel and Ray (2009) specifically addressed nursing practice as complex spiritual–ethical caring dynamics. There is a moral essence of meaning in the moral community of nursing that is reflected in the social organization of the hospital. Their insights showed that nurses in practice are ethical and are called to act responsibly for the good of the patient and the whole in the moral community. As such, the experiences of nurses are not passive; they are an intellectual and a spiritual–ethical force; "it [nursing] always conveys a moral stance" (Turkel & Ray, 2009, p. 234). Confirming Langer's (1953) ideas of a new philosophy of living form, complexity nursing science reveals the art that is already known. At its center, the art is the call, the consciousness of caring, the conscience of nursing, spiritual–ethical caring. In the network of relationships, the choice to care is the magnetic appeal that will keep the universe of nursing, medicine, and health care emerging, unfolding, and enfolding together into a new vision of relational self-organization.

Convergence of Complexity and Caring Science in the Practice Setting

Dr. Jeffrey Cohn is the Chief Quality Officer for Albert Einstein Healthcare Network. Dr. Cohn and Dr. Turkel use concepts from complexity and caring science to advance organizational transformation initiatives within the organization. The mission of Albert Einstein Healthcare Network is the following:

> with humanity, humility and honor, to heal by providing exceptionally intelligent and responsive health care and education for as many as we can reach.

The vision for Albert Einstein Healthcare Network states that it "reflects our rich past and the influence of our namesake Albert Einstein. Through that vision, we aspire to bring to our patients and community *Einstein Brilliance and Compassion in All We Touch*." The conceptual framework for the nursing service organization is Watson's Theory of Human Caring (2008), which is congruent with the mission and vision of the organization. The following edited transcript of an interview with Dr. Cohn presents his recent thinking about the relevance of complexity science to organizational quality and patient outcome initiatives.

How Did You Become Involved With Looking at Complexity Science?

When I first took on the role of Chief Quality Officer I began to do readings focused on the relationship among health care, quality and organizations. One of the books I read was *Crossing the Quality Chasm* (Institute of Medicine, 2001). I knew from reading the book that my role within the organization would be to change and improve the way care was delivered in order to improve quality and achieve desired outcomes. I was intrigued with the synthesis of ideas presented in Appendix B, Redesigning Health Care with Insights from the Science of Complex Adaptive Systems. Although I saw the relevance of the approaches from complexity science in relation to quality and organizations, I was not sure of the practical implementation within the organization. Around the same time, I became involved with our organizational transformation efforts and thought more about the readings I had done. I realized that the description of transformed organizations was what our organizational transformation was all about. As the organization became more involved with working with the Plexus Institute I went back to this work and began to understand how the concepts could be used to change the way we approached transformation and quality.

How Did You Become Involved With the Plexus Institute?

One day I received an e-mail communication related to the use of Positive Deviance (PD) as an approach to behavioral change that was sustained by individuals or groups. The story of the work was very compelling and our organization partnered with the Plexus Institute to use this approach to reduce the transmission of methicillin-resistant *Staphylococcus aureus* (MRSA).

As a result of this collaboration, we have coaches from the institute who work with me and other members of the team to use this approach for other quality initiatives. The coaches challenge us on how we are going to run a meeting. The goal is to plan meetings but to let the ideas "bubble up" from the group. An action learning framework is used with the idea of starting small when we want to initiate change. As humans, we have pattern recognition and the idea is to identify patterns of behavior. The intended result is for the actions to become the standard, which then lead to improved outcomes.

How Do You See the Concepts From Complexity Science Related to Your Role in Organizational Transformation?

Complexity science looks at systems and the interrelationship within the systems. For organizational transformation to be successful, we need to look at the work done by all employees when we are dealing with the code of conduct or specific quality initiatives. Not only do we want to change behavior but also we want to sustain the changes. Each employee needs to think, what makes my work valuable to the organization or to patient outcomes? One of the concepts from positive deviance is using a *discovery and action dialog* (DAD) to facilitate discussion with employees when you want to learn from their experiences. Employees come together as a group to share their experiences, discover solutions, and generate new ideas. Leaders need to trust the process that frontline employees will identify creative solutions and that the creative wisdom of the group allows for transformation to occur. A DAD embraces group diversity and provides an opportunity for all involved to look at how their work relates to the work of others on the team. The process is not linear; ideas emerge from the group dialog.

Another concept from the positive deviance work is *nothing about me without me* or *nothing about them without them*. This means that when we are looking at change, we do not make a decision without having received input from all those involved in the change. We will not make a decision or recommend action for a group or individual who is not at the meeting. This approach is different from the traditional approach of *just do it* because we as leaders said it should be done. The traditional approach may result in a quick change, but the change is rarely sustained over time.

Complexity Includes the Concept of a Network of Relationships. How Do You See the DADs Relating to This?

The results of the DADs have been fantastic. During the group meeting, employees from various departments are talking with each other, respecting the opinions of others, working together as a team and always asking is this enough? Their communication network extends outside of the meeting; they say hello to each other in the hall and have a sense of pride in their accomplishments. The communication and relationships from the group can be very powerful.

I would like to share an example with you from a DAD with a group of night shift nurses from the medical intensive care unit (MICU). I was facilitating a DAD related to preventing infections and we were talking about ventilator-associated pneumonia, bloodstream infections, and hand washing. We talked about why the compliance was better for hand washing when nurses came out of the room versus when they came in the room. I was comfortable with group silence during the dialog and the group really could not identify a strong answer to the question or suggest a strategy to change practice. I knew they needed time to think and reflect about this versus me saying “just start washing your hands every time you go in and out of a patient’s room.” The nurses shared the exchange with their peers. A few days later, new ideas began to emerge. The nurses began asking themselves “how can we allow this to happen?” Change started to occur. The nurses began watching their peers’ compliance with hand washing and provided feedback. They started looking at patients differently, evaluating care, and wanting to deliver proper care.

Momentum and enthusiasm resulted. The nurses spoke with the quality coordinators to have hand washing included in all the infection control bundles. The bundles were modified and the nurses continue to hold their peers accountable. To me this is an example of how a small discussion with a group of nurses resulted in a system change in practice.

Can You Give an Example of How You Use Complexity to Improve Quality Outcomes?

I think the example of a small group of nurses impacting systemwide change in several infection control bundles is an excellent example. Lately, I have been thinking about the use of various concepts from complexity science including DADs and how and when different approaches are used to improve quality outcomes. I'm relating this to the work of Ralph Stacey and the *Agreement and Certainty Matrix* (2003). This describes the relationship between certainty about the outcomes of interventions and agreement within the group about solutions of interventions to improve quality outcomes. For example, when there is a strong correlation between what should be done (group agreement) and how certain we are that this will achieve outcomes (certainty), the issue lies within the *zone of certainty* and the practice norm should be a protocol, policy, or standard of practice. An excellent example of this is fall prevention. Members of the fall committee were in total agreement when they identified evidence-based best practices to reduce falls and the research has shown that consistent use of these initiatives will reduce falls. The work of the group is to implement standards of care and hold each other accountable to change practice. Issues where there is a moderate level of group agreement and a moderate level of certainty related to interventions and outcomes are in the *zone of complexity*. In this zone, brainstorming, experimenting, having DADs, and using concepts from complexity and positive deviance are acceptable. For example, if the problem is having blood drawn at a certain time instead of creating a rigid complicated policy, bring the team members together to dialog and see what patterns of behavior begin to emerge. Begin asking "what needs to happen?" and understand how it happens may be different in different settings. In terms of quality outcomes, we need to stay away from issues in the *zone of chaos* where there is no group agreement as to interventions and no certainty of outcomes related to the interventions.

How Do You See the Concept of Network of Relationships Relating to Nursing, Patient Care, and Quality Outcomes?

Nurses are pivotal to patient outcomes and quality of care, but often the world in which we live does not allow for the networks of relationships to extend beyond a geographic area or shift. Often, care is delivered within a specific geographic location (nursing unit), and nurses only work on certain days. As a result, the nursing unit is a strong self-contained local network with strong connections to patients and weak connections to the network.

What we are doing within organizations is helping nurses become connected and intertwined with the larger social network. Ways that we have been doing this include inviting nursing leaders and frontline registered nurses to the leadership summits and brown-bag breakfast dialogs. This provides them with an opportunity to develop relationships with other disciplines and become connected to and participate in organizational initiatives.

At the unit level, we have leadership rounds where leaders in the organization schedule rounds on individual nursing units. We bring all members of the team together—nurses, medical clerks, and housekeeping—to discuss issues and collaborate on solutions. I would like to see more physician involvement in these dialogs. The goal is to break out of the silo and form that network of relationships.

As we continue to form social networks within the organization, we are able to meet the foundational needs of complex adaptive systems. This will allow for connectivity, creative problem solving, and the giving and receiving feedback and support. If we continue to build these relationships within the organization, the system will evolve and trust will result.

How Do You Use Story Telling in Your Role as a Physician and a Leader in Organizational Transformation?

Stories and storytelling have always been important to me. In graduate school I took a course on communication and speech that was part of an MBA program. The professor emphasized that people will always remember stories. I think stories are underused in both clinical practice and leadership. As a physician I use stories when I teach residents and also when I teach and talk with patients. I relate the treatment and disease in terms of a narrative or story instead of focusing on numbers, percentages, and specific outcomes. I want my patients or residents to draw their own conclusions from the story and experiences and relate the story to their world. For example, when I was working with oncology patients, I would tell a story related to the experience of chemotherapy and use the story to stimulate discussion. I wanted my patient to think about the experience and make decisions based on what she as a patient is looking for. I encouraged her to reflect and think about "what will this be like for me?" "what will my family want to know?" "how will this impact my ability to go to work?" versus focusing on "X number of patients who choose chemotherapy versus radiation" and "who will live X amount of years after the last treatment?"

As part of our organizational transformation, we start our meetings with a story or poem. The purpose is not only to recognize a difference we made in patient care but also to reflect on the meaning of the story or poem and how we can use this story to think about issues differently.

Aesthetic Expressions

Narrative stories and poetry have been used in both medicine and nursing as a way to illuminate the human side of health care, to describe the experiences of caring, and to teach and learn about patients (Bartol, 1989; Hudacek, 2000; Leslie, 2008). Sharing stories is intrinsic to all cultures as a way to share knowledge and expertise. The story comprises of an experience that is significant and meaningful to the person sharing the story with others. Stories open the opportunity to learn from each other by offering practice exemplars that illustrate hard-to-define concepts such as caring, healing, creativity, affinity, or responsiveness. Stories help convey cultural understanding and help caregivers recognize the difference they have made in another's life. Story is encouraged by asking about experiences that touched the other deeply or left a vivid impression. As humans, we can connect through the shared lessons of a story bringing harmony and healing to practice—central components of Watson's theory of human caring (Watson, 2008). Through stories, patients become real people rather than medical diseases or diagnoses.

Recently, the Institute for Healthcare Improvement has recognized the use of story as a way to demonstrate quality experiences. The following story by Dr. Cohn illustrates the power of story to capture the human side of health care and serves as an exemplar for physician–patient interactions. Reflective questions within the story provide the reader an opportunity to discover his or her own meaning and interpretation of the experience.

Moving the Moving Man

Jim, "JT" to his friends, was sitting on the side of his bed on Tower 5 when I entered his room. The IV in his neck, the ID bracelet on his ankle instead on either of his swollen arms, and the pillow he clutches to his chest when he makes any movement that stresses his chest bones (coughing, belching, etc.) indicated an eventful hospitalization. As I listened to his story, he told me of being here for nearly 3 weeks for emergency bypass surgery and its aftermath. Although he was "pretty out of it" for some of his postoperative time, he's now on his feet again and looking forward to getting home soon. JT worked for the Mayflower Company for 28 years, driving moving vans 14 times across the country. He described 16-hour days often collapsing on a bed at night and waking up to find he had fallen halfway to the floor while sleeping, but was too exhausted to have been aware of that. He'd then quickly shower, shave, and then return to the road. This schedule would continue for months, and, tiring though it was, he loved the work, the people he met, and relationships he forged, of which many continue to this day.

JT not only worked hard, he played hard too. He described lifestyle choices, which "hurt not just me but also those close to me, those that loved and supported me." He now blames these poor choices for the condition that led to his need for surgery. JT states, "I have a lot of apologizing to do when I get out of here. There still are some who have stood by me through the years even as I made these bad decisions, and I need to tell them how sorry I am. I also need to talk to the people that I've been hanging with and tell them that we were wrong. Look at what I've had to experience to learn my lesson—they have a chance to change without going through all I have and they need to hear about it."

I asked JT how his care has been and his eyes lit up. He remarked, "I am so grateful for everyone that has cared for me—everyone who walks through that doorway has been fabulous. It would take me going to Montezuma's mines and giving each person their own treasure trove of diamonds and gold to begin to try and repay everyone for how I've been treated." He noted that when staff come in and ask him why he's not reaching out more for their assistance, his response has been that they have sicker people to focus on now. He needs to begin to do more for himself, but thanks them for stopping by. He does admit that every now and then he'll ring his call bell when he's feeling lonely, just to have someone to talk to for a few minutes.

We shook hands and despite his hands being swollen, his grasp was firm and sure. Einstein's mission is ". . . to heal . . . as many as we can reach." In order to "reach," don't we need to extend our hand toward those we serve? What does it take to have our healing hand met and grasped? Clearly, this was able to happen with JT. If we can connect with someone who literally lived his life in the "fast lane," who is beyond our reach? As I left his room, I reflected on the tremendous privilege that we had been granted here. We were given the opportunity to enter JT's life, care for him during a critical time of his life, and contribute to a life experience that has left him a different person from the one who entered our institution. JT touched many people's lives as he moved their belongings around the country. I'm sure his smile and his engaging personality helped individuals and families begin to make a new life for themselves in their new homes. Similarly, we were able to play a role in helping to move JT to a new place in his life. Our caring and skills were able to support him, as he had been able to do for so many in the past. Completing the circle, now JT will work to "move" his peers from their current lifestyles to ones that focus more on love and health than thrill and abuse. We were given a gift by JT entering our institution and our lives, and it seems we've made the most of it.

Within the nursing services organization, the art of nursing is valued and honored as an expression of caring. Registered nurses were invited to share artistic expressions of caring during Nurses Week 2009. Laura Truskowsky, RN, from the Elkins Park campus was awarded first place for her poetic expression. Laura works on the spinal cord unit and was compelled to write this poem after learning that a patient with whom the staff had formed a special relationship passed away during the night.

A Day in the Life of a Nurse

I rise at the crack of dawn
with yesterday's expectation now gone.

Arriving to work at quarter to seven
I find out young John has left us for Heaven.

We console one another with tears and a hug
and try to make sense of losing the young.

Our emotions in check, we go on with our day
Compassion and caring lighting our way.

Unpredictable life is, precious nonetheless.
So let us go the extra mile to do our best.

Listen when others are feeling blue,
Provide caring and comfort, as we so often do.

Simply put, this is my verse.
A day in the life of a caring nurse.

As an artistic medium, *poetry* serves as an outlet for the expression of joy and grief associated with everyday encounters in the health care environment. Poetic expressions fabricate a tapestry of wisdom and understanding about the human experience. Through poems, health care professionals come to know self, others, the depth of human experience, and the meaning of caring (Wagner, 2000). Poems are an alternative form of reflective story and a powerful vehicle to increase self-awareness and connections to others, including coworkers, patients, and families. Poetry is a reflective process and another way to document patient encounters to teach about the art of medicine, nursing, and the human condition.

Practice Exemplar

The BOOST (Better Outcomes for Older Adults Through Safe Transition, Society of Hospital Medicine, 2010) project on Tower 6 illustrates how concepts from complexity and caring science can be integrated in the practice setting to improve patient care. BOOST is a national initiative focused on the improving discharge processes. Desired outcomes include a reduced readmission rate, increased patient satisfaction scores, and improved patient and family education practices.

One of the components of BOOST is teach back, which focuses on patient education and health care literacy. The complexity science concept of the DAD was used to facilitate dialog among members of the nursing staff, promoting quality and nursing education. Prior to starting the project, a series of DADs were held with the registered nurses to understand the most appropriate teaching strategies for the specific patient population, anticipated barriers to providing patient education, and what support would be needed to move the initiative forward. The nursing interventions provided by the registered nurses related to patient education are grounded in the tenets of Watson's theory of human caring (Watson, 2008). According to Watson, Caritas Process 4 is developing and sustaining a helping–trusting–caring relationship. Registered nurses focused on coming to know the patient as a person and what was important to the patient and his or her family during the time of illness. Understanding the person as whole and realizing the human experience allow the registered nurse to develop a caring relationship. While providing the education, the registered nurse pulls up a chair and focuses on the patient while remaining open to the patient's experience. Patients have verbalized that they learn better when they have a caring relationship with their nurses (Turkel, 1997). Caritas Process 7 is to engage in genuine teaching–learning experiences that attend to unity of being and subjective meaning—attempting to stay within the other's frame of reference. Translated into practice, this means giving patients education in a way that they can receive and understand it. For example, preprinted discharge instructions may need to be customized or pictures may need to be used to convey meaning. Registered nurses are careful not to judge or label patients as being noncomplaint. Patient outcome data are preliminary. However, the readmission rate has dropped from 24% to 18.5% and the response to the Hospital Consumer Assessment of Healthcare Providers and Systems (Center for Medicare and Medicaid Services, 2010) question "*Nurses explain things so I understand it*" has increased from 68.5% to 92.3%.

CONCLUSION

Complex relational caring dynamics in nursing and health care organizations consider the following:

- The discipline and profession of nursing in the new world of complexity sciences is a science and art grounded in the Rogerian (1970) view of the science of unitary human beings, bureaucratic caring theory, and complex relational caring theory.
- The views illuminate the role of the caring nurse as relational self-organization: generative and transformational in health and healing.
- The views identify the caring relationship as the complexity in nursing science. The view offers a challenge to the notion of self-organization, a central thesis of complexity sciences, toward the view of *relational* self-organization (Ray, 1994; Turkel & Ray, 2000, 2001, 2004, 2009).
- The theories and practice of nursing as complex caring dynamics calls for an understanding of nursing as relational caring patterns of meaning.
- The relational caring patterns of meaning include a multidisciplinary focus.

- Complex relational caring dynamics is recognized as knowledge-in-process, the epistemological basis of reality (Ray, 1994, 2010; Turkel & Ray, 2009).
- Spiritual–ethical caring is at the center of the complex relational caring practice of nursing in health care organizations.

REFERENCES

Al-Khalil, J. (2003). *Quantum: A guide for the perplexed*. London: Weidenfeld & Nicolson.

Arntz, W., Chasse, B., & Vicente, M. (2005). *What the bleep do we know!?* Deerfield Beach, FL: Health Communications.

Bartol, G. (1989). Story in nursing practice. *Nursing and Health Care, 10*(10), 564–565.

Bar-Yam, Y. (2004). *Making things work: Solving complex problems in a complex world*. Boston: NECSI. Knowledge Press.

Baum, G. (1975). *Religion and alienation*. New York: Paulist Press.

Bell, D. (1974). *The coming of post-industrial society*. New York: Basic Books.

Briggs, J., & Peat, F. (1989). *Turbulent mirror: An illustrated guide to chaos theory and the science of wholeness*. New York: Harper & Row.

Briggs, J., & Peat, F. (1999). *Seven lessons of chaos: Timeless wisdom from the science of change*. New York: HarperCollins.

Britan, G., & Cohen, R. (Eds.). (1980). *Hierarchy & society*. Philadelphia: ASHI.

Cannato, J. (2006). *Radical amazement*. Notre Dame, IN: Sorin Books.

Capra, F., Steindl-Rast, D., & Matus, T. (1991). *Belonging to the universe*. San Francisco: HarperSanFrancisco.

Centers for Medicare and Medicaid Services. Hcahpsonline.org. Baltimore, MD. Retrieved from http://www.hcahpsonline.org

Coffman, S. (2006). Marilyn Anne Ray: Theory of bureaucratic caring. In A. Marriner Tomey & M. Alligood (Eds.), *Nursing theorists and their work* (6th ed., pp. 116–139). St. Louis, MO: Mosby/Elsevier.

Coffman, S. (2010). Marilyn Anne Ray: Theory of bureaucratic caring. In M. Alligood & A. Marriner Tomey (Eds.). *Nursing theorists and their work* (7th ed., pp.113–136). St. Louis, MO: Mosby/Elsevier.

Davidson, A., & Ray, M. (1991). Studying the human-environment phenomenon using the science of complexity. *Advances in Nursing Science, 13*, 73–87.

Downs, A. (1993). *Inside bureaucracy*. Long Grove, IL: Waveland Press.

Eisler, R. (2007). *The real wealth of nations: Creating a caring economics*. San Francisco: Berrett-Koehler.

Elliston, F., & McCormick, P. (Eds. & Intro.) (1977). *Husserl*. Notre Dame, IN: University of Notre Dame Press.

Foa, U. (1971). Interpersonal and economic resources. *Science, 71*(29), 345–351.

Geertz, C. (1973). *The interpretation of cultures*. New York: Basic Books, Publishers.

Glaser, B., & Strauss, A. (1967). *The discovery of grounded theory: Strategies for qualitative research*. Chicago: Aldine.

Hudacek, S. (2000). *Making a difference: Stories from the point of care*. Indianapolis, IN: Center Nursing Press.

Husserl, E. (1970). *The crisis of European sciences and transcendental phenomenology: An introduction to phenomenological philosophy* (Trans. and Intro. by D. Carr). Evanston, IL: Northwestern University Press.

Institute of Medicine. (2001). *Crossing the quality chasm: A new health system for the 21st century*. Washington, DC: National Academy Press.

Langer, S. (1953). *Feeling and form* New York: Charles Scribner's Sons.

Leininger, M. (1978). *Transcultural nursing: Concepts, theories, and practices*. New York: Wiley.

Leininger, M. (Ed.). (1981). *Caring: An essential human need* (pp. 25–36). Thorofare, NJ: Charles B. Slack.

Leininger, M. (1991). *Culture care diversity and universality: A theory of nursing*. New York: National League for Nursing Press.

Leininger, M., & Mc Farland, M. (2006). *Culture care diversity and universality: A worldwide theory of nursing* (2nd ed.). Sudbury, MA: Jones & Bartlett Publishers.

Leslie, R. (2008). *Angels in the ER*. Eugene, OR: Harvest House.

Lindberg, C., Nash, S., & Lindberg, C. (2008). *On the edge: Nursing in the age of complexity*. Bordentown, NJ: Plexus Press.

Newman, M., Sime, M., & Corcoran-Perry, S. (1991). The focus of the discipline of nursing. *Advances in Nursing Science, 14*, 1–6.

Newman, M., Smith, M., Dexheimer-Pharris, M. & Jone, D. (2008). The focus of the discipline revisited. *Advances in Nursing Science, 31*, E16–E27.

Nightingale, F. (1992). *Notes on nursing: What it is, what it is not*. Philadelphia: J. B. Lippincott (Original Publication, 1959).

Peat, F. (2003). *From certainty to uncertainty: The story of science and ideas in the twentieth century*. Washington, DC: Joseph Henry Press.

Perrow, C. (1986). *Complex organizations: A critical essay* (3rd ed.). New York: McGraw-Hill.

Ray, M. (1981a). A study of caring within an institutional culture (PhD dissertation, University of Utah, 1981a). *Dissertation Abstracts Internationaal, 42*(06) (University Microfilms No. 81277787).

Ray, M. (1981b). A philosophical analysis of caring within nursing. In M. Leininger (Ed.), *Caring: An essential human need* (pp. 25–36). Thorofare, NJ: Charles B. Slack.

Ray, M. (1984). The development of a classification system of institutional caring. In M. Leininger (Ed.), *Care: The essence of nursing and health* (pp. 95–112). Thorofare, NJ: Charles B. Slack.

Ray, M. (1987a). Health care economics: Why the moral conflict must be resolved. *Family and Community Health, 10*, 35–43.

Ray, M. (1987b). Technological caring: A new model in critical care. *Dimensions of Critical Care Nursing, 6*, 166–173.

Ray, M. (1989a). The theory of bureaucratic caring for nursing practice in the organizational culture. *Nursing Administration Quarterly, 13*, 31–42.

Ray, M. (1989b). Transcultural caring: Political and economic visions. *Journal of Transcultural Nursing, 1*(1), 17–21.

Ray, M. (1994a). Transcultural nursing ethics: A framework and model for transcultural ethical analysis. *Journal of Holistic Nursing, 12*, 251–264.

Ray, M. (1994b). Complex caring dynamics: A unifying model of nursing inquiry. *Theoretic and Applied Chaos in Nursing, 1*, 23–32.

Ray, M. (1997a). The ethical theory of existential authenticity: The lived experience of the art of caring in nursing administration. *Canadian Journal of Nursing Research, 29*, 111–126.

Ray, M. (1997b). Illuminating the meaning of caring: Unfolding the sacred art of divine love. In M. Roach (Ed.), *Caring from the heart: The convergence of caring and spirituality* (pp. 163–178). New York: Paulist Press.

Ray, M. (1998). Complexity and nursing science. *Nursing Science Quarterly, 11*, 91–93.

Ray, M. (2001). Marilyn Anne Ray, the theory of bureaucratic caring. In M. Parker (Ed.), *Nursing theories and nursing practice* (pp. 433–444). Philadelphia, PA: F. A. Davis.

Ray, M. (2006). Marilyn Anne Ray, the theory of bureaucratic caring. In M. Parker (Ed.), *Nursing theories and nursing practice* (2nd ed., pp. 434–444). Philadelphia: F. A. Davis.

Ray, M. (2007). Technological caring as a dynamic of complexity in nursing practice. In A. Barnard & A. Locsin (Eds.), *Technology and nursing: Practice, concepts and issues* (pp. 174–190). New York: Palgrave/Macmillan.

Ray, M. (2010a). *Transcultural caring dynamics in nursing and healthcare*. Philadelphia: F. A. Davis.

Ray, M. (2010b). *A study of caring within an institutional culture*. Saabriicken, Germany: Lambert Academic Publishing.

Ray, M., DiDominic, V., Dittman, P., Hurst, P., Seaver, J., Sorbello, B., et al. (1995). The edge of chaos: Caring and the bottom line. *Nursing Management, 9*, 48–50.

Ray, M., & Turkel, M. (1996). *Econometric analysis of the nurse-patient relationship 1, 11*. Grant funded by Department of Defense, Tri-Service Nursing Research Council, Uniformed Services University of the Health Sciences.

Ray, M., & Turkel, M. (2000). *Economic and patient outcomes of the nurse-patient relationship*. Grant funded by Department of Defense, Tri-Service Nursing Research Council, Uniformed Services University of the Health Science.

Ray, M., & Turkel, M. (2004). *Economic and patient outcomes of the nurse-patient relationship*. Research Data, Final Report. Submitted to Department of Defense, Tri-Service Nursing Research Council, Uniformed Services University of the Health Science.

Ray, M., & Turkel, M. (2009). Relational caring questionnaires. In J. Watson (Ed.), *Assessing and measuring caring in nursing and health sciences*. New York: Springer.

Ray, M., & Turkel, M. (2010). Marilyn Anne Ray: The theory of bureaucratic caring. In M. Parker & M. Smith (Eds.), *Nursing theories and nursing practice* (3rd ed.), (pp. 472–494) Philadelphia: F. A. Davis.

Ray, M., Turkel, M., & Marino, F. (2002). The transformative process for nursing work force redevelopment. *Nursing Administration Quarterly, 26*(2), 1–14.

Roach, M. (2002). *Caring, the human mode of being: A blueprint for the health professions* (2nd Rev. ed.). Ottawa, ON: CHA Press.

Rogers, M. (1970). *An introduction to the theoretical basis of nursing*. Philadelphia: F. A. Davis.

Rogers, M. (1994). The science of unitary human beings: Current perspectives. *Nursing Science Quarterly, 2*, 33–35.

Society of Hospital Medicine (2010). *BOOSTing care transitions resource room*. Retrieved from http://www.hospitalmedicine.org/BOOST

Stacey, R. (2003). *Strategic management and organizational dynamics*. London: Pearson Education.

Sumner, J. (2008). *The moral construct of caring in nursing as communicative action*. Saarbrücken, Germany: VDM Verlag Dr. Müller.

Swinderman, T. (2005). *The magnetic appeal of nurse informaticians: Caring attractor for emergence.* Doctor of Nursing Science Dissertation, Florida Atlantic University, College of Nursing, Boca Raton, Florida.

Turkel, M. (1997). Struggling to find a balance: A grounded theory of the nurse-patient relationship within an economic context. *Dissertation Abstracts International, 58*(8) (UMI No. 9805938).

Turkel, M. (2001). Struggling to find a balance: The paradox between caring and economics. *Nursing Administration Quarterly, 26*(1), 67–82.

Turkel, M. (2006). Applications of Marilyn Ray's theory of bureaucratic caring. In M. Parker (Ed.), *Nursing theories & nursing practice* (2nd ed., pp. 369–379). Philadelphia: F. A. Davis.

Turkel, M. (2007). Dr. Marilyn Ray's theory of bureaucratic caring. *International Journal for Human Caring, 11*, 57–74.

Turkel, M., & Ray, M. (2000). Relational caring complexity: A theory of the nurse–patient relationship within an economic context. *Nursing Science Quarterly, 13*(4), 307–313.

Turkel, M., & Ray, M. (2001). Relational complexity: From grounded theory to instrument development and theoretical testing. *Nursing Science Quarterly, 14*(4), 281–287.

Turkel, M., & Ray, M. (2004). Creating a caring practice environment through self-renewal. *Nursing Administrative Quarterly, 28*(4), 249–254.

Turkel, M., & Ray, M. (2009). Caring for "not-so-picture-perfect patients." In R. Locsin & M. Purnell (Eds.), *A contemporary nursing process: The (un)bearable weight of knowing in nursing*. New York: Springer.

Wagner, L. (2000). Connecting to nurse-self through reflective poetic story. *International Journal for Human Caring, 4*(2), 6–12.

Watson, J. (1988). *Nursing: Human science, human care*. New York: National League for Nursing Press.

Watson, J. (2005). *Caring science as sacred science*. Philadelphia: F. A. Davis.

Watson, J. (2009). *Nursing: The philosophy and science of caring* (Rev. ed.). Boulder, CO: University Press of Colorado.

Watson Caring Science Institute-International Caritas Consortium. Retrieved April 20, 2010, from http://www.watsoncaringscience.org/j_watson/index.html

Weber, M. (1947). *The theory of social and economic organization* (Trans. by A. Henderson & T. Parsons). New York: The Free Press of Glencoe.

Wheatley, M. (2006). *Leadership and the new science: Discovering order in a chaotic world* (3rd ed.). San Francisco: Berrett-Koehler.

Response to Chapter 4

Terry Eggenberger

REFLECTIVE QUESTIONS

1. What is the relationship between theory and practice (clinical, administrative, or educational) highlighted in this chapter?

This chapter highlights clinical, administrative, and organizational relationships. Relationships are at the core of all exchanges and interactions that occur within the hospital system. The struggle for peace and harmony and finding ultimate joy in one's practice, is found in the quest to respond successfully to the unique needs of each patient, colleague, and organization. There is also fluidity in the outcome measures that are determined by the context and the perspective. As this chapter reveals, nurses are indeed attracted to nursing by the "magnetic appeal" identified as caring. Caring is a loving transpersonal relationship that illuminates harmony of body, mind, and spirit (Watson, 2005, 2008). When desired patient outcomes are linked correspondingly to desired organizational outcomes, true synergy occurs. The struggle to find a balance between patient needs and organizational values cease, and the struggle is won. This understanding is based on the recognition that hospitals exist for the delivery of nursing care. As complex systems, hospitals must be efficient, effective, and innovative to remain economically viable and create a future of sustainability for nurse caring.

Administratively, there is the importance of understanding and viewing caring relationships in context and the nurses' role within those complex relationships. Ray's theory of bureaucratic caring addresses the human caring environment, and the spiritual–ethical relationships within that environment. The focus of the discipline of nursing was first defined as "the study of caring in the human health experience" (Newman et al., 1991, p. 3). The authors raised questions regarding what it is about the relationship between nurse and patient that transforms the patient's experience. There is an acknowledgment of the feeling of interconnectedness that exists within the patient's own experience of health, within the context of his/her own understanding. Newman et al. (2008) shifted the focus of the discipline of nursing from caring to the relationship. The value of "caring presence" was highlighted as being transformative for both the "patient and the nurse" (E17). When nurses are "authentically present" (E23) patients then feel that their "concerns are being addressed" (E18).

Utilizing reflective practice in educational endeavors allows for the professional student who returns to academia to find hope for seemingly insurmountable obstacles that have been encountered. For example, as registered nurses pursuing their bachelor's degree are introduced to Margaret Wheatley's (2006) book and video related to complexity sciences, they begin to understand the complexity of interactions and the environment that surrounds them. Ray, Turkel, and Cohn (2010) stipulate that nurses then find meaning in the "cocreated" process; the pull of the "microculture" (the relationship) and "macroculture" (the hospital culture with technical, political, economic, and legal dimensions) are synthesized rather than being diametrically opposed. Instead of utilizing their energy reacting, nurses can begin to utilize their energy proactively when advocating on behalf of their patients in their professional practice, and cocreate a "synthesis" between the needs of patients and the sociocultural world of the health care system. There is an energy flow that occurs between the nurse and patient as the nurse seeks with intention to facilitate the well-being of the patient within the context of the hospital.

2. How does a nurse or physician or health care administrator learn about complexity sciences and caring through the presentation of ideas or research?

Complexity's central tenet is interconnectedness. Administrators, physicians, and nurses are interconnected with the patient and family in their efforts to preserve caring and human dignity. At the edge of chaos (a theory within complexity sciences, which illuminates the idea of a bifurcation point between disorder or disease and transformation toward health and healing), there is the need for complex clinical decision making. Today's patient care dilemmas are complex. Nurses learn about complexity sciences by pursuing continued education. Spiritual–ethical caring is the center core of the model identified as bureaucratic caring. A model informs administrators, physicians, and nurses about what they are obligated to do, and to "guide practice and decision making" (Turkel, 2006, p. 369). Ethical concepts of beneficence (doing good), nonmaleficence (doing no harm), autonomy (allowing choice), and justice (being fair) all exist in the caring relationship and should be the foundation of ethical choice making in all health care relationships.

Many times, the most effective learning takes place when modeled by those who are engaged in practice and are pursuing additional education. For example, with the issue of family presence in emergent or "code" situations, nurses or physicians who are in school or are educators, and who are members of the response team, can inform nurses, physicians, and administrators regarding research being done, which supports allowing the family members to remain present in the dying process, the final health experience. Systems should be based on what is good for the patient and the health care organization. Shared exemplars or stories, such as the examples presented by Dr. Cohn in this chapter, bring the concepts being introduced together for the other members of the health care team, and provide a framework that includes the relevance of complexity and caring theory to practice. Theory then comes alive as it functions to drive the experience through a shared understanding of complexity science and caring relationships.

In this chapter, Dr. Cohn offers one way to incorporate learning about complexity sciences and caring through practice, from the lens of quality improvement. Complexity sciences are the science of quality. Meaning is found in the shared methods and tools of quality. The concept of chaos drives the needed change in organizations. Nurses can go to the evidence to share best practices and guidelines and advocate for patients where quality of care is the focal point. Nurses can assist the team, including the patient, in moving from disorder to order, by engaging in partnership with patients and families as they make caring ethical choices. Transformation occurs as the parts are once again integrated into a whole. Patient-centered care is relationship-centered care. All members of the health care team, including the patient, need to have a voice at the table. The patient does not stand alone because he or she is in a caring relationship with the nurse. The nurse is the patient's advocate and voice.

3. What meanings are illuminated in this chapter for nursing or other health care professions? Direct your answer to education, administration, research, or practice based upon the focus of the chapter.

Implications for nursing and interdisciplinary education, practice, and research include the need to have opportunities to allow a response from the team as a whole where the contextual needs are addressed and health care members respond with both compassion and competency, including conscience, confidence, commitment, and comportment (Roach, 2002). It is important to first identify "what matters most" (Boykin & Schoenhofer, 2001) to the patient and family so that calls from the patient are understood and valued. The responses from the team of caregivers utilize the full scope of practice and their individual talents, gifts, and caring practices.

Health care providers need to understand the global mission (organization) and the unit-level mission; the missions need to be congruent. This understanding can be enhanced through the integration of theory of bureaucratic caring. Bureaucratic caring facilitates understanding families. Whether discussing family life or organizations, each is concerned with context. Economics involves an exchange. This exchange can be of ideas or of the heart, as well as of services, goods, or monetary value. Health care providers also make decisions about how they will use their time in terms of what they make or take time to do. Caring can be found in what people pay attention to. When a patient calls, nurses make the choices to leave the task they are doing in order to respond.

Every decision that the nurse makes involves ethical choice making. Nurses need to respond with compassion, competency, and ethical choice making from the perspective of what they ought to be doing. At the core, nurses make conscious decisions to humanize the system. Nurses can be supported in their choice making by leaders who share their view of the values, meaning, and importance of caring relationships.

4. What is the relevance of this work to the future of the discipline and profession of nursing, health care professions, and health care in general?

Expert practice has been previously defined as developing intuition from repetitive experiences with similar situations (Benner, 1984; Benner, Sutphen, Leonard, & Day, 2010). Clinical decision making involves being able to critically think within the patient's unique situation. Bureaucratic caring provides a framework for recognizing the institutional resources from which clinical leaders and patient care providers can draw in formulating a response. It is virtually impossible to have policies and standards that address all cases of moral dilemmas that may arise. Spiritual–ethical caring is at the center or core of the theory because caring exists across all cultures, is universal in nursing, and is contextual within specific hospital cultures. Nurses find hope within the theory because it is possible to practice excellence in nursing within the context of the current health care environments when there is appreciation and understanding of bureaucratic caring to guide clinical decision making.

Patient-centered care provides an opportunity for health care professionals to consider the role of the patient and family on the team. Programs such as "Nothing about me, without me" (The Joint

Commission, 2008) offer opportunities for cross utilization at an organizational level as well. Nurses have long understood that many of the changes they experience are nonsustainable because they or others on the health care team have not been at the table when plans were made. Patient-centered care makes the transition to relational caring complexity.

Nurses must understand that it is "caring" to be fiscally responsible. These values are not incongruent. With economic value being tied to satisfaction measures, such as the use of the Hospital Consumer Assessment of Healthcare Providers and Systems (HCAHPS) (CAHPS, 2009) survey, it is crucial that nurses and other health care providers study frameworks for caring leadership that interlink the values of the patients and families with the professional values of the caregivers and the values of the organization.

Ray and Turkel have united to come full circle with this chapter in their journey. Each of their nursing practices initially generated theory. Together they have continued to build theory and enhance other theories through their combined scholarship. Ray's theory of bureaucratic caring (1989) was visionary in its insight into the relevance of patient outcomes and satisfaction on economics through the "complex nurse–patient relational caring process" (p. 360). Context is found "where nursing is lived" (p. 361). Turkel and Ray (2000, 2001) researched the nurse–patient relationship as an economic resource. Administrators were seen as interconnected to this relationship because they control the economics and make the decisions, which impact nurses. Through research and insight, the authors (Turkel & Ray, 2000, 2001) have changed relational complexity theory to relational caring complexity theory. Knowledge of theory in hospitals facilitates initiatives to reduce costs, increase revenue, and manage the delivery of patient care in the current complex health care environment. Paradoxically, research now supports maximizing the amount of time spent on "direct nurse–patient interactions" (Turkel, 2006, p. 375) to promote safety and reduce errors. Cohn drew from this caring science and concepts from complexity science to actualize theory in practice when managing quality initiatives. The quality teams have taken an ethical perspective when focusing on what ought to happen in the relationship among the physicians, patients, and nurses. Therefore, theories generated from research in practice, such as bureaucratic caring and relational caring complexity highlight spiritual–ethical caring wisdom, furthering the idea generated by Watson (2008) of a transpersonal caritas (love) consciousness, which acknowledges the deeply human and spirit-filled loving nature of professional nursing, and other health care professionals (Watson, 2005, 2008).

REFERENCES

Benner, P. (1984). *From novice to expert: Excellence and power in clinical nursing practice*. Menlo Park, CA: Addison-Wesley.

Benner, P., Sutphen, M., Leonard, V., & Day, L. (2010). *Educating nurses: A radical transformation* (The Carnegie Foundation for the Advancement of Teaching). San Francisco: Jossey-Bass.

Boykin, A., & Schoenhofer, S. (2001). *Nurse as caring: A model for transforming practice*. Sudbury, MA: Jones & Barlett.

CAHPS Hospital Survey. (2009). *HCAHPS fact sheet*. Retrieved May 22, 2010, from http://www.hcahpsonline.org/files/HCAHPS%20Fact%20Sheet,%20revised1,%203-31-09.pdf

Newman, M., Sime, M., & Corcoran-Perry, S. (1991). The focus of the discipline of nursing. *Advances in Nursing Science, 14*, 1–6.

Newman, M. A., Smith, M. C., Dexheimer-Pharris, M. D., & Jones, D. (2008). The focus of the discipline revisited. *Advances in Nursing Science, 31*(1), E16–E27.

Ray, M. (1989). The theory of bureaucratic caring for nursing practice in the organizational culture. *Nursing Administration Quarterly, 13*, 31–42.

Roach, M. (2002). *Caring, the human mode of being: A blueprint for the health professions* (2nd Rev. ed.). Ottawa, ON: CHA Press.

The Joint Commission. (2008). *Health care at the crossroads: Guiding principles for the development of the hospital of the future*. Retrieved May 22, 2010, from http://www.jointcommission.org/NR/rdonlyres/1C9A7079-7A29-4658-B80DA7DF8771309B/0/Hosptal_Future.pdf

Turkel, M. (2006). Applications of Marilyn Ray's theory of bureaucratic caring. In M. Parker (Ed.), *Nursing theories & nursing practice* (2nd ed., pp. 369–379). Philadelphia: F. A. Davis.

Turkel, M., & Ray, M. (2000). Relational caring complexity: A theory of the nurse-patient relationship within an economic context. *Nursing Science Quarterly, 13*, 307–313.

Turkel, M., & Ray, M. (2001). Relational complexity: From grounded theory to instrument development and theoretical testing. *Nursing Science Quarterly, 14*, 281–287.

Watson, J. (2005). *Caring science as sacred science*. Philadelphia: F. A. Davis Company.

Watson, J. (2008). *Nursing: The philosophy and science of caring* (2nd Rev. ed.). Boulder, CO: University Press of Colorado.

Wheatley, M. (2006). *Leadership and the new science: Discovering order in a chaotic world* (3rd ed.). San Francisco: Berrett-Koehler.

CHAPTER 4
RELATIONAL CARING COMPLEXITY: THE STUDY OF CARING AND COMPLEXITY IN HEALTH CARE HOSPITAL ORGANIZATIONS

- *What ethical caring decisions do you make in your organization or area of practice?*
- *How are economic decisions made in your organization?*
- *The chapter discussed other bureaucratic caring dimensions, such as technological, political, and legal caring. What do they*

mean to you? How are they lived in your organization or educational facility?

- *Dr. Cohn wrote a story to describe the patient's experience and understanding of his illness. Can you think of story that you could write or share with a peer that describes a patient's experience with a caring health care provider?*
- *How is the art of nursing practiced in your organization? Is a tangible expression of this art present in the organization?*
- *In the response, Terry Eggenberger wrote about complexity science and bureaucratic caring issues in organizations and education. How could you incorporate concepts from caring science and complexity science into education related to your area of practice?*

FIVE

Why Six Sigma Health Care Is Oxymoronic in Hospitals

Bruce J. West

The Six Sigma philosophy for managing resources, so successful in product manufacturing, is counterproductive in managing a social network such as health care. The basic flaw concerns the inapplicability of normal statistics to describe the dynamics of complex social phenomena. Data show that inverse power-law distributions capture the variability of social phenomena, and we briefly discuss the modeling of complex networks necessary to understand the source of this new kind of distribution. More important, the inverse power law suggests the existence of a fundamental inequity in social organizations, a phenomenon that we describe. For health care leaders and providers, the key point is that although there is a strong body of knowledge supporting statistical methods such as Six Sigma, there is an equally powerful body of knowledge in the inverse power law that disputes the value of Six Sigma. Turkel's response addresses how caring relationships are complex social phenomena. As such, the application of Six Sigma does not sufficiently measure the value of the human-to-human interactions.

INTRODUCTION

1. Health Care and Complexity

Let me begin by stating that I am not an expert on health care. What I do know concerns the wisdom and dedication of individual health care providers working in organizations that oftentimes have different priority lists. Given this disclaimer, I address the question of how the individual health care provider, seen as one element of a network, albeit a complex element, can influence the larger and also complex organization network. Much of what I have to say in this essay is at a fairly abstract level, meaning that I do not address any specific problem of interest to the health care practitioners. This is both a strength and weakness of my comments.

As Litaker, Tomolo, Liberatore, Strange, and Aron (2006) point out, the conceptual framework for health care quality improvement is based on a mechanical view of human networks and focuses on reducing variability through incremental change. They go on to say that by not "identifying desirable variation of unique adaptation" we miss the opportunity to facilitate better health care outcomes. This view is an application of modern human resource management theory, which can be traced back to Joseph M. Juran who wrote the standard reference works on quality control (Juran, 1964). It is his work on management that evolved into the Six Sigma program, which is the basis of a worldwide quality management school. The Six Sigma program has achieved a certain currency because of its emphasis on metrics and the importance of being able to quantify the problems being addressed within an organization. What we argue is that the Six Sigma approach to management is fundamentally incompatible with health care goals, and its application is therefore counterproductive when applied to health care metrics.

2. The Gauss Worldview

The name Six Sigma is taken from the bell-shaped probability distribution (see Figure 5.1) constructed by the mathematician Johann Carl Friedrich Gauss (1777–1855), to quantify the variability observed in the results of experimental measurements taken in the physical sciences. In the Gauss worldview, the average value of an observable is the single most important quantity for characterizing a phenomenon, and the fluctuations around the average value are due to measurement errors and random environmental fluctuations. The parameter sigma (σ) quantifies these fluctuations, and in statistics, sigma is called the standard deviation; the smaller the σ value relative to the average, the better the average represents the data. Consequently, in this view, the world consists of networks made up of linear additive processes, whose variability is described by a bell-shaped curve, centered on the average value, with a width determined by sigma.

The Gauss model is very useful in manufacturing, where the specifications for the production of a widget can be given to as high a tolerance as can be achieved on a given piece of machinery. Of course, no two widgets coming off the production line are exactly the same—there is always some variability. Let us suppose that the widget can be characterized by a specified diameter and the average diameter converges on the specified value. The variation around the average diameter value is given by a bell-shaped curve. If all the measured diameters are divided by the standard deviation in a given production run, then the new variable is expressed in terms of the standard deviation σ. In Figure 5.1, the results of producing this hypothetical widget are shown. The attraction of this approach is the universality of the distribution shape when the standard deviation is the unit of measure. All processes that have Gaussian or normal statistics can be superimposed on the given curve. A well-known property of the normal distribution is that 68% of all

FIGURE 5.1 The Normal Distribution in Units of σ Centered on the Average
The typical partitioning for grading in college is indicated, with the distribution providing the relative number of students in each interval.

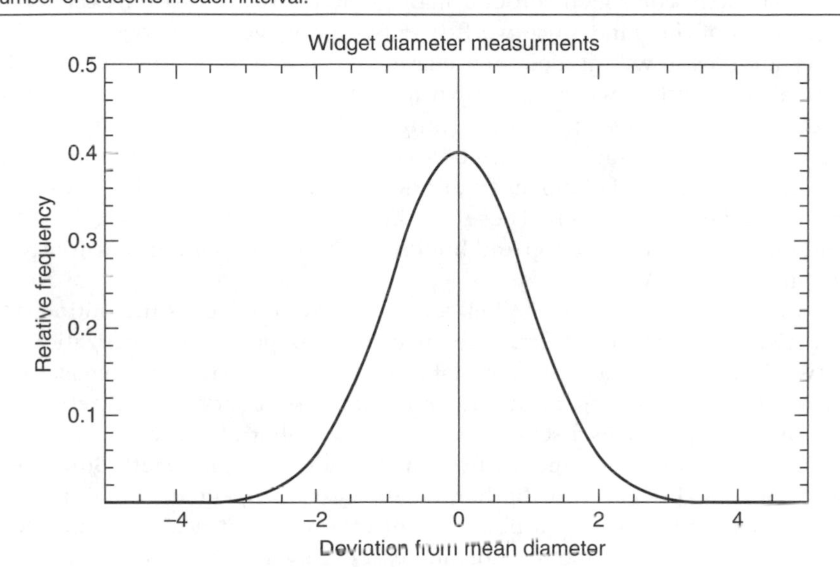

widgets produced fall between plus and minus one sigma; 95% of all widgets produced fall between plus and minus two sigma; 99.7% of all widgets produced fall between plus and minus three sigma, and so on.

The Six Sigma philosophy maintains that the variability seen in Figure 5.1 is an undesirable property in widget production and should be eliminated. Outcome uniformity at the specified tolerance level is the goal of managing the manufacturing process. To make the implications of this distribution more transparent, let us assume that one million widgets are sold and 99.7% of them are within the specified tolerance limits. On its face, this would appear to be very good, with the production line functioning at the three sigma level. On the other hand, this also implies that the company will have 3,000 unhappy customers, because they receive widgets that do not satisfy specifications. If the widgets were armor vests slated to be used by soldiers in Iraq, a three sigma level would not be acceptable. A six sigma level of 99.9997%, with its approximately 30 faulty vests out of the million shipped, would be the more acceptable number.

Therefore, this is what the Six Sigma program is all about. How to take a manufacturing plant, an organization, or any other activity whose output has an unacceptable level of variability and reduce that variability to a six sigma level; variability is assumed to be bad, and uniformity is assumed to be good.

When first introduced, the six sigma argument seems to be rather good, with simple but acceptable assumptions about the process being measured and ultimately controlled through management. We have been exposed to arguments of this kind since we first entered college and experienced that every large class was graded on a curve. That curve invariably involves the normal distribution, where as shown in Figure 5.1, between +1 and −1 lie most students' grades. This is the C range, between plus and minus one σ of the class average, which includes 68% of the students. The next range is the equally wide B and D interval from one to two σ in the positive and negative directions, respectively. These two intervals capture another 27% of the student body. Finally, the top and bottom of the class split the remaining 5% equally between A and F.

In the social sciences, the bell-shaped curve introduces the notion of a statistical performance measure relative to some goal. An organization, or network, is said to have a quantifiable goal when a particular measurable outcome serves the organization's purpose. A sequence of realizations of this outcome produces a set of data, say the number of successfully treated patients discharged per month by members of a hospital staff. Suppose a hospital's ideal successful discharge rate (*sdr*) is specified in some way. The actual *sdr* is not a fixed quantity but varies a great deal from month to month. If 68% of the time the hospital essentially achieves its ideal *sdr*, the hospital is functioning at the one sigma level. If another hospital has 95% of its realizations at essentially the ideal *sdr*, it is functioning at the two sigma level. A third hospital, one that seems to be doing extremely well, delivers 99.4% of the ideal *sdr* and is functioning at the three sigma level. Finally, a six sigma hospital delivers 99.9997% of its discharges at, or very nearly, the ideal *sdr* (Juran, 1962). What six sigma means is that the variability in the *sdr* has been reduced to nearly zero, and based on a linear additive model of the world this ought to be both desirable and achievable. The logic of assessing the hospital's quality is the same as that used to assess the students' quality and relies just as much on the bell-shaped curve.

However, does the bell curve apply in the case of complex networks?

Recently, I came across a paper where the authors analyzed the achievement tests of more than 65,000 students graduating high school and taking the university entrance examination of *Universidade Estadual Paulista* (UNESP) in the state of Sao Paulo, Brazil (Gupta, Campanha, & Chavorette, 2003). In Figure 5.2, the entrance exam results are recorded for high- and low-income students. It is clear that the humanities data in Figure 5.2a seem to support the conjecture that the normal distribution is appropriate for describing the distribution of grades in a large student population. The solid curves are the best fits of a normal distribution to the data and give a different mean and width between public and private schools. In the original publication, the data were represented in a variety of ways, such as between the rich and poor, but that does not concern us here because the results turned out to be

FIGURE 5.2 The Distribution of Grades for 65,000 Students on the University Entrance Examination Universidade Estadual Paulista (UNESP) in the State of Sao Paulo, Brazil (Gupta, et al., 2003): (a) Humanities, (b) Physical Sciences and (c) Biological Sciences.

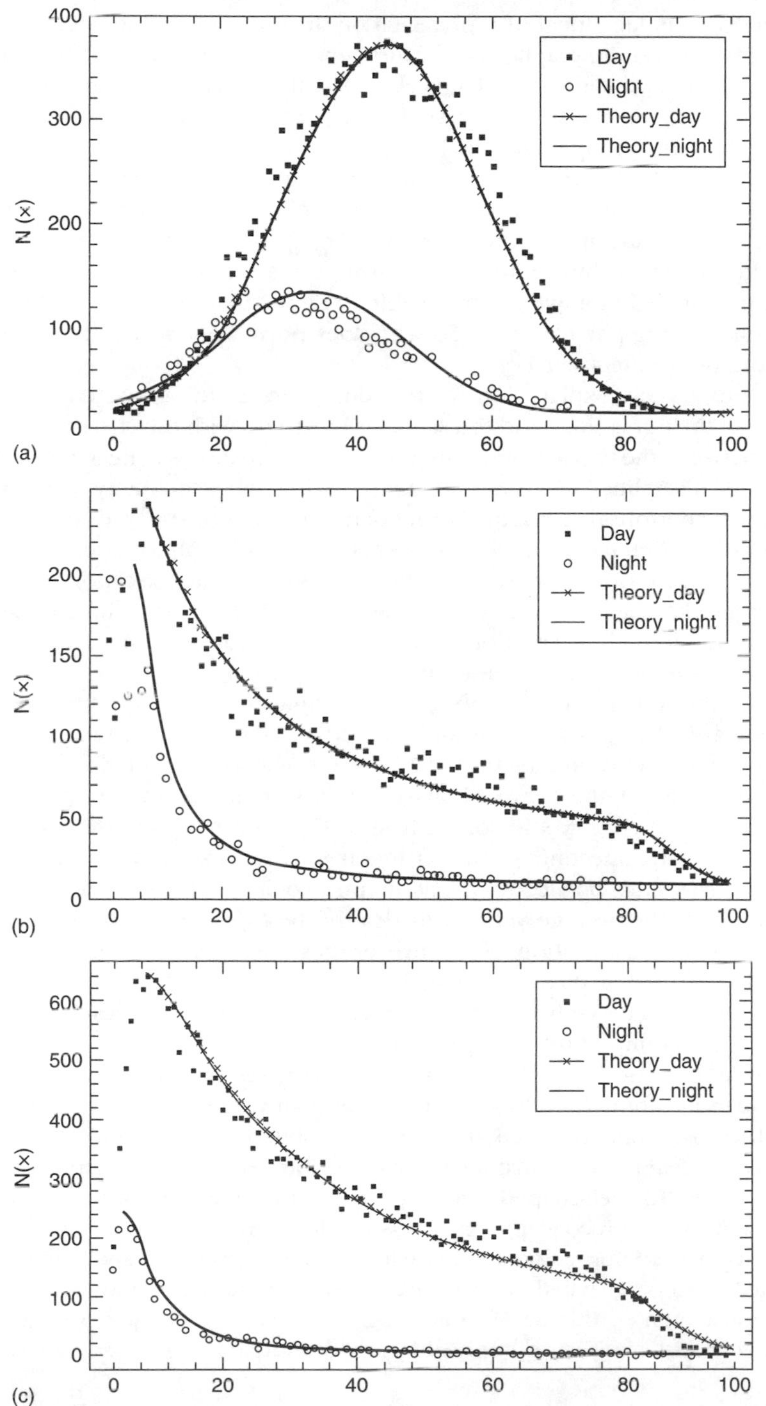

qualitatively independent of representation. In Figure 5.2b, the data from the physical sciences are graphed under the same grouping as that of the humanities in Figure 5.2a. One thing that is clear is that the distribution is remarkably different from the bell-shaped curve. Figure 5.2c depicts the grade distribution under the same grouping for the biological sciences.

The first thing to notice is that the grade distributions for the biological sciences are more like those in the physical sciences than they are like those in the humanities. In fact, the distribution of grades in the sciences is nothing like that in the humanities. The distributions are so different that if they were not labeled one would not be able to detect that they refer to the same general learning phenomenon. So why does normalcy apply to the humanities and not to the sciences?

One possible explanation for this difference in the grade distributions between the humanities and the sciences has to do with the structural difference between the two learning categories. The humanities collect a disjointed group of disciplines including language, philosophy, sociology, economics, and a number of other relatively independent areas of study. Consequently, the grades obtained in each of these separate disciplines are essentially independent of one another, or at most weakly dependent on one another, thereby satisfying the conditions of the central limit theorem for the derivation of the normal distribution. In meeting conditions originally articulated by Gauss, the humanities grade distribution takes on normalcy.

On the other hand, every science builds on previous knowledge. Elementary physics cannot be understood without algebra, and the more advanced physics cannot be understood without calculus, which also requires an understanding of algebra. Similarly, understanding biology requires mastery of some chemistry and some physics. The different scientific disciplines form an interconnecting web, starting from the most basic and building upward. This situation violates the assumption of independence and undercuts the idea that the average value provides the best description of the process. The empirical distribution of science grades clearly shows extensions out into the tail of the distribution with no clear peak and, consequently, no characteristic value with which to characterize the data. As a result, the average values, so important in normal processes, become irrelevant in complex networks. A better indicator of complex processes than the average is one that quantifies how rapidly the tail decreases in value.

The distinction between the grade distributions in the humanities and sciences is clear evidence that the normal distribution does not describe the normal situation. The bell-shaped grading curve is imposed through education orthodoxy and by our preconceptions, and is not indicative of the process by which students master information and knowledge. Thus, the pursuit and achievement of intellectual goals, whether in science or engineering, is not normal.

Consequently, the Six Sigma program, being crucially dependent on normal statistics, is not applicable to quantifying complex phenomena such as learning or health care delivery in hospitals. The variability that the Six Sigma program targets for elimination under the assumption of normal statistics is invalidated by the long-tailed distribution observed in truly complex networks. The tails indicate an intrinsic variability that is not contained in the simpler processes where normal statistics are valid, and working to reduce the tail region may in fact remove the very property that makes the process valuable. With this in mind, let us turn our attention away from the now discredited application of the Gauss distribution, at least discredited as being a viable description of the outcome for complex phenomena, and examine the first scientist's arguments to recognize the existence of the long tails depicted in Figure 5.2.

3. The Pareto Worldview

We need to have at least a preliminary understanding of how human networks operate in order to determine how to manage health care providers. In order to achieve this primitive level of understanding, let us sketch how members of a community choose from a large number of options. Suppose that a large cohort group has a hypothetical set of choices, say a large set of nodes on a computer network, to which they may connect. If we assume that the choices are made independently of the node's quality, or without regard for the selections made by other group members, the resulting distribution in the number of times a given node is selected has the familiar bell shape.[1] In this case, the selection process is completely uniform with no distinction based on personal taste, peer pressure, or aesthetic judgment, resulting in a network of random links between humans and nodes, or more generally between individuals.

A bell-shaped distribution describes the probable number of links a given node has in a random network. Barabási (2003) determined such distributions to be unrealistic; using real-world data, he was able to show that the numbers of connections between nodes on the Internet and on the World Wide Web deviate markedly from the bell-shaped distribution. In fact, he, along with others (Strogatz, 2003; Watts, 1999), found that complex networks in general have inverse power-law distributions. The inverse power law was first observed in the systematic study of scientific data on income in Western societies (Pareto, 1897) as analyzed by the engineer/economist/sociologist Marquis Vilfredo Frederico Damoso Pareto (1848–1923) in the 19th century. Subsequently, scientists recognized that phenomena described by such inverse power laws do not possess a characteristic scale and referred to them collectively as scale free, in keeping with the history of such distributions in social phenomena (West, 1999; West & Grigolini, 2010).

---

[1]The bell shape is here given by a Poisson rather than a Gauss distribution, but the two are qualitatively the same, and the former becomes the latter under certain limiting conditions.

Pareto worked as an engineer until he was middle aged, and from this perspective he gained an appreciation for the quantitative phenomena representations. Upon his father's death, he left engineering and took a faculty position in Lausanne, Switzerland. With his collection of data from various countries, he became the first person to recognize that the income and wealth distribution in a society was not random but followed a consistent pattern. This pattern could be described by an inverse power-law distribution, which now bears his name, and which he called "The Law of the Unequal Distribution of Results" (Pareto). He referred to the inequality in his distribution more generally as a "predictable imbalance," which he was able to find in a variety of phenomena, including the distribution of talent among individuals. This imbalance is ultimately interpretable as an implicit unfairness that is invariably found in complex networks.

So how do we go from random networks with their average values and standard deviations to networks that are scale free; and more important, how does this all relate to health care management?

Let us examine human achievement distribution and consider a simple mechanism that exemplifies why such distributions have long tails, as shown in Figure 5.3. In the next section, we discuss exemplars of the many inverse power-law networks in the social network of scientists and engineers as well as other professionals. Achievement is, in general, the outcome of a complex task. A complex task or project is multiplicative and not additive because an achievement requires the successful completion of a number of separate subtasks, and the failure of any one of them leads to the project's failure. As an example of such a process, consider the publication of a scientific paper. A partial list of the abilities that might be important for paper publication is (1) the ability to think up a good problem; (2) the ability to work on the problem; (3) the ability to recognize a worthwhile result; (4) the ability to make a decision as to when to stop and write up the results; (5) the ability to write adequately; (6) the ability to profit from criticism; (7) the determination to submit the paper to a journal; and (8) the willingness to answer referee's objections. If we associate a probability with each of these abilities, then to some level of approximation, the overall probability of publishing a paper would, based on this argument, be the product of the eight probabilities. The central limit theorem applied to such a process would yield a distribution for the successful publication of a paper that is log normal or inverse power law at large scales (West, 1999). Other more mathematical arguments lead to inverse power laws throughout the domain of the variate.

So how does the inverse power-law distribution affect health care provider evaluation?

Assume that a position has become available and a short list of candidates has been compiled. Suppose further that there are eight criteria that are being used in the evaluation of a group of candidates, all with ostensibly the same professional achievement level. Using the above argument on the multiplicative nature of complex processes, we see that if each of the criteria's probabilities for an individual is reduced by the same small factor, say 10%, then that person's total capability as perceived by the committee is reduced by 60%, with the probability of being promoted reduced by the same amount. In Gauss' additive world, the reduction would only be 10% because this is the average reduction; in fact, this is typically how such small influences are dismissively discussed. In the inverse power law world, the Pareto world we live in, small changes in specific attributes can result in large changes in one's career trajectory.

This achievement distribution model is not a Gauss additive situation, where a strong individual can oppose a promotion and bully others on the committee into going along with him/her. This is a more subtle, multiplicative situation, where each rumor, innuendo, and slur can detract in a cumulative way from a person's potential being fully recognized. Moreover, this suggests that significantly different evaluations of individuals or organizations may not be attributable to any single cause, but may be the result of an uncounted number of overlooked and often forgotten impressions that do not contribute on their own merits but do color the overall evaluation.

4. Distributions in Complex Networks

The understanding of inverse power laws in the context of social networks began with small-world theory, a theory of social interactions in which social ties can be separated into two primary kinds: strong and weak. Strong ties exist within a family and among the closest of friends, those that you call in case of emergency and contact to tell when you get a promotion. Then there are the weak ties, such as those with many of the colleagues at work, friends of friends, business acquaintances, and most of our teachers, with whom you chat, but never reveal anything of substance.

Clusters form among individuals with strong interactions, forming closely knit groups in which everyone knows everyone else. These clusters are formed from strong ties, but then the clusters are coupled to one another through weak social contacts. The weak ties provide contact from within a cluster to the outside world. The weak ties are all important for interacting with the world at large, say for getting a new job. A now classic paper by Granovetter (1973) explains how the weak ties to near strangers are much more important in getting a new job than are the stronger ties to one's family and friends. In this "small world," there are shortcuts that allow for connections between one tightly clustered group and another tightly clustered group very far away. With relatively few of these long-range random connections, it is possible to link any two randomly chosen individuals with a relatively short path. This has become known as the *six-degrees-of-separation* phenomenon (Barabasi, 2003; Strogatz, 2003;

Watts, 1999). Consequently, there are two basic elements necessary for the small-world model, clustering and random long-range connections.

Small-world theory is the conceptual precursor to understanding inverse power-law networks. Recent research into the study of how networks are formed and how they grow over time reveals that even the smallest preference introduced into the selection process has remarkable effects. Two mechanisms seem to be sufficient to obtain the inverse power-law distributions that are observed in the world. One of the mechanisms is contained in the principle that the rich get richer and the poor get poorer (Barabasi). In a computer network context, this principle implies that the node with the greater number of connections attracts new links more strongly than do nodes with fewer connections, thereby providing a mechanism by which a network can grow as new nodes are added.

The inverse power-law nature of complex networks affords a single conceptual picture spanning scales from those in the World Wide Web to those within an organization. As more people are added to an organization, the number of connections between existing members depends on how many links already exist. In this way, the status of the oldest members, those that have had the most time to establish links, grows preferentially. Thus, some members of the organization have substantially more connections than do the average, many more than predicted by any bell-shaped curve. These are the individuals out in the tail of the distribution, the gregarious individuals who seem to know everyone and are involved in whatever is going on. In a health care context, these are the individuals that members of the hospital seek out for discussion and support.

5. The Pareto Principle

In itself, determining that a given data set has a distribution of the inverse power-law type would not necessarily be important. What makes the Pareto (inverse power-law, scale-free) distribution so significant are the sociological implications that Pareto and subsequent generations of scientists were able to draw from its form. For example, he identified a phenomenon that later came to be called the Pareto Principle, that being that 20% of the people owned 80% of the wealth in Western countries. It actually turns out that fewer than 20% of the population own more than 80% of the wealth, and this imbalance between the two groups is determined by the Pareto index. The actual numerical value of the partitioning is not important for the present discussion; what is important is that the imbalance exists.

In any event, the 80/20 rule has been determined to have application in all manner of social phenomena in which the few (20%) are vital and the many (80%) are replaceable. In the late 1940s, Juran coined the phrase "vital few and trivial many," and he is the person who invented the name Pareto Principle and attributed the mechanism to Pareto (Pande & Holpp, 2002). The 80/20 rule caught project managers' and other corporate administrators' attention, who now recognize that 20% of the people involved in any given project produce 80% of all the results; that 80% of all the interruptions come from the same 20% of the people; resolving 20% of the issues can solve 80% of the problems; that 20% of one's results require 80% of one's effort; and so on. Much of this is recorded in Richard Koch's book *The 80/20 Principle* (Koch, 1998). This principle is a consequence of the inverse power-law nature of complex social networks.

Power laws are the indicators of self-organization, so that inverse power laws are apparently ubiquitous, with order emerging from disorder in complex networks. In the transition from disorder to order, networks give up their uncorrelated random behavior, characterized by average values, and become scale free, where the network is dominated by critical exponents. It is the power-law exponent that determines the inverse power law falloff rate and captures the network's global character, the imbalance between the rich and poor, between those that publish rarely and those that are prolific, and between those whose work goes unread and those cited by everyone. This behavior is implicit in Figure 5.3, where we schematically show the dramatic difference between the bell-shaped curve of Gauss and the inverse power law of Pareto.

It is evident from Figure 5.3 that all the action takes place in the vicinity of the bell curve's peak (the average value), as a random network has no individual

FIGURE 5.3 A Schematic Comparison Is Made Between the Bell-Shaped Distribution of Gauss and the Inverse Power-Law Distribution of Pareto.
The vertical axis is the logarithm of the probability and the horizontal axis is the variable divided by σ.

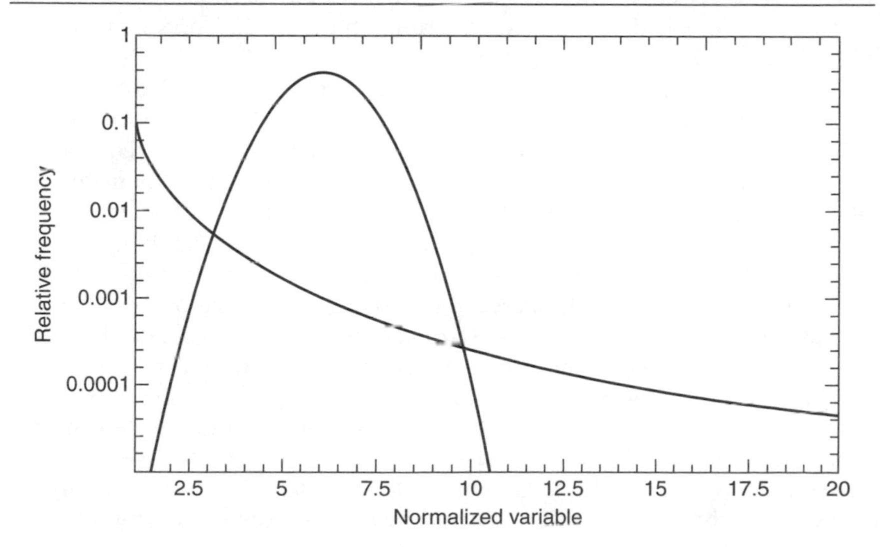

preferences. In the inverse power-law network, the action is spread over essentially all the available values to which the process has access, but in direct proportion to the number of links already present. This preference of a new member hooking up with the oldest member, having established connections, is what makes a star. A luminary in health care (given a certain level of technical ability) is socially determined in substantially the same way as is a luminary in Hollywood (given a certain level of acting ability).

From Figure 5.3 we see how unfair social phenomena are in the real world. If events happened according to Gauss, then everyone would be similarly situated with respect to the average value of the quantity being measured. If the normalized variable were a measure of income, then everyone would make approximately the same amount of money. If, rather than money, the distribution described the number of patients, then the Gauss world would have most nurses treating the average number of patients, with a few treating more and a few treating less than the average. However, the health care world would be basically fair. Applying Six Sigma to this world would have every health care practitioner caring for exactly the same number of patients and, because there is little variability in the resultant measures, this would appear to be a desirable goal.

However, that is not the world in which we live. In the real world, the Pareto world, there may not even be an average value and people may have almost none of the variable, or they may have a great deal of it. Therefore, we find people at the level of poverty and people making millions of dollars a year and all those in between. The number of patients is no exception in this regard; we (Philippe, Garcia, & West, 2004) determined that the distribution in the number of hours a patient occupies a bed in an emergency ward is, in fact, inverse power law, just like that of income. The difference between the two distributions is the parameter determining the inverse power-law falloff, the power-law index or Pareto index.

The Pareto law was derived for income distribution, but there are other laws with the same inverse power-law form: the Auerbach law on the distribution of the sizes of cities, the Zipf law on the relative frequency of words in language, the Richardson law for the distribution of the magnitude of wars, the Lotka law on the distribution in the number of scientific citations, and many others (West & Grigolini, 2010). All these distributions stem from the implicit multiplicative nature of the underlying phenomena that make them amenable to the Pareto Principle. The essence of this principle is that *a minority of input produces a majority of the results*, and this is the key to understanding the management of health care providers. The Pareto Principle results in the few being important, and in the present context, these few determine the methods, goals, and direction of complex health care teams, the health care direction of hospitals, and in fact, the overall health care direction of the country. The principle also maintains that the many are irrelevant in the sense that they are not the health care leaders, but this does not imply that they are not valuable. Most of health care is accomplished by the faceless majority that refines, improvises, and brings to fruition the often nearly unintelligible ideas of the few.

In order to develop an intuition about the inverse power law, let us consider the number of citations to published scientific papers in any given year. The cumulative values yield (de Solla Price, 1986): 35% of all papers published in the sciences in a given year have no citation; 49% of all such publications have one citation; 9% have two citations; 3% have three citations; 2% have four citations; 1% have five citations; and 1% have six or more citations. The average number of scientific citations per year is 3.2, and the inverse power-law distribution implies that 95% of all papers published are below average in the number of times they are cited. The manager who uses the number of citations as a direct measure of the researcher's quality is using a linear worldview and is not being fair. Most of the research published is in alignment with the 95% (Redner, 1998). If the number of citations to a paper is average, then that research falls in the upper 5% category. How a manager ought to use the information on citations and other such criteria for purposes of evaluation is therefore strongly nonlinear.

In the case of health care, using the properties of inverse power-law distributions it is possible to say that, surprisingly, most nurses care for fewer than the average number of patients. It is definitely not the case that half the nurses care for more than average and half care for less than the average, but rather that the vast majority of nurses care for less than the average number of patients. Consequently, if the number of patients were the metric of quality used by the hospital to evaluate its health care providers they would not fare well. However, this would be a distortion of the true quality of the workers and the health care being done, as this application of the metric presupposes a linear worldview as the basis for comparison among health care team members.

Suppose we applied Six Sigma to the "number of patients cared for" metric. Reducing the variability in the metric would imply eliminating 90% of the nurses (based on an analogy with citations), leaving only those 5% above and below the average value out in the tail. The rationale for this action would be that most nurses were dead wood because they are not caring for enough patients and therefore ought to be replaced. A less Draconian response might be to give the 90% a substandard evaluation. Neither of these last two alternatives properly accounts for the nonlinear character of the process, and consequently both are not only inappropriate but insulting, demoralizing, and just plain wrong.

6. Conclusions

The evaluation of a health care provider is, all too often, a consequence of the manager being the product of our competitive education system. There is, of course, no a priori reason why everyone taking a college course, which has absolute criteria for mastery of the material, cannot receive an A. The competition that arises in classes is all too often the result of the restriction in the number of top slots in the class, treating students as if they were independent entities, all struggling for the top grade, as one does with a Gauss model. The Gauss worldview enforces a self-fulfilling prophecy of human behavior. This educational model does not serve the reality of the multiply connected health care world into which the student will graduate. On the other hand, working with students and promoting their interactions with one another enhances their interdependence and facilitates collaboration. This latter model better prepares them for the future world of health care. Individuals' interdependence in both science and the learning of science suggests a different way of thinking about how to facilitate the complex social interactions among health care providers.

The inverse power laws that measure the quantities of interest to the health care communities suggest that any attempt by management to equalize the reward structure within a hospital, such as imposing a normal distribution in the evaluation process, will necessarily be counterproductive and disrupt the interactive modes that would ordinarily, adaptively develop. This "unnatural" equalization process is a consequence of the failure to recognize the Pareto nature of the world in which we live. Consequently, with the 80/20 rule in mind, the data coming out of the analysis and understanding of complex networks suggest that managers, evaluating the overall quality of health care, should concentrate on the 20% that truly make a difference and only lightly monitor the 80% that do not significantly influence the hospital's operation. Using this principle for guidance, the health care professionals should take cognizance of what motivates these individuals. For example, health care providers are stimulated by interesting and rewarding work, but only the top 20% are sufficiently talented to lead the hospitals into the frontier areas. The other 80% can often be guided to transition methods toward both long- and short-term applications. The two groups are complementary to one another, and a successful hospital requires both in order to generate innovative ideas, to carry to completion the necessary testing of these ideas, and to give talks at national and international meetings to disseminate knowledge obtained from such testing.

Successful health care management can only be accomplished by enabling the 20% to focus on the development of new health care methods while unreservedly supported by the 80%. Management must also bear in mind that this 20% does not always consist of the same individuals. From one year to the next any individual can be one of the 20% or one of the 80%, but only if the hospital environment is structured in such a way as to facilitate that mobility. Both health care leadership and support are important in an effective hospital, and enabling one at the detriment of the other, in the end, serves neither.

In closing, I reiterate that I know very little about health care, and any conclusions I have drawn are speculations based on extrapolations I have made using the general properties of complex networks. However, I hope that these comments will help guide the health care expert, if not the speculations themselves then at least the reasoning based on inverse power law rather than the normal distribution.

REFERENCES

Barabási, A. L. (2003). *Linked*. New York: Plume Publishing Company.

de Solla Price, D. (1986). *Little science, big science—and beyond*. New York: Columbia University Press.

Granovetter, M. (1973). The strength of weak ties. *American Journal of Sociology, 78,* 1360–1380.

Gupta, H., Campanha, J., & Chavorette, F. (2003). *Power-law distribution in high school education: Effect of economical, teaching and study conditions*. (arXir.0301523v1).

Juran, J. M. (1962). *Quality control handbook*. New York: McGraw-Hill.

Juran, J. M. (1964). *Managerial breakthroughs*. New York: McGraw-Hill.

Koch, R. (1998). *The 80/20 principle: The secret of achieving more with less*. New York: Nicholas Brealey.

Litaker, D., Tomolo, A., Liberatore, V., Strange, K., & Aron, D. (2006). Using complexity theory to build interventions that improve health care delivery in primary care. *Journal of General Internal Medicine, 21*, S30–S34.

Pande, P., & Holpp, L. (2002). *What is six sigma?* New York: McGraw-Hill.

Pareto, V. (1897). *Cours d'economie politique*. Lausanne: F. Rouge.

Philippe, P., Garcia, M., & West, B. (2004). Evidence of "Essential Uncertainty," Emergency-Ward length of stay. *Fractals, 12*, 197–209.

Redner, S. (1998). How popular in your paper? An empirical study of the citation distribution. *The European Physicals Journal. B, 4,* 131–134.

Strogatz, S. (2003). *SYNC*. New York: Hyperion Books.

Watts, D. (1999). *Small worlds*. Princeton, NJ: Princeton University Press.

West, B. J. (1999). *Physiology, promiscuity and prophecy at the millennium: A tale of tails*. Singapore: World Scientific.

West, B. J., & Grigolini, P. (2010). *Complex webs: Anticipating the improbable*. New York: Cambridge University Press.

Response to Chapter 5

Michael Brooks Turkel

Early in my undergraduate education, I studied engineering. My training in engineering introduced me to two of West's points: the measurement of variation as a critical tool for control in manufacturing systems and the use of the Gaussian or Normal distribution in grading tests. As undergraduates, we all knew that test results slightly above the mean for the class resulted in a grade of B minus and test results slightly below the mean resulted in a grade of C plus. I still remember a particularly grueling calculus final examination where the mean score was 50 out of a possible 200. The high degree of difficulty on the test was specifically designed to result in a broad grade distribution.

Over the last 25 years I have been directly involved in health care management. I have witnessed, complained about, and led organizational change in response to changing reimbursement models, increased competition among providers and greater regulation. Along the way, I have seen health care embrace tools and techniques from other industries to improve care delivery. The abbreviations TQM (total quality management) and CQI (continuous quality improvement) seemed almost as prevalent in hospitals as CHF (congestive health failure) and EKG (electrocardiogram). Tools from other industries seem beneficial but, in practice, are difficult to apply in the complex, chaotic health care environment.

In 2003, Anthony Benedetto published in the *Journal of Healthcare Management* an article from his Fellow project. Benedetto described the application of Six Sigma to a radiology film library. Critical in the application of Six Sigma was the definition of a defect. As West points out, Six Sigma was designed to measure how many of the widgets were out of tolerance. For the film library, not having the film available at the agreed upon time became a defect. However, the agreed-upon time was inherently arbitrary and based on the custom for having films available not an objective measure of when they needed to be available. Additionally, there was no measurement of when the department anticipated patient needs and got the film to a caregiver when the film was needed in advance of the predetermined deadline. It is evident that the measurements associated with Six Sigma and the discipline to reduce variation could assure that films were more likely to be available at the predetermined time, but it was also apparent that in the health care setting, defining that standard was inherently difficult and fulfilled only part of the department's purpose.

West describes the health care environment as a series of complex social phenomena. Accordingly, in order to apply Six Sigma we must define defects in these social phenomena. If, as part of a Six Sigma approach, we define patients that stay in the hospital one or more days longer than the number of days paid for by the insurer as a defect, then how do we account for an extra day that results in a substantial benefit to the patient? What if the extra day resulted from a nurse's intuition and the nurse advocating for an additional diagnostic test for the patient? What if the caring interaction between nurse and patient led to the early discovery, diagnosis, and treatment of a condition that if otherwise untreated would have resulted in a substantial decline in the patient's autonomy? I am not arguing for extended lengths of stay. I am far too familiar with nosocomial infections and other negative outcomes (clinical and financial) associated with unnecessary days spent in the hospital. I use longer length of stay only as an example. In the complex social environment of health care, variations from the mean results are just that. We have very limited tools to measure the effectiveness of what caregivers do. In fact, because the duration of a patient's stay in the acute care hospital is so short, it is likely that the caregivers will never know the implications of their work.

Perhaps Six Sigma could be applied to improving the efficiency of transporting patients throughout the hospital? However, we are again confronted with the problem of defining a defect. We could measure transport effectiveness based on the number of transport runs that are completed within the allowed time. For example, let's say that the standard transport time, from the time of dispatch to when the patient is brought from the inpatient room to the radiology department for testing, is 15 minutes. Is it a defect if when the transporter is ready to bring the patient to radiology the patient shares with the nurse that he is afraid, and if the nurse (following Sr. Roach's principle [2002] of compassion) facilitates the patient by talking to his wife on the phone so that the patient can share his anxiety, and if by taking the time to truly care for the patient the patient is now more comfortable in getting the test and is less anxious about his environment, but as a result of these caring interactions the transport time for this run is 10 minutes longer than the standard. Instead, what if the transporter stopped on the way to get the patient and helped an elderly visitor find his loved one's room—is this a defect? Applying Six Sigma within the social context of the health care environment is at best difficult. Although the principles are useful, the application is problematic.

For the health care leader, measurement and management of the quantitative and the qualitative aspects of health care delivery are crucial in assessing staff and departmental performance. Today's environment demands that we measure, monitor, and manage. It is imperative that we constantly improve health care delivery, and we must do so with an understanding of the complex social environment of health care.

REFERENCES

Benedetto, A. (2003). Adapting manufacturing-based six sigma methodology to the service environment of a radiology film library. *Journal of Healthcare Management, 48*(4), 263–280.

Roach, M. (2002). *Caring, the human mode of being* (2nd rev. ed.). Ottawa, ON: CHA Press.

CHAPTER 5
WHY SIX SIGMA HEALTH CARE IS OXYMORONIC IN HOSPITALS

- *Does your organization use Six Sigma, the quality improvement tool? Have you found instances where trying to apply Six Sigma is harder than expected?*
- *Are there examples in your organization where the distribution of results does not seem to follow the bell-shaped curve?*
- *Have you heard of the Pareto Principle or 80/20 rule? What examples are there in your organization of the Pareto Principle?*
- *Have you seen forced distributions in school or at work that are counterproductive? Given Dr. West's conclusions, how should your organization adapt to the realities of complex human networks?*
- *Does your organization struggle in measuring the "right thing" as part of its improvement projects? How can your organization better use data to improve patient outcomes while recognizing the underlying complex social phenomena?*

SIX

Entropy as Information Content in Medical Research

Nikhil S. Padhye

There is a wide variety of definitions of entropy in scientific literature. It is not the aim of this chapter to provide a comprehensive treatment of all entropies, but rather to maintain focus on the entropies that are of most interest in medical research. The discussion of entropy in this chapter is primarily from the point of view of application to time-series data. For example, a patient's temperature taken throughout the hospitalization period is time-series data. Entropy is closely tied to disorder and information content of a system. An example of entropy in physical law is that of the second law of thermodynamics, which states that the entropy of a system always increases in time. Imagine that we inserted a smoke ring into a sealed box and then let the closed box evolve for some time. Over time, the smoke ring would disappear and all molecules would mix throughout the box. This represents the loss of order, or increase in disorder, of the system over time. Entropy definitions that involve physical laws are closely tied to the workings or mechanisms of those laws. Dr. Rapp's response discusses the relationship of entropy to health care situations. She notes that complexity is a measure of uncertainty where the degree of uncertainty is associated with health status. Complexity implies that all systems work harmoniously responding at the appropriate time and with sufficient intensity to adjust and adapt to physiological, psychological, and social changes.

INTRODUCTION

There is a wide variety of definitions of entropy in scientific literature. It is not the aim of this chapter to provide a comprehensive treatment of all entropies but rather to maintain focus on the entropies that are of most interest in medical research. It is assumed that the reader is familiar with discrete probabilities and with probability distributions of continuous variables. Knowledge of concepts from nonlinear dynamics is helpful but not necessary. The same is true of information theory. The discussion of entropy in this chapter is primarily from the point of view of application to time-series data rather than motivated by mathematical models or maps for deterministic systems, because the latter are usually unknown in medical research. The use of mathematical expressions will be avoided as much as possible, but not entirely so, and familiarity with logarithms is assumed.

Entropy is closely tied to disorder and information content of a system. Although most definitions of entropy are related to each other to various extents, it is helpful to make a primary distinction between entropies that involve physical laws and those that involve information theory. An example of entropy in physical law is that of the second law of thermodynamics, which states that the entropy of a system always increases in time. We have here in mind a system like a gas that is composed of very large numbers of molecules in motion. Imagine that we created an orderly structure in the gas, such as could be done by inserting a smoke ring, and then left the system in a closed box to evolve for some time. A little while later, the structure of the smoke ring would be gone and all molecules would mix homogeneously throughout the box. This represents the loss of order, or increase in disorder, of the system over time. Entropy definitions that involve physical laws are closely tied to the workings or mechanisms of those laws. Our interest in this chapter is tied more closely to entropy in its more general sense as it pertains to information theory, than in physical entropy that is of interest to physicists and chemists.

What Is Information Theory?

Information theory is a field of study that had its beginnings after the advent of large telegraph networks in the early part of the 20th century. There was a need to determine error rates and the fastest possible rates of information exchange across telegraph networks. Although some of the issues involved were linked intimately to physical law related to transmission of electric current through conducting wires, information theory evolved more purely to address those questions that were not related to the physical processes. An information source can have a certain range of possible messages that can be created from symbols, such as 0s and 1s or letters of the English alphabet, for example. One may therefore view the message as being uncertain and randomly distributed, at least to an extent. Information theory concerns itself primarily with the resolution of this uncertainty.

Claude Shannon (1949) is widely credited with launching information theory with a classic paper in which he introduced two remarkable ideas. One is the so-called sampling theorem that specifies the minimum rate at which measurements must be made to avoid loss of information about the signal that is being measured, and another was the concept of entropy as a measure of information content. An excellent account of the history and basic ideas of information theory is provided by Pierce (1980).

Our interest in information theory revolves around questions such as, "How much do we learn about the system when we make one or more measurements?" If the system were unchanging, for example, a homeostatic system, one measurement is sufficient to tell us all that there is to know about it. For a periodic system, for example, circadian rhythmic system, measurements over one period are sufficient to know system behavior over all subsequent (and precedent) periods. On the other hand, for a string of 0s and 1s produced randomly by a computer, we cannot predict the next outcome with any confidence greater than 50% even if we knew 10,000 previous outcomes (Kantz & Schreiber, 2004). It turns out that entropy is a powerful tool in answering such questions. This chapter begins with a discussion of *Shannon entropy*, its interpretation, and the closely related variant of *Kolmogorov–Sinai entropy* (K–S). The following section builds on these ideas to introduce modern definitions of *approximate entropy* and *sample entropy* statistics, which have been found to serve well in biomedical applications.

SHANNON ENTROPY

The introduction of the concept of entropy by Shannon (1949) was an important milestone in information theory. It is sometimes referred to simply as *information theoretical entropy*, although we will see that it is not the only definition of entropy that is used in information theory.

Shannon Entropy of a Discrete Variable

We start with the mathematical expression of entropy, as defined by Shannon (1949), and then proceed to interpret it.

$$H = \sum_{i=1}^{N} p_i \log_2\left(\frac{1}{p_i}\right)$$

In the above formula, H is the entropy and p_i is the probability of occurrence of the ith event. For instance, p_i could be the probability of occurrence of the letter "e" of the alphabet in English literature of the 17th century. The logarithm to base 2 is denoted by $\log_2$ and the summation sign (Σ) denotes summation over all events. The events are labeled from 1 to N, so i ranges over those values in order to account for all events. From analogy to weighted sums, the above formula may also be understood as being the expectation value (mean) of the logarithm of the quantity $1/p_i$, which is called the *surprise*. It is so named because it is inversely related to the probability of occurrence of an event. The smaller the probability of an event, the higher is the surprise of its occurrence (Duda, Hart, & Stork, 2001). If an event were guaranteed to occur, that is, $p_i = 1$, the surprise would be zero. Thus Shannon entropy is a measure of surprise. Note that the expression for the entropy involves only the probabilities. The actual events or symbols that might appear in a message are irrelevant in the calculation of entropy. Next, we proceed to examples that help to understand how the entropy is a measure of information content.

Consider the case that the four digits 0 to 3 could appear in a message with equal likelihood for each digit, so $p_i = ¼$ for $i = 1$ to 4. The Shannon entropy evaluates to $H = 2$. If you repeat this calculation for the case in which eight digits from 0 to 7 appear with equal likelihood, you will get $H = 3$. Notice that our two answers represent the maximum length of the binary number that is required to represent the digits. The digits 0 to 3 can be represented by the binary numbers 00, 01, 10, and 11. These numbers have a maximum length of two binary digits. The numbers 0 to 7 can be represented by the same binary numbers up to 3, followed by 100, 101, 110, 111. In other words, the entropy provided us with the number of *bits* (binary digits) required to represent all of the information in the system. It is in this sense that Shannon entropy is a measure of information content of a system. It is the use of the logarithm to base 2 that results in the entropy yielding the number of bits. Logarithm to a base other than 2 could also be used in the expression for entropy, but it would result in entropy that is not measured in bits. The unit of measurement is called *nats* when the natural logarithm is used.

It can be shown that the entropy is highest when events are equally likely. A simple practical application of this is the roulette wheel. Each sector on the wheel is of the same size so that the probability of any one sector being the winning one is the same. It requires no profound knowledge of entropy to recognize that if some of the sectors on the roulette wheel were larger than others, it would pay to bet on the larger sectors.

The next example shows how entropy as a measure of information content can be put to practical use (Van den Eijkel, 2003) in matters that are of more interest to us. At the outset, it may be helpful to realize that a bit can also be interpreted as the resolution to a *yes/no* question. Our task is to pose questions that can be answered either *yes* or *no* to find out which of four nurse practitioners were chosen for primary care by a new patient. Let us label these nurse practitioners A, B, C, and D. One way to find the answer is to ask questions that either confirm or eliminate each possibility. We would have to ask a maximum of three questions because after elimination of three possibilities the answer would have to be the only remaining (fourth) possibility. However, there is a more efficient way to do it. We can arrive at the answer by asking no more than *two* questions.

The first question would ask whether the chosen nurse practitioner was among *A* or *B*. If the answer to the first question is *yes*, we then ask whether it was *A*. The *yes/no* answer to this second question determines whether the chosen nurse practitioner was *A* or *B*. Had the answer to the first question been *no*, we proceed to ask whether it was *C*. The *yes/no* answer to this second question determines whether the chosen nurse practitioner was *C* or *D*. Now, let us see how the Shannon entropy helps to measure the efficiency of our questions (see Figure 6.1).

The total information content of the system is given by the Shannon entropy, assuming that each possibility is equally likely, that is, $p_i = 1/4$ for $i = 1$ to 4 (corresponding to choices *A* to *D*). The Shannon entropy evaluates to $H = 4 \times 1/4 \log_2 4 = 2$. Next, we compute the conditional entropy for the first question "Was it *A* or *B*?" in the two-question system. The probability of the answer being yes, p(yes), equals ½, as is the probability of the answer being no, that is, $p(\text{no}) = 1/2$. If the answer is yes, the conditional probability $p(i\,|\,\text{yes})$ is ½ for either nurse practitioner *A* or *B* and 0 for options *C* and *D*. If the answer is no, the conditional probability $p(i\,|\,\text{no})$ is ½ for either nurse practitioner *C* or *D* and 0 for options *A* and *B*. Thus, $H = \Sigma\, p(\text{yes})\, p(i\,|\,\text{yes}) \log_2 [1/p(i\,|\,\text{yes})] + \Sigma p(\text{no})\, p(i\,|\,\text{no}) \log_2 [1/p(i\,|\,\text{no})] = 2 \times 1/2 \times 1/2 \log_2 (2) + 2 \times 1/2 \times 1/2 \log_2 (2) = 1/2 + 1/2 = 1$. The conditional Shannon entropy evaluates to $H = 1$, which is a measure of information content left in the system after we know the answer to the first question.

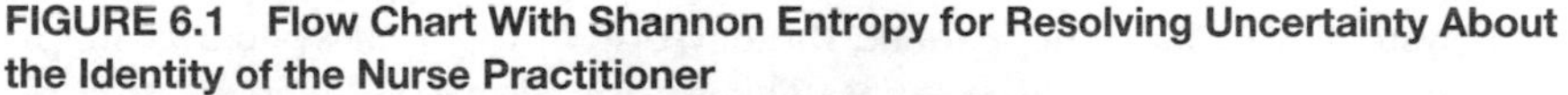

FIGURE 6.1 Flow Chart With Shannon Entropy for Resolving Uncertainty About the Identity of the Nurse Practitioner

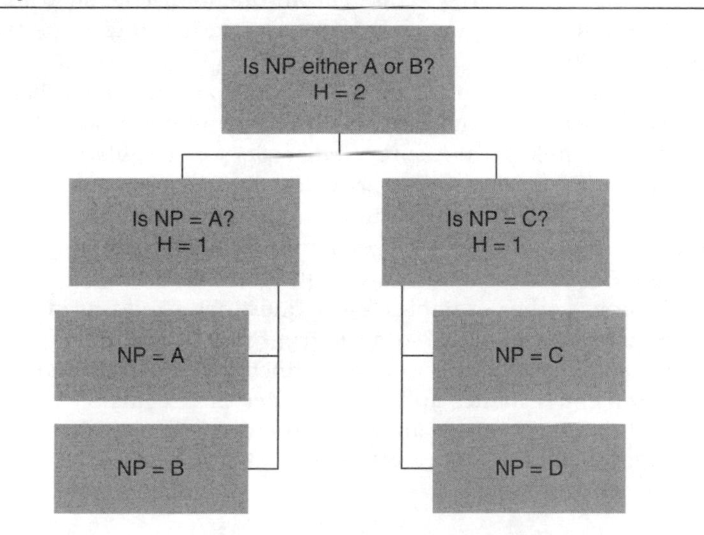

Thus, the gain in information that we obtained from the first question is given by the difference between the total information at the outset ($H = 2$) and the information content left after knowing the answer to the first question ($H = 1$), which is 1 bit. Our second question helps to resolve this remaining 1 bit of uncertainty.

On the other hand, consider the possibility that we had chosen to ask a less efficient first question: "Was it *A*?" The yes/no answer to this question leaves us with a conditional entropy of $H = \Sigma\, p(\text{yes})\, p(i|\text{yes}) \log_2 [1/p(i|\text{yes})] + \Sigma\, p(\text{no})\, p(i|\text{no}) \log_2 [1/p(i|\text{no})] = 1/4 \times 1 \log_2 (1) + 3 \times 3/4 \times 1/3 \log_2 (3) = 0 + 1.19 = 1.19$. The advantage gained by the question is therefore 0.81 bit, which is the difference between the total information ($H = 2$) and the remnant information ($H = 1.19$). Comparing the information gained by the questions in our two strategies clearly shows that asking "Was it *A* or *B*?" is more efficient than asking "Was it *A*?" when there are four possible options.

In this manner, Shannon entropy is a useful measure in decision trees and other learning algorithms. In our simple example, it may seem as if the use of entropy to decide the best questions to ask is like using dynamite to move a pebble out of the way. The real power of the entropy techniques becomes clear in more realistic and complex applications. The interested reader is referred to the text by Breiman, Friedman, Olshen, and Stone (1984).

Note that the use of logarithm to base 2 in the definition of entropy has been critical for the interpretation in terms of bits of information. There is another nice feature that results from the use of the logarithm. If there are two systems that are independent, the entropy of the joint system is simply the sum of the individual systems. The information contained in multiple independent systems can therefore simply be added to give the total information of the joint system.

Although we have seen in our discussion up to this point that Shannon entropy is a measure of information, entropy is often also described in the literature as being a measure of disorder or irregularity of a system. Equivalently, it is a measure of surprise contained in a system. These are not independent interpretations but can be understood in the terms of our preceding discussion. An orderly or regular system has less surprise and less information. Fewer observations are required to characterize such systems. In regard to our earlier example, fewer questions would need to be asked to resolve any answers that we seek from an orderly or regular system. As the entropy and information content increase, the potential surprise in any measurement is higher and the disorder or irregularity of the system increases. This will be clearer after we extend the Shannon entropy to continuous variables and introduce other entropies later in the chapter, such as the *approximate entropy*.

Shannon Entropy of a Continuous Variable

The extension of the definition of Shannon entropy from discrete probabilities to the case of a continuous variable is accomplished by means of the probability distribution $p(x)$.

$$H = \int_{-\infty}^{+\infty} p(x) \log\left(\frac{1}{p(x)}\right) dx$$

The sum over discrete cases is replaced by integration on the entire range of real numbers from negative infinity ($-\infty$) to positive infinity ($+\infty$). The integration may be understood as an operation that computes the area under the curve of the quantity $p(x) \log (1/p(x))$. Because most physiological variables are continuous, this extension of the definition of entropy is of particular interest to us. Nevertheless, entropy for discrete events continues to be relevant for applications to time-series data. As there are only a finite number of observations in a time series, it is not possible to truly have continuous data that can be resolved to any desired precision. In practice, we simply get a sum over a large number of discrete events even when the variable is continuous in principle.

The normal curve or Gaussian probability distribution occupies a prominent place in statistics. This is primarily due to the property that sample means tend to be normally distributed even when the data distribution is not normal. The mathematical proof of this property is called the *central limit theorem*. As it turns out, the normal distribution has special status accorded to it by entropy considerations, too. For any given mean and variance of a variable, the distribution that has maximum entropy is the normal distribution. Thus, among all possible probability distributions with a specified mean and variance, the normal distribution has maximum information content.

The reader would be justified in wondering why it is not the uniform distribution that has maximum entropy, because we noted earlier that the entropy is maximum for discrete events that are equally likely. The problem is that a uniform distribution has infinite variance unless it is bounded, that is, there is a minimum and a maximum value that the continuous variable can take. For such bounded probability distributions, it is in fact true that entropy is maximized when the distribution is uniform. Upon removing the restriction of being bounded, the normal distribution has maximum entropy for a continuous variable.

In the case of discrete events, switching labels of the events does not alter the entropy because the entropy depends only on the probabilities of events and not on the labels of events or on symbols used. For a continuous variable, we could ask whether the entropy changes when the variable is transformed. A simple transformation could be just a change in units of measurement, for example, changing from centimeter to millimeter would result in variable x being changed to $10x$. A slightly more involved transformation might change x to x^2 or to $\log(x)$. We would like it if the entropy would be invariant under such transformations because there is no new information being created or destroyed. Unfortunately, our wish is not granted. Shannon entropy does change under transformation of the underlying variable, even if it is a simple linear change of units. Nevertheless, entropy *differences* continue to be valid measures of information gained or lost. Therefore, in making comparisons of entropy, it is important to focus on relative levels of entropy rather than the base values.

Kolmogorov–Sinai Entropy

Shannon entropy assumes that the probability of each event is independent of other events. Clearly, this is not always true. For instance, the probability that you would find the letter *U* in an English word that also has the letter *Q* is 100% because *Q* is always followed by *U*. In time-series data, consecutive measurements are often correlated, which implies that the probabilities of measured values are not independent of each other. Kolmogorov (1958) and Sinai (1959), working independently, modified Shannon entropy to accommodate systems with serial correlation.

Kolmogorov and Sinai introduced two innovations. First, the probability distribution of a single data point was replaced by the joint probability distribution of m consecutive data points. This allows for serial correlation to exist between the m data points. Second, m is made very large by virtue of a mathematical limiting process that allows m to approach infinity. The resolution of measurement simultaneously approaches zero. It should be noted that Kolmogorov and Sinai were interested in theoretical principles and entropy calculations for well-defined mathematical and physical systems. The advent of computers since their work has made computations possible for large time series. Even so, it is not possible to allow m to be infinite. In practice, m is chosen to be large enough so that there is no serial correlation beyond m points. Doing so allows us to assume that one block consisting of m points is independent of a neighboring block of m points. In summary, the K–S entropy replaces individual events in Shannon entropy with blocks of events. The events within blocks can be correlated, but each block is independent of other blocks.

K–S entropy is usually interpreted as the rate of creation of information in a system. This is to be understood in the following sense. Due to limitation of the resolution of measurement, two nearby trajectories may initially appear to be the same. As time proceeds, the trajectories move far enough apart that they can be distinguished. For a deterministic system, this observation then allows us to determine the initial condition (earlier measurement) to greater precision by reverse calculation or propagating backward along the trajectory.

Our observations, therefore, yield more and more information about the trajectory that was initially restricted by the resolution of measurement (Ott, 1994). The advanced reader is referred to Kantz and Schreiber (2004) for a discussion of Bernoulli shift and its relationship to understanding the meaning of K–S entropy as being the rate at which bits are shifted to the left.

Readers familiar with Lyapunov exponents that measure the rate of divergence of trajectories might wonder if there is a connection between K–S entropy and Lyapunov exponents. In fact, there is such a relationship. Pesin's identity connects Lyapunov exponents, partial dimensions, and K–S entropy. A corollary is that K–S entropy can never be larger than the sum of all positive Lyapunov exponents of the system (Kantz & Schreiber, 2004).

Computation of K–S entropy requires lengths of time series that are usually prohibitive. In applications in which it is possible to compute with good reliability, it is typical to plot the K–S entropy versus changing resolution of the data for various block sizes. A plateau in the figure, that is, K–S entropy that is independent of resolution, is considered to be evidence of deterministic chaos. In the next section, we discuss *multiscale entropy* that is similarly plotted but has more robust statistical properties that make it of greater practical value in measuring complexity. The work of Grassberger and Proccacia (1983) and Eckmann and Ruelle (1985) to modify K–S entropy for improved computational ease served as a stepping stone for the modern definitions of *approximate entropy* and *sample entropy* that we discuss next.

APPROXIMATE ENTROPY AND SAMPLE ENTROPY

The difficulty of reliably computing K–S entropy and its misuse in certifying the existence of chaos prompted Pincus (1991) to introduce *approximate entropy* (ApEn) statistics. Pincus (1991) made it clear that approximate entropy cannot be used to certify chaos in a system, but it can serve as a measure of complexity and it can be computed more reliably than K–S entropy. ApEn is to be thought of as a family of parameters that measure the complexity of a system. The corresponding estimates of ApEn on a sample of data obtained from the system are called ApEn statistics.

ApEn is characterized by block size (or pattern size), m, and resolution, r. ApEn parameters are therefore denoted ApEn(m, r). In a departure from K–S entropy, the blocks in ApEn do not have to be large. They are parameters that can be varied to advantage in measuring the complexity of the system. The resolution refers to the metric that is used to decide whether patterns of data are sufficiently close to each other. ApEn(m, r) is defined as

$$\mathrm{ApEn}(m,r) = \mathrm{mean}\left[\log\left(\frac{p(m,r)}{p(m+1,r)}\right)\right].$$

In this expression, $p(m, r)$ is the probability of observing a pattern of m consecutive data points that matches a chosen template sequence of m data points. The mean is taken over all such template sequences in the data. A pattern is considered to match if the maximum distance between data points is less than r, as explained in more detail below. Similarly, $p(m+1, r)$ is the probability of observing a pattern of $m+1$ consecutive data points that matches the chosen template sequence of $m+1$ data points. Notice the similarity to Shannon entropy: Instead of the probability of the ith event, we have here the conditional probability, $p(m+1, r)/p(m, r)$, that a pattern that matches for m consecutive data points will continue to match at the $(m+1)$th data point. ApEn(m, r) is the expectation value (mean) of the *surprise* corresponding to this conditional probability.

Although K–S entropy accounts for serial correlation in data by using large blocks of data and joint probabilities of observing data values within blocks, ApEn also accounts for serial correlation but does so in a different manner that was proposed by Eckmann and Ruelle (1985). The correlation is handled by consideration of probabilities of observing patterns of consecutive data points. The probability of observing a matching pattern of m consecutive data points, $p(m, r)$, is calculated by counting all such instances in the data. Because the data consist of a continuous variable, we must specify what is meant by a match. This is accomplished by setting a resolution (tolerance), r, for defining a match. If data values that are being compared lie within the tolerance band, they are counted as a match. The resolution r is typically set as a fraction (10–25%) of the standard deviation of the data. A small value of the resolution results in lower numbers of matches and generally higher value of ApEn. Conversely, a high value of resolution leads to larger numbers of matches and generally lower value of ApEn. The resolution should be ideally as low as possible but not lower than the expected level of noise in the data. If the resolution is lower than the typical level of noise in the measurements, matches that would otherwise have been counted (had the measurements been noise free) would fail to be counted as such due to interference from noise.

Take the case in which $m = 2$, so that we are looking for matching pairs of data points. The first template sequence consists of the first two data points in the series, shown as columns A and B in Figure 6.2. Our task is to compare all other pairs of consecutive data points in the series to the template pair and count the number of matches. The dashed lines in Figure 6.2 show the tolerance bands around the values of columns A and B that are defined by the value of the resolution r. If there is a pair of data points in which the first column falls within the tolerance band for column A and the second column falls within the tolerance band for column B, it is considered to be a matching pair. In Figure 6.2, columns I and J represent a pattern that matches columns A and B. Once all matching pairs are found for the first template, the probability of matches is expressed as a fraction of the matches to the total number of available pairs.

FIGURE 6.2 Matching Patterns in the Calculation of Approximate Entropy
Each column labeled with a letter represents a data value. The pair of consecutive data points *AB* is under consideration for matching patterns here. The dashed lines represent the tolerance bands around each value. Data values (or column endings) that lie within these tolerance bands match one of *A* or *B*. When two consecutive data points such as *I* and *J* match *A* and *B*, it is counted as one matching pattern.

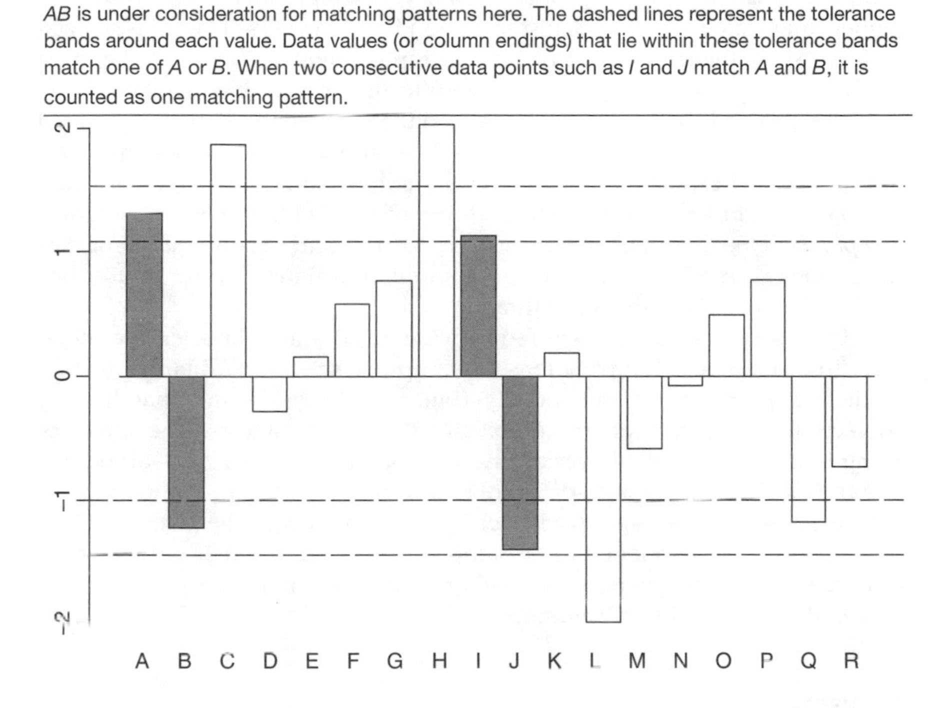

This process is repeated for each available template sequence. In Figure 6.2, the next template sequence is represented by columns *B* and *C*. Estimates of the probability of a matching pair are obtained for all available template sequences. For computing the value of ApEn(2, *r*), it is necessary to compute all available values of *p*(2, *r*) and repeat the procedure for sequences of three consecutive data points to get *p*(3, *r*). We can then estimate the conditional probability that a pattern that matches at two consecutive points will continue to match at the next (third) point. There are as many estimates of the conditional probability as there are template sequences. A mean of these estimates is computed after taking the logarithm of the inverse conditional probability, that is, *surprise*, to obtain ApEn(2, *r*).

ApEn(*m*, *r*, *N*) are the statistics that are estimates of the parameters ApEn(*m*, *r*). Because ApEn(*m*, *r*, *N*) do not require a large *m* or that *r* be close to zero, they can be reliably computed on time series with as few as $N = 100$ data points for $m = 2$. The wider tolerance band in ApEn, moreover, allows for its application to data that have presence of noise. Pincus (1991) used data generated from theoretical models of varying levels of complexity to establish that ApEn provided measures of complexity. Data from a system that has more complex behavior tends to produce fewer repeating patterns and results in higher ApEn values. Pincus, Gladstone, and Ehrenkranz (1991) used ApEn statistics to show that complexity or information content is reduced in the heart rate variability of a sick neonate compared to that of a healthy neonate.

Richman and Moorman (2000) realized that ApEn had a flaw that resulted from counting self-matches of patterns. The counting of self-matches produced a biased estimate of entropy under some conditions. They corrected this flaw and, in keeping with the tradition of this field, gave it a new name: *sample entropy* (SampEn). SampEn(m, r, N) has better statistical properties than ApEn(m, r, N) as well as faster computation times due to an advance made in the design of the algorithm.

ApEn and SampEn quickly found wide applications in scientific literature, from biomedical applications to computer science to finance. In biomedical applications, it is generally (but not always) found that healthy subjects have greater system complexity than sick subjects. The effect of aging is also to generally lower the system complexity (Lipsitz & Goldberger, 1992). In principle, ApEn and SampEn can measure system complexity on any timescale by varying m, without being restricted to the one timescale at which the data happened to be sampled. However, the data series length requirements rapidly increase for higher m, a problem that is handled more efficiently using *multiscale entropy*.

Multiscale Entropy

The pattern size m in ApEn(m, r, N) and SampEn(m, r, N) can be increased to get estimates of entropy on different timescales. This strategy runs into a practical problem: the length of time series required for reliable estimation of the entropy increases exponentially with m. Costa, Goldberger, and Peng (2002) suggested a different approach to estimate the entropy across timescales. Instead of increasing the pattern size, the time series is coarse grained at different timescales and the SampEn is computed for the new series. The estimates of SampEn at various timescales are collectively referred to as *multiscale entropy* (MsEn).

Coarse graining is a procedure that increases the timescale (see Figure 6.3). The shortest timescale that is available in a data series equals the time between each observation. For simplicity, we assume that data have been collected uniformly, that is, measurements are equispaced in time. Thus, we have a series of N observations: $x_1, x_2, x_3, x_4, x_5, x_6, \ldots, x_N$. These observations were made at times: $0, t, 2t, 3t, 4t, 5t, \ldots, (N - 1)t$, where t is the time between each measurement. Coarse graining at scale 2 has the following meaning: Each pair of data points is replaced by its average value. The series coarse grained at scale 2 results in: $(x_1 + x_2)/2$, $(x_3 + x_4)/2$, $(x_5 + x_6)/2$, and so on. The timescale corresponding to the new

FIGURE 6.3 Example of Coarse Graining a Time Series at Scales 5 and 10
Note that the sampling interval effectively increases with coarse graining.

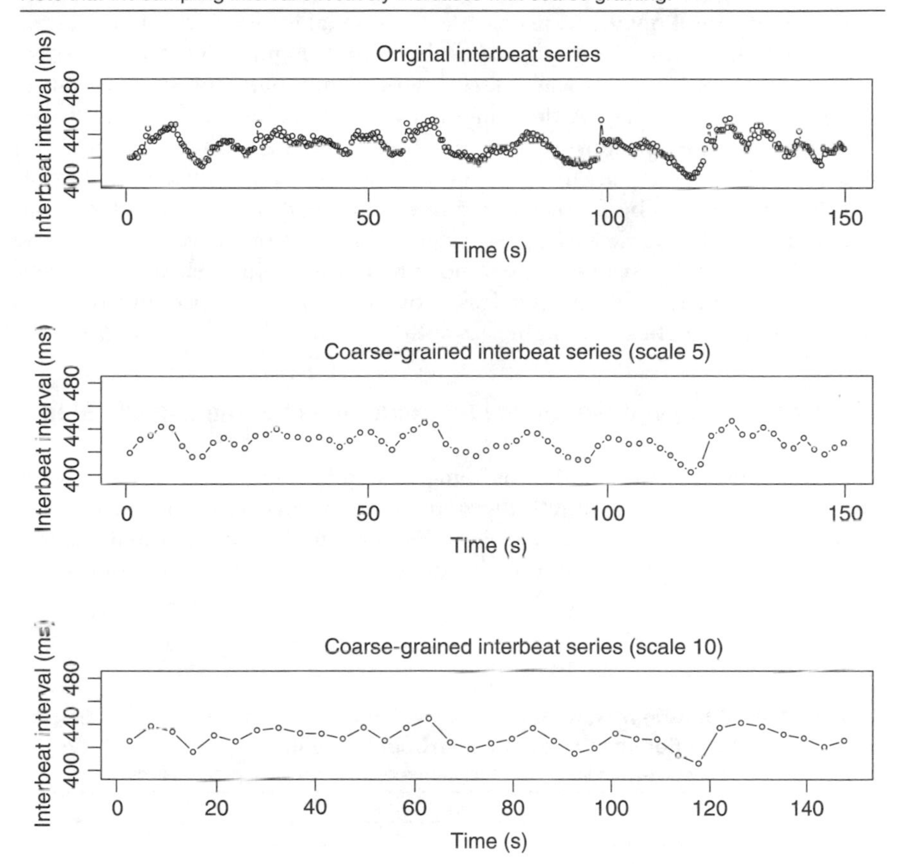

series is $2t$ because each pair of observations is separated by a time interval of $2t$. The length of the new series is approximately half that of the original series. (It is exactly half if N is even; otherwise, it is one short of half.)

Similarly, coarse graining at scale 3 results in a series with timescale $3t$, with values: $(x_1 + x_2 + x_3)/3$, $(x_4 + x_5 + x_6)/3$, and so on. The length of the series is approximately a third of the original length. As the timescale is increased, the coarse-grained series is shortened, which reduces the statistical reliability of the estimated entropy of the coarse-grained series. However, the loss of reliability is still lower than the loss that would result if we had increased the timescale by increasing the pattern size.

The sample entropy is typically computed for $m = 2$ for the coarse-grained series at each timescale. It is convenient to present MsEn as a plot of SampEn(2, *r*, *N*) versus timescale. The timescale is usually replaced simply by the scaling factor, for example, timescale 3*t* is simply represented as 3. Variation of SampEn with scale provides additional information about system complexity that SampEn at a fixed scale cannot. Figures 6.4 and 6.5 show MsEn of a neonatal interbeat interval series computed for various values of *m* and *r*, respectively. A drop in SampEn at a higher scale indicates the presence of more structure or order at that scale. The reduction of SampEn at scale 2 in some of the curves is likely to be due to a tendency for a short interbeat interval to be followed by a long interbeat interval, thereby resulting in a smoother series when averaged over pairs of heartbeats. Because this tendency is not strong, it does not present itself in each of the curves. Lower values of resolution offer less protection against noise and result in fewer pattern matches and higher SampEn.

Multiscale Entropy of the Normal Distribution and of the Logistic Map

It is illustrative to study MsEn for some standard data series. The following two data series were synthetically generated: normal distribution and the logistic map. The normal data series was generated with a normal random number generator. (See Figure 6.6 for the data series and the distributions of the generated series.)

FIGURE 6.4 SampEn(*m*, *r*, *N*) at Different Values of *m* (Pattern Size), Holding *r* and *N* Fixed, for a Series of Neonatal Interbeat Intervals
In this example, there is more complexity at higher timescales for each pattern size.

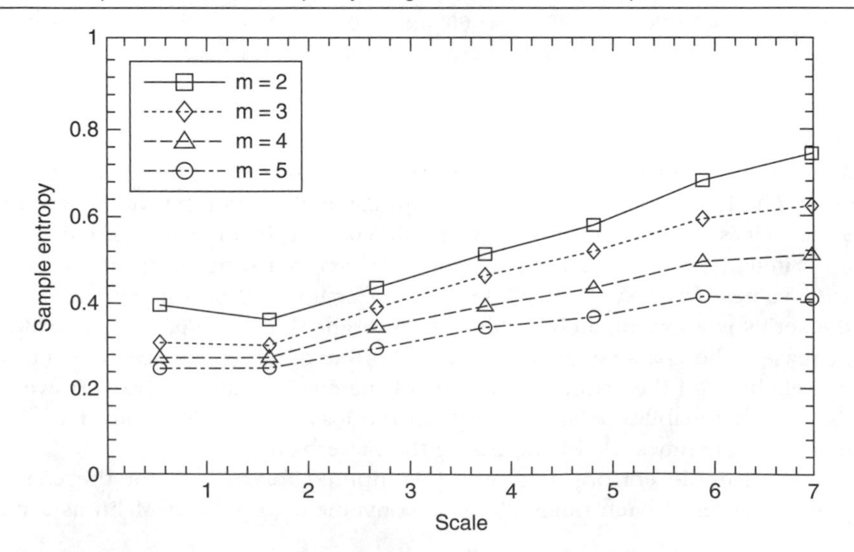

FIGURE 6.5 SampEn (*m, r, N*) at Different Values of *r* (Tolerance), Holding *m* and *N* Fixed, for a Series of Neonatal Interbeat Intervals

In this example, there is more complexity at higher timescales and the entropy is lower for higher tolerances.

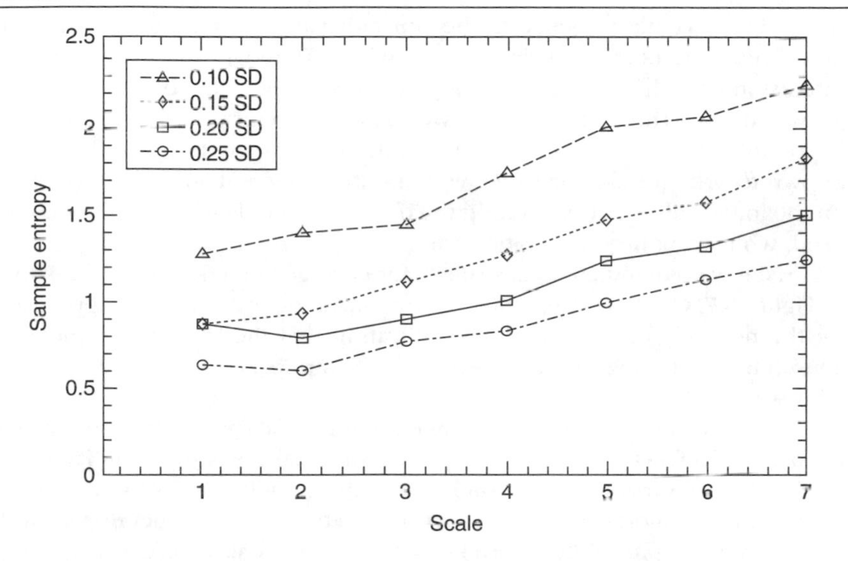

FIGURE 6.6 Examples of Data for a Normal Distribution and for a Chaotic Logistic Map

Below each data set is shown their observed distributions over large data sets that contain 50 times as many data points as in the top panels.

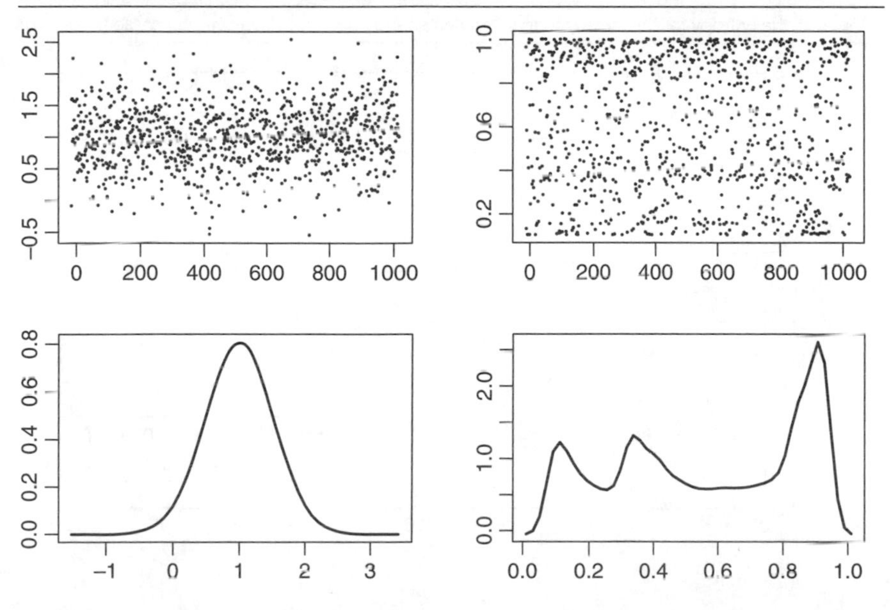

The logistic map is given by

$$X_{n+1} = Rx_n = Rx_n(1 - x_n),$$

which is a formula to generate the $(n+1)$th value of x from the nth value of x. The value of x is restricted to lie within 0 and 1. The map is of wide interest in growth curves because it generates a new data point based on the growth up to a certain time (x_n) as well as the amount of further growth that is possible $(1 - x_n)$. For some values of the multiplying factor R, this map is known to produce chaotic series with a critical dependence of future values on the initial value. For generating the data shown in Figure 6. 6, $R = 3.9$ was used, which produced a chaotic series.

MsEn was computed up to scale 20 for each generated series and is shown in Figure 6.7. Over this range of scales, SampEn does not change appreciably for the normally distributed series, indicating that the level of complexity is constant. For the logistic map series, the SampEn and complexity increase with scale and then saturate.

The application of MsEn to heart rate variability data from a prematurely born neonate and a fetus that were close in postmenstrual age is shown in Figure 6.8. Fetal heart rate variability was found to be more complex across timescales, and other measures indicated that the nervous system was better developed in the fetus than in the prematurely born neonate (Padhye et al., 2008).

FIGURE 6.7 Multiscale Entropy for a Normal Distribution and for the Logistic Map

Multiscale entropy for a normal distribution shows a small decrease over the timescales shown. However, the entropy for the logistic map increases sharply and stabilizes over the same timescales.

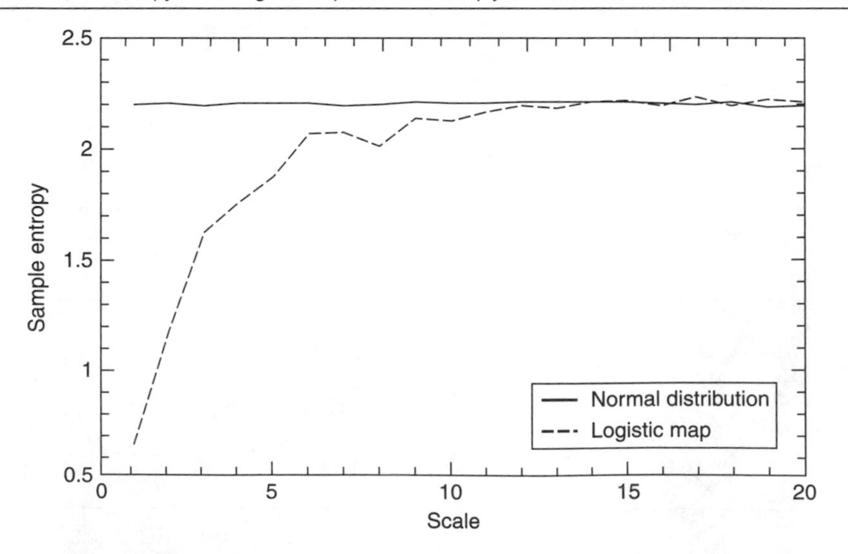

FIGURE 6.8 Examples of Prematurely Born Neonatal (top panel) and Fetal (middle panel) RR Series of Heart Rate Variability

The multiscale entropy computed from full data sets for the neonate (N) and the fetus (F) are shown in the bottom panel.

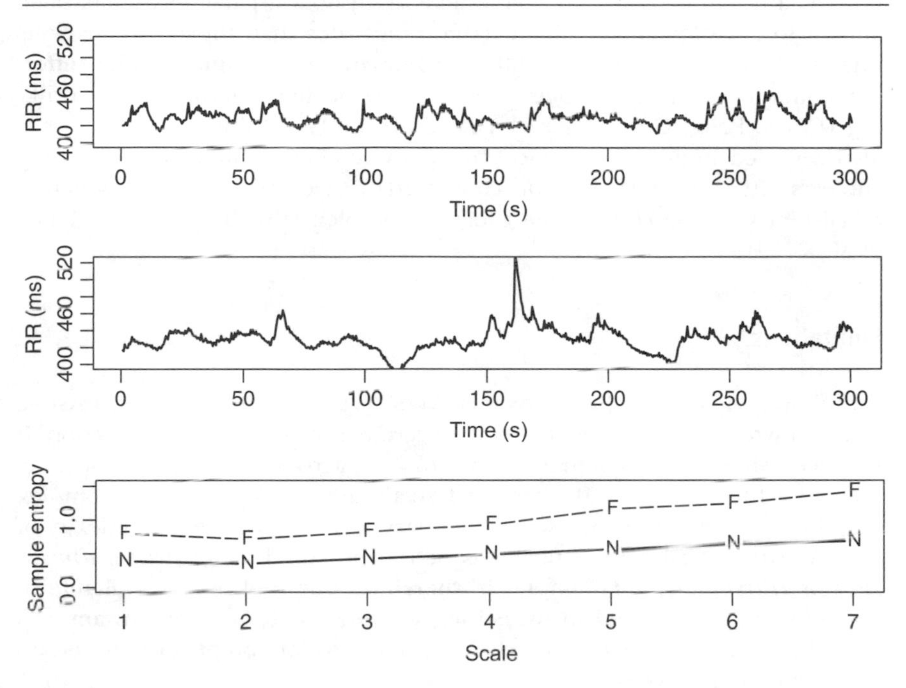

Multiscale entropy has wide applicability, ranging from its original use in heart rate variability (Costa et al., 2002), to its ability to distinguish between types of fluid flows (Zheng & Jin, 2009), to image segmentation (Leung, Lam, & To, 2002), to mechanical fault diagnosis in the production process (Zhang, Xiong, Liu, Zou, & Guo, 2009), and to detection of anomalous traffic on computer networks (Riihijärvi, Wellens, & Mähönen, 2009). MsEn consistently yields higher values when there are long-range correlations in the time series (Yoo & Yi, 2006), as are typically found in data from healthy subjects or even in "healthy" computer networks.

In the health sciences, multiscale entropy has established a growing record of separating healthy and pathologic groups. Costa et al. (2002) showed that the pattern of SampEn variation with scale was distinct for patients with normal hearts, with atrial fibrillation, and with congestive heart failure. Although SampEn for atrial fibrillation was higher at small timescales than in normal heart rate variability, at larger timescales SampEn was highest for healthy subjects. This is in keeping with a widely held conjecture in the theory of complex systems that complexity is higher in the healthy state. Reduced level of MsEn of heart rate variability has been used to predict mortality in trauma patients (Norris, Stein, & Morris, 2008). A variation of MsEn has been shown to be reduced in subjects with aortic stenosis (Valencia et al., 2009). Renal nerve activity in anaesthetized rats has been shown to have lower SampEn at large timescales than those in conscious rats (Li, Yan, Yang, & Zhang, 2008). In human gait dynamics under different conditions, MsEn has been observed to be highest for normal walking (Costa, Peng, Goldberger, & Hausdorff, 2003). Multiscale entropy has also revealed higher electroencephalographic (EEG) complexity in young subjects after visual stimulation compared with elderly subjects, which is consistent with the conjecture of loss of complexity with aging (Takahashi et al., 2009).

Limitations

SampEn(m, r, N) are a family of measures of complexity, but it is possible to reach wrong conclusions if adequate care is not taken in interpretation. It has been shown that SampEn is close to zero across a wide range of timescales for the Weirstrass–Mandelbrot fractals despite the relatively complex self-similar behavior of such series (Riihijärvi et al., 2009). The complexity of the Weirstrass–Mandelbrot fractals is better measured using *fractal dimension* or *Hurst exponent*. SampEn is, therefore, one tool that is available to the researcher interested in measuring complexity. It is an important and versatile tool, but it is known to fail as a measure of complexity for certain kinds of complex systems.

SampEn does succeed in measuring the relative complexity of processes that share the same underlying theoretical structure. If processes have different underlying theoretical structures, it is generally not possible to claim that the process with higher SampEn has higher complexity. In practical applications, this implies that comparisons must be restricted to multiple series of the same variable because they arise from a common physiological control mechanism. Therefore, comparison of SampEn of two series of interbeat intervals of the heart obtained from two subjects is straightforward and justified, but comparison of SampEn of a series of interbeat intervals to SampEn of a temperature series could be problematic because the control mechanisms are likely to be different.

SampEn and MsEn also suffer from some technical flaws that can be important in certain conditions. Thuraisingham and Gottwald (2006) have discussed the impact of sampling time on MsEn and the confusion that it can potentially create in entropy signatures of complex physiological systems. Variants of MsEn have been proposed to reduce problems arising from the coarse-graining procedure (Valencia et al., 2009) and from nonstationarity of the time series (Aziz & Arif, 2005).

REFERENCES

Aziz, W., & Arif, M. (2005). *Multiscale permutation entropy of physiological time series.* In 9th International Multitopic Conference, IEEE INMIC 2005, Karachi, Pakistan, 1–6.

Breiman, L., Friedman, J. H., Olshen, R. A., & Stone, C. J. (1984). *Classification and regression trees.* Belmont, CA: Wadsworth International Group.

Costa, M., Goldberger, A. L., & Peng, C. K. (2002). Multiscale entropy analysis of complex physiologic time series. *Physical Review Letters, 89* (068102).

Costa, M., Peng, C.-K., Goldberger, A. L., & Hausdorff, J. M. (2003). Multiscale entropy analysis of human gait dynamics. *Physica A: Statistical Mechanics and Its Applications, 330,* 53–60.

Duda, R. O., Hart, P. E., & Stork, D. G. (2001). *Pattern classification.* New York: Wiley.

Eckmann, J. P., & Ruelle, D. (1985). Ergodic theory of chaos and strange attractors. *Review of Modern Physics, 57,* 617–654.

Grassberger, P., & Procaccia, I. (1983). Estimation of the Kolmogorov entropy from a chaotic signal. *Physical Review A, 28,* 2591–2593.

Kantz, H., & Schreiber, T. (1997). *Nonlinear time series analysis.* Cambridge, MA: Cambridge University Press.

Kolmogorov, A. N. (1958). New metric invariant of transitive dynamical systems and endomorphisms of Lebesgue spaces. *Doklady of Russian Academy of Sciences, 119,* 861–864.

Leung, C. K., Lam, F. K., & To, J. T. P. (2002). *Entropy-based multiscale image segmentation with edge refinement.* In Proceedings of the First International Conference on Information Technology and Applications (ICITA 2002), Bathurst, Australia, 143–148.

Li, Y.-T., Yan, R., Yang, Z., & Zhang, T. (2008). Multiscale entropy analysis of renal sympathetic nerve activity in conscious and anaesthetized Wistar rats. *Chinese Journal of Biomedical Engineering, 27,* 498–501.

Lipsitz, L. A., & Goldberger, A. L. (1992). Loss of 'complexity' and aging. *Journal of the American Medical Association, 267,* 1806–1809.

Norris, P. R., Stein, P. K., & Morris, J. A., Jr. (2008). Reduced heart rate multiscale entropy predicts death in critical illness: A study of physiologic complexity in 285 trauma patients. *Journal of Critical Care, 23,* 399–405.

Ott, E. (1994). *Chaos in dynamical systems.* New York: Cambridge University Press.

Padhye, N. S., Verklan, M. T., Brazdeikis, A., Williams, A. L., Khattak, A. Z., & Lasky R. E. (2008). *A comparison of fetal and neonatal heart rate variability at similar postmenstrual ages.* In Proceedings of the 30th Annual International Conference of the IEEE Engineering in Medicine and Biology Society, Vancouver, BC, Canada, 2801–2804.

Pierce, J. R. (1980). *An introduction to information theory symbols, signals and noise.* New York: Dover Publications.

Pincus, S. M. (1991). Approximate entropy as a measure of system complexity. *Proceedings of the National Academy of Sciences, 88,* 2297–2301.

Pincus, S. M., Gladstone, I. M., & Ehrenkranz, R. A. (1991). A regularity statistic for medical data analysis. *Journal of Clinical Monitoring, 7,* 335–345.

Riihijärvi, J., Wellens, M., & Mähönen P. (2009). *Measuring complexity and predictability in networks with multiscale entropy.* In Proceedings of IEEE Conference on Computer Communications (INFOCOM), Rio de Janeiro, Brazil, 1107–1115.

Shannon, C. E. (1949). Communication in the presence of noise. *Proceedings of the Institute of Radio Engineers, 37,* 10–21. (Reprint as classic paper in *Proceedings of the IEEE, 86,* 1998.)

Sinai, Ya. G. (1959). On the notion of entropy of a dynamical system. *Doklady of Russian Academy of Sciences, 124,* 768–771.

Takahashi, T., Cho, R. Y., Murata, T., Mizuno, T., Kikuchi, M., Mizukami, K., et al. (2009). Age-related variation in EEG complexity to photic stimulation: A multiscale entropy analysis. *Clinical Neurophysiology, 120,* 476–483.

Thuraisingham, R. A., & Gottwald, G. A. (2006). On multiscale entropy analysis for physiological data. *Physica A: Statistical Mechanics and its Applications, 366,* 323–332.

Valencia, J. F., Porta, A., Vallverdu, M., Claria, F., Baranowski, R., Orlowska-Baranowska, E., et al. (2009). Refined multiscale entropy: Application to 24-h Holter recordings of heart period variability in healthy and aortic stenosis subjects. *IEEE Transactions on Biomedical Engineering, 56,* 2202–2213.

Van den Eijkel, G. (2003). Information-theoretic tree and rule induction. In M. Berthold & D. Hand (Eds.), *Intelligent data analysis* (Appendix B, 465–473). Berlin: Springer-Verlag.

Yoo, C.-S., & Yi, S.-H. (2006). On the physiological validity and the effects of detrending in the multiscale entropy analysis of heart rate variability. *Journal of the Korean Physical Society, 48,* 670–676.

Zhang, L., Xiong, G., Liu, H., Zou, H., & Guo, W. (2009). An intelligent fault diagnosis method based on multiscale entropy and SVMs. In H. Yu & N. Zhang (Eds.), *Lecture notes in computer science* (Vol. 5553, pp. 724–732). Berlin: Springer-Verlag.

Zheng, G.-B., & Jin, N.-D. (2009). Multiscale entropy and dynamic characteristics of two-phase flow patterns. *Wuli Xuebao/Acta Physica Sinica, 58,* 4485–4492.

Response to Chapter 6

Mary Pat Rapp

The physiological changes with aging, chronic disease, prevention of functional decline, and the care of older adults involve a complicated and dynamic interplay of biological, social, and cultural processes. Traditional descriptions of physiological and functional decline in aging suggest a linear downward trajectory based on identifying the cause, or trigger, of a cascade leading to functional dependency and death (Covinsky, Eng, Lui, Sands, & Yaffe, 2003). Challenges related to the functional decline and care for older adults possibly stem from thinking in this linear fashion.

Linear explanations fail to incorporate the constantly changing dynamics of aging, disease-related physiology, and the interface with the external systems of care including primary care practices, acute care, long-term care, and informal or formal social networks. When understood in this manner, health systems function in a closed loop identifying the malfunctioning parts, developing a plan of action, studying the reaction, and if necessary, refining the intervention leading to a revised plan of action. This may explain that although processes in screening, preventing, and treating chronic disease in older adults are known (Wenger, Roth, & Shekelle, 2007), progress in preventing functional decline remains problematic.

In this chapter, Dr. Padhye presents a nonlinear method to measure probabilities in health status. Using time series data, entropy offers an alternative to homeostatic or periodic measures and may improve our understanding of physiological changes in disease and health. As explained by Dr. Padhye, nonlinear methods measure the element of surprise where small changes can have exaggerated and unanticipated effects.

In order to link theory to method, the practical application of entropy in nursing practice is illustrated here by using substruction as suggested by Dulock and Holzemer (1991). Table 6.1 illustrates how information can be organized from abstract concepts through proposed outcomes. Note that entropy is on a micro scale as compared to the macro scale concepts of complexity and health.

COMPLEXITY

As explained by Dr. Padhye, information theory is concerned with resolving uncertainty. Complexity is a measure of uncertainty where the degree of uncertainty is associated with health status. Complexity implies that all systems work harmoniously, responding at the appropriate time and with sufficient intensity to adjust and adapt to physiological, psychological, and social changes. The process of senescence and disease weakens coordination among systems, undermining complexity and making the old person frail (Fried et al., 2009).

TABLE 6.1 Information Collection

TERM	DEFINITION	EXAMPLE
Construct	Abstract idea	Complexity, health
Axiom	Relational statement between constructs	Complexity is an indicator of health
Concepts	Word expressing a mental image of some phenomenon	Variability, frailty
Proposition	Relational statement between two concepts	Frailty is an indicator of the loss of variability
Postulate	Relational statement between construct and concept	Frailty is characterized by a change in complexity. Variability may increase or decrease as health status changes
Transformational statement	Logically derived links between the concepts and empirical indicators	Thermistors measure variability in temperature
Empirical indicators	Instruments or experimental conditions	Thermistor
Hypothesis	Relational statement between two empirical indicators	Temperature variability changes with aging and disease
Measurement	Act of assigning units, timescales	Times-series data examples: temperature measured in preset intervals
Scores/values	Unit of measure	Entropy
Descriptive or inferential statistics	Data analysis techniques	Means, standard deviations
Outcomes	Goal of measuring and describing complexity	Interventions, single, multifactorial, or extrinsic

HEALTH

Defined by the World Health Organization, health is characterized as a state of physiological, social, and psychological well-being. The definition recognizes that health includes persons with acute and chronic disease as well as functional disabilities (WHO, 1948). Complexity implies that health is dynamic as opposed to static. Adaptation, adjustment, coping, and change are measures of complexity that explain the dynamic state of health for older adults. Atypical presentation of disease is the norm, and alterations in psychological and mental status are common with disruptions of the physiological system. In sum, complexity is an indicator of health.

Dr. Padhye makes an important distinction between complexity, homeostasis, and periodic rhythms where complexity is more closely related to the behavior of the system. Temperature is commonly measured as a dichotomous homeostatic variable high or low, where either is associated with a variable state of health. When associated with infections, temperatures at least one degree above normal suggest a healthy immune system. The return to a stable temperature near 98.6°F assumes a healthy recovery. A grave sign of morbidity and mortality is an infection associated with a low temperature. Although the implications are unclear, institutionalized older adults, most of whom are frail, have constant lower body temperatures (Aung, Darvesh, Alam, & Auerbach, 2004). Circadian temperature rhythms in a periodic system (usually 24 hour), where higher or lower temperatures are compared around the referent mean (98.6°F), are preserved in healthy older adults (Czeisler et al., 1999; Monk, Buysse, Reynolds, Kupfer, & Houck, 1995; Monk & Kupfer, 2000). Investigators have shown that chronic disease and dementia alter circadian rhythms when compared with older adults without those conditions (Davis & Lentz, 1989; Harper et al., 2001; Nakamura, Tanaka, Motohashi, & Maeda, 1997; Okamoto-Mizuno, Yokoya, & Kudoh, 2000; Sund-Levander & Wahren, 2002). Like homeostatic systems, periodic systems are dependable and predictable.

Frailty is a constellation of signs and symptoms characterized by unplanned weight loss, low strength, low energy, slowed motor performance, and low physical activity (Fried et al., 2001). Frailty is the result of age and disease-related changes occurring at microscopic levels and ultimately affects the overall function of the individual. The risk of frailty increases nonlinearly with the number of biological system alterations. The odds ratio for older adults with one or two system alterations is 4.8 and rises to 26 with six system alterations (Fried et al., 2009).

Lipsitz suggests that greater variability in physiological systems is observed in functional decline and frailty (Lipsitz, 2002). Using the temperature example, older individuals exposed to heat or cold demonstrate greater and faster fluctuations in body temperature and are susceptible to hyperthermia and hypothermia out of proportion to younger persons. Equating to a loss of complexity, the body is incapable of making micro adjustments to restore normal body temperature. Visually, high complexity with low variability is a rippling wave of minor adjustments, whereas high variability with low complexity is the exaggerated fixed response of a seesaw (Lipsitz, 2002).

Lipsitz's concept was observed in older nursing facility residents whose skin temperature was measured in response to bathing, a commonly occurring event. Residents at higher risk for pressure ulcers, a proxy measure for frailty, demonstrated a longer mean time for temperature return to baseline, that is, 153 minutes (SD = 102) in the high-risk group versus 136 minutes (SD = 100) in the lower-risk group. Additionally, skin temperature mean multiscale entropy was lowest among those who developed pressure ulcers, $F(1, 15.78) = 35.14, p < 0.001$ (Rapp, Bergstrom, & Padhye, 2009).

Frailty is associated with the degree and direction of change in complexity and variability (Vaillancourt & Newell, 2002). There is evidence to suggest that temperature dysregulation in old age is concomitant with chronic disease and frailty, and may be associated with excess morbidity and mortality. A means to quantify temperature as a continuous variable to aid in diagnosis or prognosis may be clinically important (Cuesta et al., 2007; Varela et al., 2009). Measuring complexity is a relatively young science, particularly when applied to temperature data (Varela, Jimenez, & Farina, 2003). Measures of central tendency and differences are inadequate to evaluate the dynamic nature of the temperature regulation. Entropy, using time-series analysis, allows us to learn more about the system.

The ultimate goal of measuring and describing entropy of any system is the exciting possibility that the knowledge may lead to more complex interventions, matching the complexity of the individual. As a single intervention, exercise is unlikely to reverse the trend of functional decline in frailty. It is more likely that a multisystem, interdisciplinary approach will be more effective.

REFERENCES

Aung, M. M., Darvesh, G. M., Alam, S., & Auerbach, C. (2004). *Older is colder: Temperature range and variation in the elderly*. Paper presented at the American Geriatrics Society 2004 Annual Scientific Meeting, Las Vegas, NV.

Covinsky, K. E., Eng, C., Lui, L. Y., Sands, L. P., & Yaffe, K. (2003). The last 2 years of life: Functional trajectories of frail older people. *Journal of the American Geriatrics Society, 51*(4), 492–498.

Cuesta, D., Varela, M., Miro, P., Galdos, P., Abasolo, D., Hornero, R., et al. (2007). Predicting survival in critical patients by use of body temperature regularity measurement based on approximate entropy. *Medical and Biological Engineering and Computing, 45*(7), 671–678.

Czeisler, C. A., Duffy, J. F., Shanahan, T. L., Brown, E. N., Mitchell, J. F., Rimmer, D. W., et al. (1999). Stability, precision, and near-24-hour period of the human circadian pacemaker. *Science, 284*(5423), 2177–2181.

Davis, C., & Lentz, M. J. (1989). Circadian rhythms: Charting oral temperatures to spot abnormalities. *Journal of Gerontological Nursing, 15*(4), 34–39.

Dulock, H. L., & Holzemer, W. L. (1991). Substruction: Improving the linkage from theory to method. *Nursing Science Quarterly, 4*(2), 83–87.

Fried, L. P., Tangen, C. M., Walston, J., Newman, A. B., Hirsch, C., Gottdiener, J., et al. (2001). Frailty in older adults: Evidence for a phenotype. *The Journals of Gerontology Series A Biological Science and Medical Science, 56*(3), M146–M156.

Fried, L. P., Xue, Q. L., Cappola, A. R., Ferrucci, L., Chaves, P., Varadhan, R., et al. (2009). Nonlinear multisystem physiological dysregulation associated with frailty in older women: Implications for etiology and treatment. *The Journals of Gerontology Series A Biological Science and Medical Science, 64*(10), 1049–1057.

Harper, D. G., Stopa, E. G., McKee, A. C., Satlin, A., Harlan, P. C., Goldstein, R., et al. (2001). Differential circadian rhythm disturbances in men with Alzheimer disease and frontotemporal degeneration. *Archives of General Psychiatry, 58*(4), 353–360.

Lipsitz, L. A. (2002). Dynamics of stability: The physiologic basis of functional health and frailty. *The Journals of Gerontology Series A Biological Science and Medical Science, 57*(3), B115–B125.

Monk, T. H., Buysse, D. J., Reynolds, C. F., III., Kupfer, D. J., & Houck, P. R. (1995). Circadian temperature rhythms of older people. *Experimental Gerontology, 30*(5), 455–474.

Monk, T. H., & Kupfer, D. J. (2000). Circadian rhythms in healthy aging—effects downstream from the pacemaker. *Chronobiology International, 17*(3), 355–368.

Nakamura, K., Tanaka, M., Motohashi, Y., & Maeda, A. (1997). Oral temperatures of the elderly in nursing homes in summer and winter in relation to activities of daily living. *International Journal of Biometeorology, 40*(2), 103–106.

Okamoto-Mizuno, K., Yokoya, T., & Kudoh, Y. (2000). Effects of activity of daily living and gender on circadian rhythms of the elderly in a nursing home. *Journal of Physiological Anthropology and Applied Human Science, 19*(1), 53–57.

Rapp, M. P., Bergstrom, N., Padhye, N. S. (2009). Contribution of activity and skin temperature to the risk of developing pressure ulcers in nursing facility residents. *Advances in Skin Wound and Care, 22*(11), 506–513.

Sund-Levander, M., & Wahren, L. K. (2002). The impact of ADL status, dementia and body mass index on normal body temperature in elderly nursing home residents. *Archives of Gerontology and Geriatrics, 35*(2), 161–169.

Vaillancourt, D. E., & Newell, K. M. (2002). Changing complexity in human behavior and physiology through aging and disease. *Neurobiology of Aging, 23*(1), 1–11.

Varela, M., Cuesta, D., Madrid, J. A., Churruca, J., Miro, P., Ruiz, R., et al. (2009). Holter monitoring of central and peripheral temperature: possible uses and feasibility study in outpatient settings. *Journal of Clinical Monitoring and Computing, 23*(4), 209–216.

Varela, M., Jimenez, L., & Farina, R. (2003). Complexity analysis of the temperature curve: New information from body temperature. *European Journal of Applied Physiology, 89*(3–4), 230–237.

Wenger, N. S., Roth, C. P., & Shekelle, P. (2007). Introduction to the assessing care of vulnerable elders-3 quality indicator measurement set. *Journal of the American Geriatrics Society, 55*(2, Suppl. 2), S247–S252.

WHO. (1948, June 19–22). Preamble to the Constitution of the World Health Organization as adopted by the International Health Conference, New York, 1946; signed on 22 July 1946 by the representatives of 61 States (Official Records of the World Health Organization, no. 2, p. 100) and entered into force on 7 April 1948.

CHAPTER 6
ENTROPY AS INFORMATION CONTENT IN MEDICAL RESEARCH

- *Does the idea of entropy (things going from order to disorder as in the smoke ring example) fit what you see in patients?*
- *As explained by Dr. Padhye, nonlinear methods measure the element of surprise where small changes can have exaggerated and unanticipated effects. What are some examples where small changes have had a profound effect on a patient's or the organization's well-being?*
- *What are some examples where small changes by managers or nurse leaders have significantly changed health care delivery?*
- *Dr. Rapp points out that entropy theories break from a traditional view of linear decline in patient function. How can care delivery systems take into account nonlinear changes in patient well-being?*
- *How does your organization balance the concepts of homeostasis and circadian rhythms in the assessment of patients or the work life of nurses?*

SEVEN

Caring Science and Complexity Science Guiding the Practice of Hospital and Nursing Administrative Practice

Mary Beth Kingston
Michael Brooks Turkel

This chapter contains practical examples of how a Chief Nurse Executive and Chief Executive Officer from different health care facilities use tenets from complexity and caring sciences as a framework to guide leadership styles and practices in nursing and health care. Examples include changes in one organization's performance management approach, implementation of Schwartz rounds, and collaboration between organizations to address a crisis in the provision of obstetrical care. A subsequent section highlights how concepts from complexity and caring frame health care economics and financial decision making within complex organizations. Respondent Alfonso discusses Caring and Complexity Science approaches to health care challenges. He presents the differences between customer service and professional practice grounded in theory, values and ethics.

INTRODUCTION

Nursing practice and health care leadership is driven by a complex system of economics, technology, regulations, and multimodal organizations (Turkel & Ray, 2000). Because of the need to balance economic realties with the tenets of caring science, the cultures of nursing, medicine, and health care organizations are in a constant state of chaos. The focus on costs is not a transient response to decreasing reimbursement; instead, it has become the catalyst for change at the national level within health care organizations and with health care financing. At the same time, sustaining human caring ideals and a caring ideology is what patients, nurses, and physicians need and value (Watson, 2005).

Sister Simone Roach (2002) wrote that caring is the human mode of being and that the attributes of caring can be described in the six *C*s of commitment, compassion, competence, confidence, conscience, and comportment. Drawing on the work of Roach, health care delivery systems can be understood with another group of *C*s. According to Turkel (1993), the six *C*s of health care systems are complicated, counterintuitive, convoluted, confusing, chaotic, and complex. Caring and complexity are inextricably bound, recognizing that living systems cannot be viewed as segmented pieces, but rather as interactive and whole. Integrating caring and complexity theory into administrative practice is essential for leaders in the discipline of nursing. However, many health care leaders, including nurses, have "grown up" with a leadership and interpersonal style that is focused on linear thinking and linear problem solving. (Lindberg, Herzog, Merry, & Goldstein, 1998). This approach typically involves identifying the issue or problem, considering potential strategies, implementing interventions, and evaluating the results. Clinicians often employ this method to address clinical problems. It is also the preferred strategy for many in health care administration and leadership roles. Breaking down problems or issues into discrete parts and attempting to remove variation in processes works well for certain tasks, but health care leaders may increase innovation and creativity by integrating tenets from caring science and examining issues through the "lens" of complexity.

Other chapters in this book have described caring science and complexity science in detail. This chapter's focus is to illustrate how concepts from caring science and complexity science provide the framework for examining leadership styles and practices in nursing and health care leadership. An example of leadership innovations that are based on an understanding of these principles include (a) changes in one organization's performance management approach, (b) implementation of Schwartz Center Rounds (Schwartz Center, 1995), and (c) collaboration between a community hospital and major teaching hospital to address a crisis in provision of obstetrical services. The common theme in each example is that the approaches changed the way that individuals related to, communicated, and interacted with each other. These practical applications advance the theory of relational complexity where relationships are viewed as a cocreative process as organizational systems respond to the changing health care environment (Ray, Turkel, & Marino, 2002). A subsequent section highlights how concepts from complexity and caring can frame financial decision making within complex organizations and health care economics.

KEY COMPLEXITY CONCEPTS RELATING TO HEALTH CARE LEADERSHIP

Complexity science is the study of living systems and examines the manner in which these systems behave. It is based on the belief that the whole of a system is greater than the sum of its parts. Health care organizations are examples of a complex adaptive system (CAS) that is nonlinear, dynamic, unpredictable, and interactive. A CAS is self-organizing; self-organization is a process where agents (elements, staff) adjust their behaviors to cope with the demands of internal and external environments. Self-organization has the potential to be a source of creativity and innovation—a process termed "emergence." A balance of variability (not too much, not too little) is viewed as positive; indeed, creative ideas arise when one begins to break away from the norm. Complicated plans and forecasts are replaced with minimum specifications providing general guidance and vision to promote self-organization and successful outcomes (Lindberg, Nash, & Lindberg, 2008). Stacey (2007) applied complexity theory to human interactions proposing that patterns emerge through self-organization, which can lead to widespread change in behavior and interactions. It is the relationships between the agents, elements, and/or individuals that serve as the primary forces that change an outcome or environment.

To further awaken this creative leadership and caring spirit, nurse leaders need to recreate the organizational culture. Watson's (2008) theory of human caring serves as the professional model for nursing practice within the Albert Einstein Healthcare Network (AEHN). Nursing meetings open with a caring story where members are invited to share a caring moment. Unit-based bulletin boards celebrate lived examples of nurse-to-nurse caring and examples of where caring made a difference for the patient and/or the patient's family. Monthly caritas forums provide an opportunity for nurses to dialogue about caring and to reflect on how caring has transformed their practice of nursing and their understanding of the patient experience.

Consider the groundbreaking research that Anderson, Issel, and McDaniel (2003) and Anderson et al. (2005) conducted in nursing homes, examining the relationship between management practices and nursing home resident outcomes. Nursing homes, or long-term care facilities, are one of the most heavily regulated areas of health care with prescriptive and detailed standards developed to ensure quality of care for all residents. As in other health care organizations, there has been a tendency to employ a hierarchical leadership structure, in part due to the significant regulatory oversight and high numbers of unlicensed personnel. The study was conducted at 164 nursing homes and the results suggested that resident outcomes are heavily dependent on the quality of the relationships among nursing home caregivers and the existing management practices. Significant findings included (a) lower use of restraints when staff could speak freely without fear of reprisal, (b) lower levels of resident aggression when registered nurses (RNs) participated in decision making, and (c) lower prevalence rates of fractures and complications from immobility with a relationship-oriented leadership style.

As the complexity of an issue increases, and the leader must focus on relationships, coalition building, and negotiation skills or strategies to allow innovation, creativity, and new ways of operating emerge (Anderson & McDaniel, 2000; Stacey, 1996). Nursing executives need to foster a practice environment where building caring relationships is valued as part of the culture. An example of this was the decision to expand visiting hours within AEHN. Although the evidence clearly shows that increased visitation translates into increased patient satisfaction, the concept was challenging to integrate into the culture of an urban hospital practice environment. The vision was to use a caring framework for the creation of guiding principles related to visitation that balances and respects the needs and safety of patients, families, visitors, and employees. Discovery action dialogues were held with direct care RNs, physicians, family members, and patient safety officers. As a result of dialogue and collaboration, extended visiting hours were implemented with overnight requests and children visitation being honored.

CARING INFORMING HEALTH CARE LEADERSHIP

The nurse leader, functioning within a dynamic CAS, can strengthen relationships and promote self-organization using the unifying framework of caring science. Focusing on the patient as the basis for decision making, creating caring and trusting relationships with staff, devoting time for self-reflection and renewal, and reflecting on the holistic nature of nursing are congruent with caring and complexity. Boykin and Schoenhofer (2001) noted that all nursing administrative activities are ultimately directed to the patient or "person(s) being nursed." Modeling and demonstrating that this truly informs administrative decisions provides all staff members with guiding principles on which to base their practice and serves to connect the leader with all nurses in the organization. O'Connor (2008) described the need for the nursing leader to focus on staff (the most important resource in the health care setting) by creating trusting and caring relationships. There are daily opportunities to build and strengthen these relationships with intention. Difficult situations, such as budget issues or individual performance, are then approached with mutual respect and understanding, even though differences of opinion exist.

Pipe (2008) and Turkel and Ray (2004) highlighted the need for the nurse leader to devote time for reflection and renewal as a means of exerting positive influence on individual staff and colleagues, as well as groups and teams. External influence in the chaotic health care environment is enhanced by strengthening one's inner journey. This is an area often overlooked by individuals with competing priorities and mounting pressures; however, caring for oneself is essential to facilitate growth and intentional presence. Nursing leadership mirrors the holistic and dynamic nature of nursing and demonstrates flexibility, inclusiveness, and adaptability (Jackson, Clements, Averill, & Zimbro, 2009). As noted earlier, these are also primary characteristics of CAS (leadership, caring, complexity) that are indeed interwoven in the health care environment.

APPLICATION TO HEALTH CARE LEADERSHIP PRACTICE

Providing Direction: Changing Performance Management Approach

Like many organizations, health care tends to have a very prescriptive approach to addressing poor performance in the work setting. Over the years, program titles have changed from discipline to counseling to performance management, but the basic tenets remained—a very detailed program with progressive steps and punitive consequences. The underlying philosophy of "discipline" is that the more one is "punished" the better one will behave. It is based on a hierarchical, parent–child model and is usually a one-sided view of a particular situation.

AEHN became inspired to change this approach after reading Grote's (2006) book, *Discipline Without Punishment*. The nurse managers and directors met with representatives from the human resource department to discuss development of a new program that incorporated a change in the way managers and staff interacted when a performance issue was identified. The change in focus was clearly not a directive from senior management.

Although the book outlined a template for changing the current performance management program, those involved quickly realized that importing a system of change would not result in success. Discovery action dialogues were held with nursing staff to discuss the issue and to involve them in the design. It became apparent that staff was often caught off guard when called to a meeting where formal discipline was presented. Ideas that emerged were simple, yet effective:

- Begin with informal interaction, that is, everyday conversation to discuss feedback or concerns.
- If issues continue, schedule a structured conversation to align the goals of the organization with performance.
- During two-way conversation, determine barriers in the environment or work setting and collaborate to create a "Success Plan."
- Continual feedback about progress or improvement.

All of the above items fall in the realm of informal action and do not follow the staff member through their employment with the organization. Staff also identified that feedback from peers had, at times, a greater influence on day-to-day behavior. Because some staff felt uncomfortable providing feedback, opportunities, such as walking rounds, were structured to promote these conversations. Staff nurses were encouraged to use walking rounds as an opportunity for nurse-to-nurse caring by providing positive feedback with peers. Managers also began to recognize that a degree of tension is often helpful in creating change and were beginning to allow the conversations to occur.

In a perfect world, all of these conversations would result in changed performance. There continues to be a more formal process if behaviors that are not aligned with the Mission Statement and Code of Conduct persist. However, a few subtle changes were incorporated. A meeting is scheduled to discuss gaps, to review the Success Plan, and to determine if additional support is needed. A written summary of the meeting replaces the prior standardized form. In the past, if issues remained unresolved, staff members were suspended for 3 days without pay. This practice produced anger and rarely resulted in behavior change. Yet there was agreement that time to reflect was an important component of any process requiring change. In *Discipline Without Punishment*, Grote (2006) recommended a paid day off as a step before leaving an organization involuntarily. This was adopted and termed a "decision-making day" where a staff member decides whether or not he or she can or wants to be guided by the organization's general principles. Additionally, the staff member submits a written reflection discussing what actions he or she will take to promote his or her alignment with the organization's values.

Initially, the program was piloted in nursing and as one might suspect, the paid day off was met with concern from a number of organization leaders. Were we rewarding staff with a paid day off? Would staff members truly reflect and develop their plans? Would informal conversations have an impact? Did our managers have the skills needed for this approach? The answers vary, but the results have been positive. The change promoted a shift in management style from authoritarian to an interactive and relationship-oriented approach. Grievances about performance management actions have decreased due to increased staff involvement in determining the course of action. This example illustrates (a) the use of relationship-oriented management, (b) staff participation in decision making, (c) the emergence of ideas fostering creativity, and (d) the utilization of minimum specifications rather than detailed progressive steps.

Although this was a change in one particular process, nursing leaders and staff had been utilizing complexity concepts in other forums without realizing it. A primary example is self-scheduling. Most of the nursing units have utilized self-scheduling for many years. Guidelines regarding numbers of personnel required, mix of RNs to assistive staff, and weekend/holiday parameters are provided. Staff members then self-organize to determine the schedule and, on most occasions, produce balanced staffing while meeting personal needs. It is often helpful to identify examples of CAS that already exist in order to stimulate use in other situations.

Inspiring Dedication and Commitment: Schwartz Center Rounds

Being a health care provider is a rewarding experience, but often takes an emotional toll on the caregiver due to the complexity of the environment and multiple stressors. Caregivers often do not have a forum to express the impact of the many interactions they experience on a daily basis. Debriefing for specific incidents, educational offerings, and individual counseling are appropriate in many situations, but do not provide an ongoing forum. At AEHN, a very committed board member suggested implementing Schwartz Center Rounds (Schwartz Center, 1995) as a way to support caregivers. As defined by the Schwartz Center, the "Rounds" provide a multidisciplinary forum where caregivers discuss difficult emotional and social issues that arise in caring for patients. The process of Schwartz Rounds differs dramatically from traditional education practices. A case or topic is briefly presented, and the majority of the program time is dedicated to audience response and sharing.

The Rounds exhibit many qualities of complexity and caring science. The planning committee provides a framework, but aside from gentle facilitation no one really controls what happens or what is said during Rounds. Distributed control, or control shared by many agents or individuals, is central to the process. Rounds are self-organizing and unpredictable, promoting "emergence" of new structures or interactions.

At the first Rounds, the case focused on caring for a colleague who was dying. After the brief case presentation, a physician leader asked, "How do nurses handle the stress of caring for someone who is dying?" followed by the statement, "Physicians are only with the patient for such a short time period, but nurses are present for hours." A nurse in the audience simply stated, "That is what nurses do—we support and care for patients and families even though cure is not possible. That is the privilege of nursing." Those words had what complexity science denotes as a "butterfly" effect (Zimmerman, Lindberg, & Plsek, 1998, p. 261). An individual observation or sharing touches a chord in others, and they leave, changed in some way, sometimes profoundly and sometimes not immediately cognizant of the impact.

Research findings from the Schwartz Center (Schwartz Center, 1995) identified that 51% of caregivers at experienced Round sites and 40% of caregivers at new Round sites report that Rounds have prompted patient-centered changes in practices, or policies within their departments, or in their hospitals at large. Additionally, caregivers reported improved communication and teamwork within and between departments and that Rounds had made unique and profound contributions to their organizations (Schwartz Center, 1995).

Rounds can only have this impact when leaders are comfortable allowing the process to unfold. Initially, a number of participants were uncomfortable with silence while waiting for someone to share his/her thoughts or feelings, but this was typical when a particularly impactful story was shared. Recognizing that caring and complexity principles are present in Rounds has promoted widespread application throughout the organization.

PROVIDING ASSISTANCE IN FACING ADAPTIVE CHANGE

An Exemplar: Crisis in Obstetrical Care Services

The delivery of perinatal services in the Philadelphia, Pennsylvania, region is an excellent example of how concepts from caring and complexity impact care delivery. Over the last 11 years, 16 hospitals have discontinued maternity services in Philadelphia, stretching the capacity of the remaining programs. In a traditional market environment one might expect a change in demand or some technological change underlying this significant shift in the delivery system. Instead, the reality is the six *C*s of health care delivery (complicated, counterintuitive, convoluted, confusing, chaotic, complex) at work. Health care leaders need to integrate Roach's (2002) six *C*s of caring (commitment, compassion, competence, confidence, conscience, comportment) into the innovative solutions. The reality of health care economics is not the focus for patients. What matters most to patients is the presence of a caring, compassionate, competent RN or physician.

The current system is *complicated*. At its heart, obstetrics is the most fundamental aspect of health care delivery—birthing. Every human has experienced birth. It is the most fundamental aspect of the human condition; all of us have been born. However, science, cultural preference, and other factors have resulted in an elaborate set of options for the delivery of babies that include

- At home;
- At birthing clinics;
- At community hospitals;
- At teaching hospitals;
- Mother alone;
- Mother with the assistance of family;
- Support from a non-licensed practitioner, licensed midwives, or certified nurse midwives;
- Family practice physician-supervised delivery; and
- Obstetrician-supervised delivery.

The system is further *complicated* by resource availability. It was once common, in the hospital setting, for physicians to meet their patients at the hospital when called. Today a "laborist" model of dedicated physicians based in the hospital has grown common in large urban hospitals. Caregiver availability also varies. Specialist nurses or physicians on call come in when needed or on-site specialized physicians and nurses (neonatologists, neonatal nurse practitioners, neonatal trained nurses) provide staffing for nurseries and for level I, level II, or level III neonatal intensive care units.

The choices continue for newborn care. These include rooming in with mom, a dedicated nursery, and transfer of sick babies to highly specialized centers or on-site specialized nurseries.

Another dimension of complexity is physician training and the role of the trainees in care delivery. The presence or absence of family practice residents, obstetrics residents, pediatrics residents, and neonatology fellows shapes the delivery care. In many cases, the mother-to-be makes a preference for the setting of delivery as part of picking a practitioner. Risks to mother's the health that are recognized early in the pregnancy may guide decisions toward institutions with higher technology and specialized staff. Preference for a different setting may suggest home birth with assistance in order to preserve a more natural experience. No early decision at all may result in a random assignment of a delivery setting based on proximity. A normal and necessary human event can occur in many different ways and in a planned or unplanned fashion. It's *complicated!*

The economics of maternity services are *counterintuitive*. The care of high-risk patients serves as an example of the *counterintuitive* nature of the economics in play. It would be rational to expect that the system of obstetrics hospitals would sort itself out like other hospital structures. In most urban communities there is a general system of community hospitals, major teaching hospitals, and academic medical centers. Community hospitals serve the core health care needs of the community through emergency departments, common surgical interventions, primary care, and basic medical care. The major teaching hospitals offer a quantum increase in sophistication. The faculty and staff have greater expertise in rarer conditions, and the facilities offer access to more sophisticated technology. The academic medical centers provide expertise for the most rare and most complicated conditions and, along with the major teaching hospitals, provide a full array of core services. This somewhat ordered structure is challenged amidst the uncertainties of obstetric services. The desire to address unforeseen complications to mother or baby creates an incentive to provide higher degrees of services. However, the economics of a fee-for-service health care system only pay for these services when used. It is analogous to only paying fire fighters when there is a fire. The result is that community hospitals struggle with balancing the level of expertise and the intensity of services delivered. Because some mothers will only choose the hospital if the high technology is available, the community hospital must have highly skilled professionals and capabilities. At the same time, the facility needs to not stretch beyond its capabilities and thereby increase the likelihood of poor outcomes. The scenario is a classic example of catch-22 (Heller, 1961).

If a community hospital does not offer costly high-technological services, mothers will not choose the hospital and the hospital cannot afford to have the obstetrical service. If the hospital chooses to have high-technological services, but does not attract women at risk for needing those services, the fee-for-service aspects of the health care system make the provision of obstetrical services too expensive. If the hospital chooses to provide high-risk services, the number of poor outcomes increases. This increase in poor outcomes occurs particularly as women without prenatal care are directed to the hospital by an emergency management system of ambulances. The legal system then looks to the hospital as the "deep pocket" to pay for the impact on mother and child, regardless of whether the hospital was the proximate cause of the incident. The tort system does so primarily because there are real expenses that need to be paid and the hospital is the agent most readily tapped for the funds. Thus, a basic human function of delivering babies escalates toward provision in only high-technology centers at high cost.

Adaptation of Leaders to the Complex Economic Environment

It is crucial for nurse leaders to understand how their organization is adapting to the environment. It is easy to become frustrated by the economics of a challenging situation. One of the first steps to avoid frustration is to understand that health care economics and, hence, health care decision making is *counterintuitive*. From the perspective of the nurse leader, the delivery of competent and compassionate care to patients in a caring, healing environment should be enough to define program excellence. However, health care delivery system economics may create results that are exactly the opposite of what is intended. For example, looking at the delivery of obstetrics services from the patient perspective, a nurse leader may arrive at the conclusion that, in order to maintain the appropriate level of competency for the staff in a 1,000-deliveries-a-year program, the program will only focus on delivery of noncomplicated patients and partner with tertiary centers to provide higher levels of acuity. This would appear to be a sustainable model and, for many communities, it is. If the incidence of mothers without adequate prenatal care choosing a hospital is more than rare, if the nearby facilities can provide tertiary services, if the legal environment punishes the facility for not having the same level of tertiary services, and if the insurers do not preferentially pay for the less efficient hospitals, then the environment will not sustain what is a very good program.

Rather than be frustrated, the nurse leader needs to adapt practice to the *counterintuitive* nature of health care systems where complexity can overwhelm caring. Caring and complexity science must be integrated. The opportunity is to bring the caring practices to the higher volume setting and, in so doing, have a nurse–patient relationship that transcends the challenges of a high-volume facility. In other words, 6,000 deliveries a year need not feel like a factory if tenets from the philosophy of caring science are valued and integrated within the complexities of the professional practice model.

The *convolution* of maternity services can be seen in the way medical education shapes delivery systems. Babies typically arrive after 40 weeks of gestation. However, there is enough variation that an obstetrics practice cannot schedule office visits and time to deliver with the precision of a general surgery practice. With increasing demands on their time, pregnant women, more of whom are in the workforce than ever before, are less tolerant of prenatal care visits being cancelled or delayed by obstetricians leaving the office to deliver babies (the same applies to pediatricians attending deliveries). As a result, a system of laborists attending deliveries on a dedicated or rotating basis has been introduced in many settings. The benefit is an orderly prenatal system. But now, the laborist is on his/her own for all deliveries. For example, in a community hospital with 1,000 annual deliveries, this means that an 8-hour shift averages 1 delivery, but has a range of 0–3 deliveries, and perhaps an unscheduled operating room procedure. The result is that, despite an average of one delivery per shift, there is enough variation to make the demands too much for a single physician.

The graduate medical education of residency programs in the United States is funded primarily by the Medicare and Medicaid programs. The teaching hospitals use residents to provide additional support. At one time, urban community hospitals had access to residents; however, with the reduction in allowed work hours, a limitation on the number of residency slots, and no payment by commercial insurers for teaching programs, urban community hospitals find themselves without residents. The choices for coverage for unanticipated increases in patient demand for a community hospital are (a) to ask obstetricians to leave their offices to come in and assist (impacting the patient's perception of these practices), (b) to hire extra physicians without reimbursement, and (c) to hire advanced practice nurses or nurse midwives to address the need. Each of these choices results in higher costs with no incremental reimbursement and, in urban settings, may lead to the delivery of babies only at institutions with teaching programs where residents can provide the services and thus avoiding a negative financial impact on the hospital.

What Is the Solution?

For the nurse leader, the challenge is to look beyond the hospital's four walls. This orientation is important for all services, not just obstetrics. In order to anticipate the *convoluted* shape of the health care system, nursing leaders should look to the patient and physician needs outside of the hospital. Patients using emergency departments for primary care are a good example for nurse leaders to consider. It is widely acknowledged that the emergency department is a very expensive setting for primary care services. Yet every day across the United States this occurs. Why? One reason is because emergency departments are open and available 24 hours a day and will see everyone regardless of the ability to pay. According to the most recent published report by the United States Census Bureau, in 2008 there were 46.3 million people in the United States (15.4% of the population) who were without health insurance for at least part of that year (DeNavas-Walt, Proctor, & Smith, 2009). For those without insurance the network of private primary care providers is largely unavailable. Few private practices are available to those who cannot pay. The next layer is the free clinics operated by local governments, by charitable organizations, or by hospitals as part of their teaching and community service mission. However, the free clinic often fails the working poor. Scheduling challenges in free clinics often result in very long wait times. It is not uncommon for a patient to spend more than a day to have the same patient experience that in a private office may take 1 hour. The impact on the working poor is that often times they most forego a day's wages or risk loss of employment for this free care. The result is that care is not free at all. In fact, to the working poor, a day in a free clinic may be a very significant part of their household income. The result is that, despite the best efforts of federal and state governments, charitable organizations, and hospitals, people without insurance will see the emergency department as their best and perhaps only choice for primary care. For the nurse leader appreciating the *convoluted* nature of health care systems is an initial step to understanding the needs of patients.

In a state of confusion, an expectant mother asks the hospital administrator, "Why can't I deliver at home?" followed by "And if I need your services I will then use them." Her question seems reasonable, except it touches on the economic "free-rider" problem. The name *free rider* comes from a common economic textbook example; someone using public transportation without paying the fare. If too many people do this, the system will not have enough money to operate. Under a fee-for-service system for health care delivery, each user is paying for services that he or she wants available should they be needed—so each delivery pays the hospital to have staff and services available 24 hours per day. If deliveries shift to the home setting then hospital reimbursements diminishes. The costs to the hospital are exacerbated by the phenomenon of home delivery settings "risking out" deliveries to hospitals increasing the likelihood of higher costs per patient cared for without a change in reimbursement to the hospital by third-party payers. Why are patients confused by hospitals not reaching out to birthing centers and midwives? The reason is often because payment systems and tort systems punish them for doing so. Nurse leaders should be aware of patient confusion generated by the health care delivery system. Hospitals, like public transportation systems, have a very high degree of fixed costs; the day's costs to run the hospital changes very little regardless of the number of patients seen. This high fixed-cost structure adds to confusion in understanding health care economics. For the nurse leader it is important to understand that small changes in volume, up or down, have a profound impact on hospital finances.

The Answer: Understanding Hospitals as Complex Systems in Partnership

Chaos theory, one of the theories in complexity science, has investigated the sensitivity of systems to variations in initial conditions as one cause of complex behavior. In a classic economic model, individuals acting in their own self-interest should create an environment of improved efficiency and greater choice. For example, health care delivery in an urban setting is predicated on patient prenatal choice, resident availability, and demand for high-technology services. Patients value predictable office visits, residents are scarce, neonatal intensive care unit (NICU) services are in demand, and the legal system punishes undesired outcomes rather than inappropriate practices. As a consequence, obstetrics services will shift to large teaching settings as community hospitals give up the service. However, small variations in these preferences may result in the sustainability of these services. For instance, unusual local demand for NICU services, unrelated to the general population of patients, can create sufficient economic strength to allow for a NICU and, thus, a community obstetrics provider. The presence in the community of a drug treatment center and the higher number of babies suffering from the results of maternal substance abuse can create a sufficiently large number of babies in a NICU for a community hospital to sustain a service. Also, a facility may be sufficiently important to third-party payers for other procedures and be able to command a reimbursement premium to offset the high cost of an infrequently used program. The disproportionate impact of the absence or presence of key variables results in very different behaviors. As we have learned, maternity services represent both complex behavior and a complex system.

Complexity theory has long been concerned with the study of complex systems. The behavior of a complex system is often said to be due to emergence and self-organization. These systems can be biological, economic, social, or technological. Recently, complexity is a natural domain of interest in real world socio-cognitive systems and emerging systems research. Complex systems tend to be high dimensional, nonlinear, and hard to model. In specific circumstances they may exhibit low-dimensional behavior (Lindberg et al., 2008). This complexity carries over to the other demands on teaching hospitals. Teaching hospitals are facing economic challenges for maternity services. They have an ethical responsibility to provide care while bearing a high burden of the tort system. Yet they sustain the service because of the need to provide obstetrics training for their other teaching programs and their ethical caring responsibility to the community served. However, as part of the chaos and complexity innovation, creativity and new ways of operating emerge. An adaptation to the complexity in obstetric services in Philadelphia was a unique nonlinear collaboration between a major teaching hospital and a community hospital. In this arrangement services were neither bought nor sold. Instead, each organization began with a focus on patient and community needs. Physicians, administrators, and RNs made the decision to work in collaboration to balance economic realities with the need to deliver compassionate and competent care.

What were the community needs? Deliveries in Philadelphia were flat whereas capacity was diminishing. Mothers valued highly competent clinicians, high technology, and a caring practice setting for delivery. What emerged was an innovative partnership between the clinicians at a major teaching hospital with the resources to provide the care, and a community hospital discontinuing obstetric services that had been in place for 100 years. There was an intentional will, a focus, and a commitment by leaders in both organizations to collaborate. The major teaching hospital made available by phone on-call specialists to the emergency department of the community hospital to consult on patients. These consults were particularly critical during the months immediately following program closure. In the event a delivery happened in the community hospital's emergency department, a commitment was established to accept transfers of mothers and babies once stable. Staff from the community hospital trained with the staff at the major teaching hospital. This training increased both awareness of capabilities and created a personal relationship among providers. The obstetricians from the community hospital actively transitioned patients to the major teaching hospital. The clinicians improved communication of clinical information and bridged the patient's needs from one organization to another. The major teaching hospital literally saved the community hospital's family practice training program by allowing the family practice residents to train at the major teaching hospital. In a time of a growing shortage of primary care physicians, the family practice program is a treasured community asset. The community hospital in turn provided a gynecology rotation from the major teaching hospital, including access to advanced robotic surgery that was not available at the teaching hospital.

This practice exemplar integrated tenets from complexity and caring science and illustrates how systems self-organized to cocreate order at the edge of chaos. Two organizations that compete for patients in order to survive a perilous financial environment were willing to believe that partnership and collaboration would result in better outcomes for both the human (patient) system and the organization (hospital) system. The organizations realized that the complex nature of obstetric services required flexibility, partnership, and commitment to patient-centered care over traditional structured clinical processes.

HEALTH CARE REFORM

As discussed in the introduction of this chapter, health care economics and health care systems are at the edge of chaos. Significant choices for change need to occur for self-organization to emerge. Over the last 20 years, attempts at radical system change have not worked and with an increased number of uninsured, closures of hospitals, increased health care costs, and insurance companies controlling decisions, one could say the system is destined to deteriorate. Making a bad situation better through *incremental change* may be the only practical solution. Attempts at wholesale reform appear to be stymied by their impact on those who prefer the status quo or who are concerned that the "cure is worse than the disease." From an ethical caring perspective, the system needs to self-organize. Perhaps the butterfly approach of small incremental change will allow for self-organization, leading to transformation.

In 2010, what may be better described as "health insurance reform" became law in the United States. Some provisions of the new law relating to how health insurance is sold have an immediate impact, but those provisions that substantially increase access to care are years away. There is not universal support for these new laws, and the delay in implementation for much of the law creates uncertainty over whether the new rules will ever take effect. Accordingly, the health care economic environment is turbulent. Organizations face significant economic uncertainty with respect to federal, state, and commercial reimbursement for health care services.

It is possible that out of the political chaos which is U.S. health care policy making, there will be a search for an innovative set of options to reform health care delivery. One example of an alternative reform plan can be referred to as "bottom up and top down." Envision this working. Expand Medicaid coverage to more of the poor without health insurance, thus using an existing system to provide needed primary and complex care from the bottom up. Individuals who lose their coverage due to job loss, preexisting conditions, or lifetime maximums on employer-provided insurance would be enrolled in Medicare as a top-down plan. There are many details that would still need to be addressed, including better use of the health care system resources as a whole, but at least bottom up and top down could be another opportunity to improve access to care and reduce the cost of providing care.

CONCLUSION

Using caring and complexity frameworks to examine health care leadership and administrative issues may promote innovation and creative strategies for long-standing problems. Strategies that increase the quality of interactions and intentionally solicit a variety of opinions are much more likely to promote positive outcomes in health care organizations. Although many of the caring and complexity applications may appear to be a "common sense" approach, leaders often revert to more traditional management styles in difficult times. Burns (2001) conducted a survey of 59 leaders in health care (supervisors, managers, directors, administrators) to identify if complexity principles resonated or made sense. There was general agreement that cooperation and building relationships are essential leadership components, and, indeed, the majority of health care leaders embraced relationship building as a key competency. However, only 29% of the respondents supported the concept of providing minimum specifications rather than detailed plans, and 36% of the leaders believed that the focus should be on design first and then action rather than trying multiple actions while waiting to see what emerges. Health care leaders learn by experience and by expanding the repertoire of possible vantage points or "mental models" (Senge, 1990/2006) from which to approach problems have the potential to find new solutions to recurrent and long-standing issues. Practicing from a caring framework in complex organizations provides unifying structure that guides choices for action (Ray et al., 1995). The health care leader can focus on the principles of providing direction, inspiring dedication and commitment, providing assistance in facing emergent change, and unleash the creativity of caring individuals in health care organizations.

REFERENCES

Anderson, R. A., Ammarell, N., Bailey, D. E., Colon-Emeric, C., Corazzini, K., Lekan-Rutledge, D. et al. (2005). The power of relationship for high-quality long-term care. *Journal of Nursing Care Quality, 20*(2), 103–106.

Anderson, R. A., Issel, L. M., & McDaniel, R. R., Jr. (2003). Nursing homes as complex adaptive systems: Relationships between management practice and resident outcomes. *Nursing Research, 52*(1), 12–21.

Anderson, R. A., & McDaniel, R. R. (2000). Managing healthcare organizations: Where professionalism meets complexity science. *Healthcare Management Review, 25*(1), 83–97.

Boykin, A., & Schoenhofer, S. O. (2001). *Nursing as caring: A model for transforming practice*. Sudbury, MA: Jones & Bartlett.

Burns, J. P. (2001). Complexity science and leadership in healthcare. *Journal of Nursing Administration, 31*(10), 474–482.

DeNavas-Walt, C., Proctor, B., & Smith, J. (2009). *U.S. Census Bureau, Current population reports, income, poverty, and health insurance coverage in the United States: 2008*, U.S. Government Printing Office, Washington, DC, P60-236(RV).

Grote, D. (2006). *Discipline without punishment: The proven strategy that turns problem employees into superior performers* (2nd ed.). New York: Amakon.

Heller, J. (1961). *Catch-22*. New York: Simon & Schuster.

Jackson, J. P., Clements, P. T., Averill, J. B., & Zimbro, K. (2009). Patterns of knowing: Proposing a theory for nursing leadership. *Nursing Economics, 27*(3), 149–159.

Lindberg, C., Herzog, A., Merry, M., & Goldstein, J. (1998). Healthcare applications of complexity science. Life at the edge of chaos. *Physician Executive, 24*(1), 6–20.

Lindberg, C., Nash, S., & Lindberg, C. (2008). *On the edge: Nursing in the age of complexity*. Bordentown, NJ: Plexus Press.

O'Connor, M. (2008). The dimensions of leadership: A foundation for caring competency. *Nursing Administration Quarterly, 32*(1), 21–26.

Pipe, T. B. (2008). Illuminating the inner leadership journey by engaging intention and mindfulness as guided caring theory. *Nursing Administration Quarterly, 32*(2), 117–125.

Ray, M., DiDominic, V., Dittman, P., Hurst, P., Seaver, J., & Sorbello, B. (1995). The edge of chaos: Caring and the bottom line. *Nursing Management, 26*(9), 48–50.

Ray, M., Turkel, M. C., & Marino, F. (2002). The transformative process for nursing on workforce redevelopment. *Nursing Administration Quarterly, 26*(2), 1–14.

Roach, S. (2002). *Caring, the human mode of being: A blueprint for the health professions* (2nd ed.). Ottawa, Ontario: Canadian Healthcare Association Press.

Schwartz Center, Kenneth B. (1995). *Schwartz Center Rounds*. Retrieved 2010 from http://www.theschwartzcenter.org/programs/rounds.html

Senge, P. M. (1990/2006). *The fifth discipline: The art and practice of the learning organization* (Rev. ed.). New York: Random House.

Stacey, R. D. (1996). *Complexity and creativity in organizations*. San Francisco: Berrett-Koehler.

Stacey, R. D. (2007). *Strategic management and organizational dynamics* (5th ed.). Upper Saddle River, NJ: Prentice Hall.

Turkel, M. B. (1993). *Healthcare economics for nursing administration*. Presentation. Christine E. Lynn College of Nursing, Florida Atlantic University, Boca Raton, FL.

Turkel, M. C., & Ray, M. (2000). Relational complexity: A theory of the nurse-patient relationship within an economic context. *Nursing Science Quarterly, 13*(4), 307–313.

Turkel, M. C., & Ray, M. (2004). Creating a caring practice environment through self-renewal. *Nursing Administration Quarterly, 28*(4), 249–254.

Watson, J. (2005). Caring science as sacred science. Philadelphia: F. A. Davis.

Watson, J. (2008). *Nursing: The philosophy and science of caring* (Rev. ed.). Denver, CO: University Press of Colorado.

Zimmerman, B., Lindberg, C., & Plsek, P. (1998). *Edgeware: Insights from complexity science for healthcare leaders*. Irving, TX: VHA, Inc.

Response to Chapter 7

Jim D'Alfonso

ECONOMICS, CARING, AND COMPLEXITY

The emerging theme in the chapter was how economic changes related to reimbursement and financing at the national and local levels are driving shifts in health care. As health care reform and financing are being debated at the national level, there is uncertainty as to how these changes will impact the delivery of health care within organizations. I am reminded of the words of Saraydarian (1992, p. 327–328): "It is very interesting that, in general, spiritual (caring) ideas have been thrown out of hospitals and only science and money rule there. In most cases a patient is a source of income for doctors and hospitals or he or she is a case for study or research but he or she is not a soul with hopes and dreams." Financial challenges are an undeniable force in our health care world, and there is no straightforward solution or "quick fix"; yet, change remains necessary, unavoidable, and ever present. Sustainable transformation of our healing environments is a complex, interconnected, and organic process that must be fully embraced and lovingly nurtured if we hope to find creative, innovative, and lasting changes for a new millennium. Leaders at all levels of the health care organization must awaken to new ways of leading lasting change, remaining continually aware of and seeking to balance the fiscal and often dehumanizing aspects of the health care business debate with the ethical–moral demands to care for the whole person (body–mind–spirit) who remains central to our "raison d'être" ("reason to be" in French). Top-down strategies and prescriptive mandates are targeted toward sweeping reform and blind compliance, whereas a revolution or renewal from within our organizations is what poses the most exciting possibilities for meaningful change. Authentic dialogue and a healing approach to leading change for self and others offers us the best opportunity to reconnect and reaffirm our deeper core values of universal love, caring, and healing connections that are both timeless and at the heart of our shared covenant with society and humanity as a whole.

However, as illustrated in this chapter, external and economic changes can serve as a catalyst for true innovation, creativity, and collaboration. Yes, the environment is unequivocally complex and complicated, but at the same time there is a simultaneous spark of the human spirit, which allows for both innovative and creative change from within. Through the integration of tenets from caring science, care providers are once again returning to caring and our shared humanity as core values of professional practice and to the universal virtues of ancient wisdom that is integral to all healing traditions. The Caritas processes or caring practices that are part of Watson's Theory of Human Caring (Watson, 2008) are helping to bridge and giving voice to the often elusive and perhaps more implicit concepts of caring, making caring more explicit and real to everyday caregivers. Health care leaders are using caring science concepts to restore human caring to the core of leadership practices and to help facilitate cocreative and dynamic approaches to the innumerable challenges facing nursing and health care today.

CARING SCIENCE LEADERSHIP

There is an interconnectedness of concepts from caring science and complexity science. Physics, social theory, complexity theory, chaos theory, human caring, and other theories help inform the work of postmodern leaders. As research and outcomes begin to demonstrate meaningful shifts, caring science leaders are seeking new ways to move beyond industrialized "fix it" approaches and seeking out more authentic, collaborative, and engaging relationships based on trust and shared values. Accelerated innovation is occurring within health care organizations where leaders already have a view on the future and value the integration of relational ethics, support collaborative policies at all levels of the organization, and foster authentic caring-loving cultures that are rich with cocreative possibilities. In organizations that have helped to pioneer and foster caring science cultures, the spirit of authenticity, dialogue, partnership, creativity, transparency, trust, and love are not only welcome and refreshing concepts to caregivers, but those same tenets are also emerging as a powerful transformational strategy with incredible benefits to practitioner, organization, and community. The external customer service models, focused primarily on driving up patient satisfaction scores, have repeatedly failed to touch the hearts of practitioners and are often met with caregiver resistance or lukewarm compliance at best. What caring science does help us to understand is that transdisciplinary caregiver teams intuitively know, but often struggle giving voice to the fact, that healing our hospitals requires much more than a steady stream of "one-size-fits-all" prescriptive programs or never-ending trail of service initiatives. Customer service models, often borrowed or adapted from hotel or non–health care service settings, are a poor match for empowering caregivers and helping them reconnect with authentic professional practice. Scripting caregiver connections is not the same as helping caregivers find their own voice or fulfill their deeper longings to deliver care and be an integral part of a healing organization that is aligned with their personal and professional values of human caring. The healing journey for both hospital and caregiver is all about shared consciousness, intentionality, presence, and striving toward caring-healing moments with patients, families, peers, and self.

Leaders must actively unlearn or let go of ineffective industrialized management tactics that are often ego based, polarizing, competitive, controlling, and simply out of step with the caring imperative in health care. The beautiful convergence of caring science and complexity science is the shift that occurs when postmodern leaders place the highest value on "igniting the inner spirit" of both individual and team versus forcing change or reacting to crisis by deliberately setting out to light a fire underneath caregivers. A caring science leader takes responsibility to engage people in a shared journey of collaboration, continually advancing win–win goals with meaningful solutions, while continually seeking to help caring practitioners reconnect or stay connected with their true purpose or "calling" as healers. The resulting transformation is a culture beyond blind compliance, where practitioners and teams are both renewed and fully capable of revolutionizing health care cultures from the inside out. Essential caring science leadership literacy helps make explicit the necessary shift away from the technological, military-based, and common managerial competencies of another generation. The term *literacy* is used deliberately in caring science, which makes more explicit the dynamic and lifelong pursuit of "being and becoming" (Figure 7.1) and encompasses the ongoing development of key attributes based on being, knowing, and doing required for a new generation of caring leaders.

When caring becomes a central part of a culture, there is an opportunity for all to reconnect with their true purpose as a healer and to openly share and give voice to experiences that celebrate caring and healing. An important part of this culture shift is providing forums for nurses and all caregivers to name, claim, articulate, and act upon the phenomenon of caring (Watson, 2008). As discussed in this chapter, Schwartz Rounds provide a safe, interactive, and educational forum for ethical discussions to occur and allow for the expressions of positive and negative feelings by all members of the care team. As employees become more trusting with dialogue, they begin to value opportunities to share their voices and will ultimately embed the practice of valuing story and giving voice to caring in new and innovative ways, both personally and professionally. The resulting outcome is human flourishing, which is a new perspective and a different measure of caring economics and resource success within complex organizations.

FIGURE 7.1

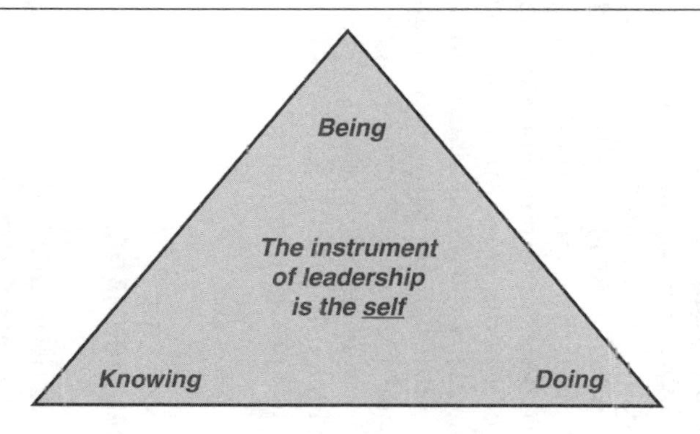

Being

- Loving - Caring
- Respectful
- Calm – Steady
- Supportive
- Grateful
- Ethical
- Curious
- Joyful
- Resilient
- Informed - Aware
- Knowledgeable
- Teacher / Mentor
- Selfless
- Disciplined - Responsible
- Courageous
- Inclusive - Sharing
- Hopeful
- Spiritual
- Inspiring
- Reflective

Knowing In Light of Noble Values

- Integrate a vision of Love, Compassion, Caring and Wisdom
- Cooperate with your conscience and remain in right-relation with your inner truth
- Avoid the devastating effects of ego, vanity and pride on self / others
- Discern when to intervene and when best to coach / support others in co-creating or resolving new challenges
- Model the importance of self-care in sustaining caring cultures
- Transform judging or labeling to enable self/others to relate more effectively with the true or deeper needs of others
- Value the beauty and gifts that emerge from conflict and strife
- Honor authenticity of being and becoming in all your relationships, communication and leadership actions

Doing

- Protect the joy and spirit of others
- Spread optimism & positive energy
- Be a leading voice of Love and Caring
- Create a safe space for all
- Anticipate needs of self and others
- Work for the success of others
- Practice self-mastery & support others to do the same
- Dedicate time to reading & life-long learning
- Lead by continuous living-example
- Uphold excellence in stewardship of resources and a climate of caring economics

CARING ECONOMICS

As leaders, we need to remain aware that what we carry in our hearts really does matter. There is always a need to prioritize initiatives, so we need to embrace a caring science approach and awareness of complexity science as we "connect the dots." Caring cultures do not happen by accident, but through the awareness and commitment of leaders willing to take the risks and actively move toward more inclusive, open, and flexible leadership models that advance a new level of connectedness between practitioner and organization. The words we choose also give voice to our intentions, and what we say is equally as important as what we do and model as leaders. When the leadership voice, language, and focus is borrowed from industrialized settings and focused on budgets and the "bottom line," what caregivers hear in the mixed messages can be dispiriting, dehumanizing, and incongruent with caring, particularly when finances appears to be the primary driving force and first consideration in all decision making. Creativity and innovation thrive most in climates of mutual respect for caring economics combined with solid ethical decision making, reflected in the caring practices of the people, processes, and choices made daily across the organization. Caring economics places a healthy and congruent focus on mission before margin and not the other way around (Eisler, 2008). Caregivers need to value the importance of a strong financial position led by good caring business practices, but finance must likewise be placed in proper context and never in a way that gives the impression that budgets trump caring values. The shortest distance between inspired transformation and failure of a new idea is a narrow or reductionist view, which continually places budget facts before authentic dialogue, actively listening, seeking to understand, and exploring all options and feasibility for implementation together as a team. New ideas and dreams are fundamental to overcome financial challenges and find new ways of solving old problems. Ongoing research informs us that innovation will diminish or disappear in a climate where caring and authentic dialogue is not modeled or valued in leaders' hearts and minds.

ORGANIZATIONAL TRANSFORMATION

Organizations are making strides to move away from fear-based cultures toward love-based caring cultures. Because the organization is a living, breathing, organic system, change impacts all within the system. For nursing, this is the opportunity to transform our cultures from a biocidic or life-destroying toxic environment to more biogenic life-giving and life-receiving environments (Halldorsdottir, 1991).

Administrators need to realize that there is no magic fix, no fast solution, and no secret recipe to instantly transform a culture. There needs to be a continual building of trust (Ray, Turkel, & Marino, 2002) and a realization that integration of caring science and caring theory is an ongoing journey more than a short-term destination to improve scores. Caring theory is the philosophy, values, traditions, and the ethics that guide professional practice. In caring science, leaders listen to their hearts and intuitions and value, honor, and reward those same virtues in others. External service programs help inform our practices, but they are not a theory, ethical value, our true voice, or the foundation for the discipline of nursing or caring professions. Popular industrial leadership models may offer added insight, helpful advice, and even new ways of seeing, being, and doing as caring leaders. The following Caring Science and Leadership Crosswalk (Table 7.1) explores the relationship between the caritas processes of Watson's Theory of Human Caring (Watson, 2008) and various leadership models. There are common threads that we can connect with and benefit from in other leadership models, including many outside of health care. What is most important is not to apply any direct transfer from any industry or external source directly into your caring–healing environment, but to first actively engage practitioners and members of the team in helping to identify "how" we might learn from and incorporate some new, innovative, and perhaps appropriate concepts from external service models into our caring cultures.

One example is the simple tweaking of a scripted service initiative phrase, by which care providers are often instructed to end each patient's encounter with the statement "Is there anything else I can do for you? I have the time." One caring science leader and a team of caritas nurses struggled with this statement's wording, not so much with the patient-centered intent of the service initiative. Caring science nurses felt it was not authentic or appropriate to say, "I have the time," because in their very real and often chaotic environments the one thing they do not have to freely extend to each patient is more time! Background that can be shared on this scripted phase may include the fact that patients frequently believe nurses are too busy, and they really don't want to be a bother or become labeled as "needy" or demanding, so often that their needs may go unexpressed and may even go unmet. By asking this simple statement at the end of each encounter, the nurse (or anyone on the care team) truly has the opportunity to improve communication, connection with the patient, and ensure legitimate needs are prioritized and met. In reality, nurses also have become skeptical and aware that cleverly crafted service-oriented initiatives have a more operational intent and nonclinical goal in mind, which may be to improve patient's perceptions and drive up specific patient score measures for the unit, hospital, or administrator's dashboard. The way caring science leaders resolve these gaps is the differentiating factor. Complex theory informs us that there are many different agendas at play in all things and on all levels, both seen and unseen. Caring

TABLE 7.1 Crosswalk of Caring Science Concepts Within Sample Leadership Sources

FIVE CONSTITUENT MEANINGS OF CARING AND RELATED CARITAS PROCESSES (CPs)	LEADERSHIP CHALLENGE	RELATIONSHIP-BASED CARE FIVE CARING PRACTICES	LEADER'S DIGEST	THE POWER OF FULL ENGAGEMENT	CARING LEADERSHIP MODEL©
Watson, Jean. (2008). *Nursing: The philosophy and science of caring* (Rev.). Boulder, CO: University Press of Colorado. Smith, M. (1992). Caring and the science of unitary human Beings. *Advances in Nursing Science, 21*(4), 14–28.	Kouses, J. M., & Posner, B. Z. (2007). *The leadership challenge.* San Francisco: Wiley.	Koloroutis, Mary. (Ed.). (2004). *Relationship-based care.* Minneapolis, MN: Creative Health care Management.	Clemmer, J. (2003). *The leader's digest.* Toronto, Canada: TCG Press.	Loehr, J., & Schwartz, T. (2003). *The power of full engagement.* Paperbacks, NY: Free Press.	McDowell, J., & Williams, R. (2009). The caring leadership model, Created by Caritas Coaches, Contact: rlwillia@wfubmc.edu.
1. Manifesting Intention **CP 1:** Cultivating the practice of loving kindness and equanimity toward self and other **CP 8:** Create a healing environment for the physical and spiritual, which respects human dignity	**Model the Way** • Find your voice by clarifying your personal values • Set the example by aligning actions with shared values	**Maintaining Belief** • Hope-filled attitude • Sees possibilities • Respects and values all people • Does whatever it takes to do the right things • Follows through on commitments	**Authenticity** • Know thyself • Be true to me and live creatively • "I'll go first"	**Fully Engaged: It's About Managing Energy and Not Time** • Full engagement is the energy state that best serves performance • Define purpose • Face the truth • Take action	Always live with kindness, compassion, and equality

2. **Experiencing the Infinite** **CP 2:** Being authentically present **CP 4:** Developing and sustaining a helping–trusting authentic caring relationships **CP 10:** Open to mysteries and allow for the unknown to unfold	**Inspire a Shared Vision** • Envision the future by imagining exciting and ennobling possibilities • Enlist others in a common vision by appealing to shared aspirations	**Knowing** • Seek to understand others • Avoid assumptions • Seek to understand others' experiences	**Focus and Context** • Search for meaning not certainty • Imagine the possibilities • Predict the future by inventing it	**Principle 1** • Full engagement requires drawing on four separate but related energy sources: physical, emotional, mental, and spiritual	**G**enerate hope and faith through cocreation
3. **Creative Emergence** **CP 6:** Creatively using self and all ways of knowing as part of the caring processes **CP 9:** Reverentially and respectfully assisting with basic physical, emotional, and spiritual human needs	**Challenge the Process** • Search for opportunities by seeking innovative ways to change, grow, and improve • Experiment and take risks by constantly generating small wins and learning from mistakes	**Doing For** • Resolves problems • Gets things done • Models desired behaviors	**Responsibility for Choices** • Find reasons why it can be done and just do it • Know the difference between what can and cannot be changed • Optimism—Accentuate the positive	**Principle 2** • Because energy diminishes both with overuse and with underuse, we must balance energy expenditure with intermittent energy renewal	**A**ctively innovate with insight, reflection, and wisdom

(*Continued*)

TABLE 7.1 Crosswalk of Caring Science Concepts Within Sample Leadership Sources ***(Continued)***

FIVE CONSTITUENT MEANINGS OF CARING AND RELATED CARITAS PROCESSES (CPs)	LEADERSHIP CHALLENGE	RELATIONSHIP-BASED CARE FIVE CARING PRACTICES	LEADER'S DIGEST	THE POWER OF FULL ENGAGEMENT	CARING LEADERSHIP MODEL©
4. Attuning to Dynamic Flow **CP 2:** Being authentically present (in the moment) **CP 3:** Cultivation of one's own spiritual practices and transpersonal self, going beyond ego-self **CP 6:** Creatively using of self and all ways of knowing as part of the caring processes **CP 7:** Engage in genuine teaching–learning experiences that attend to unity of being and meaning, attempting to stay within the others' frame of reference **CP 8:** Create a healing environment for the physical and spiritual which respects human dignity **CP 10:** Open to mysteries and allow for the unknown to unfold	**Enable Others to Act** • Foster collaboration by promoting cooperative goals and building trust • Strengthen others by sharing power and discretion	**Being With** • Listens • Suspends judgment • Promotes healthy productive interactions • Consciously monitors own reactions • Models healthy personal boundaries • Offers support	**Growing and Developing** • Create leaders • Treat people as if they were what they ought to be and help them become what they are capable of • Coach and mentor people • Trust people	**Principle 3** • To build capacity we must push beyond our normal limits, training in the same systematic way that elite athletes do	**P**urposely create protected space founded upon mutual respect and caring

5. **Appreciating Pattern** **CP 4:** Developing and sustaining a helping–trusting authentic caring relationships **CP 5:** Promote and accept positive and negative feelings; authentically listen to another's story CP **CP 6:** Creatively using self and all ways of knowing as part of the caring processes **CP 7:** Engage in genuine teaching–learning experiences that attends to unity of being and meaning, attempting to stay within the others frame of reference **CP 9:** Reverentially and respectfully assisting with basic physical, emotional, and spiritual human needs	**Encourage the Heart** • Recognize contributions by showing appreciation for individual excellence • Celebrate the values and victories by creating a spirit of community	**Enabling and Informing** • Articulates expectations • Seeks and supports opportunities for growth and development • Serves as a teacher, coach, and/or mentor • Leads with purpose and integrity	**Mobilizing and Energizing** • Help people draw energy out of themselves • Motivation is a myth • Be a servant leader • Know the power of teams • Recognize and celebrate people	**Principle 4** • Positive energy rituals—highly specific routines for managing energy—are the key to full engagement and sustained high performance	**E**mbody an environment of caring–helping–trusting for self and others

science leaders approach the challenge of implementing a service initiative with mixed goals by first opening up an authentic dialogue and sharing the core caring rationale, which includes inviting patients to share needs without fear of labels or reprisal and ultimately demonstrating caring while ensuring patient needs are fully met. The caritas nurse teams came back with a solution versus resisting or sabotaging the planned initiative; they simply asked to change the script to "Is there anything else I can do for you? I will take the time." They felt this change was real, authentic, and something they would absolutely feel good about sharing with patients. Patients would know without a doubt that the nurses would always "take the time" to help them with any needs they may have. As a plan for success, there were never any compliance issues with different service initiatives for this team, because the caring science leaders and teams sought first to connect with the core of caring and then everyone was invited to find solutions together to improve care, support caring values, and cocreate a caring culture. It was never about driving up satisfaction scores or moving benchmarks, which could have created more tension, chaos, dissatisfaction, and offered a lukewarm compliance at best. In this approach, all team members are offered a win–win solution, and this caring approach has become an exemplar for guiding future successes with new initiatives and more importantly to sustaining a trusting caring–healing culture.

As caring science becomes integrated within organizational cultures innovative caring, economic outcome measures need to become the measures of success. For example, leaders need to seek evidence of human flourishing, purposive involvement, building capacity, honoring unique individual talents for advancement, and allowance for creative emergence. There is a focus on relationships, connectedness, and work groups. As caring science standards guide administrative practices, there is a move beyond traditional economic indicators toward humanistic values, ethics, morality, and evolved standards of caring science. The health care organization being an institution with an industrial model approach is now a transformed caring–healing system.

Integration of caring science into an organizational culture will continue to be measured and will demonstrate a return on investment. Traditional outcomes measures, such as registered nurse satisfaction and patient satisfaction are increased, and turnover, patient errors, sick calls are decreased. Leaders need to actively reframe values and actions and model or "live out" caring behavior. For example, the traditional response to a low volume is "we need to get more patients to justify staffing." This can be reframed to "how can we let patients and our community know we care?" As patients receive care centered in loving kindness they make choices to come to the hospital for care, and census becomes a lower-level concern than actively sustaining caring–healing environments.

Fluctuating census is a practice reality for all nurse leaders. The complexity science concept of pattern recognition can guide choice making and inform decisions, which are congruent with a culture of caring for patients, registered nurses, and all staff. Having an understanding of "the pulse" of the complex organization is an important literacy for a caring science leader. Looking at trends in high volume, specialty procedures and developing innovative staffing patterns can result in meeting patients' needs and reducing the practices of overtime, reduced hours, and floating for registered nurses.

Nyberg (1998) encouraged hospital administrators and nurse leaders to "reach for what might be, rather than succumbing to what already is," sage advice for any leader facing the challenges of transforming health care environments in the new millennia. Incorporating chaos theory and caring science concepts requires leaders to stretch, explore, and connect with new ways of knowing, being, and doing. "You've got to read to lead," is a mantra frequently shared with students and colleagues, which encourages others to actively seek out new learning and expanded networks with other caring science leaders. The first and most profound step for the journey of transforming our caring and compassionate environments actually begins with us, wherein all leaders embrace the caring science concept that "the instrument of leadership is the self." Translated literally, it means we must stop looking outside of ourselves for solutions and realize we are indeed the leaders we have been waiting for! It is clear we cannot solve or heal current systems with yesterday's "quick fix" solutions. Solutions borrowed in the past from techno-military-industrial models are vastly different than the philosophical, theory-guided, ethical and moral demands of delivering and administering human caring services. Time has demonstrated that these external and typically economic-focused models may have actually done more harm than good, leaving both providers and receivers of care dispirited, disconnected, and out of control. Renewal from the inside out is the greatest hope for true transformation, not forced by external reforms, new rules, or added bureaucracy, but led by authentically inspired and informed administrators and leaders who acquire a higher evolved understanding for and passionate commitment to the complex and deeper human dimensions of caring.

REFERENCES

Eisler, R. (2008). *The real wealth of nations: Creating caring economics*. San Francisco: Berrett-Koehler.

Halldorsdottir, S. (1991). Five basic models of being with another. In D. A. Gaut & M. Leininger (Eds.), *Caring: The compassionate healer*. New York: National League for Nursing.

Nyberg, J. (1998). *A caring approach in nursing administration*. Boulder, CO: University Press of Colorado, Colorado.

Ray, M., Turkel, M., & Marino, F. (2002). The transformative process for nursing in work force redevelopment. *Nursing Administration Quarterly*, *26*(2), 1–14.

Saraydarian, T. (1992). *New dimensions in healing*. Cave Creek, Arizona: T.S.G. Publishing Foundation.
Watson, J. (2008). *Nursing: The philosophy and science of caring* (Rev. ed., pp. 187–189). Boulder, CO: University Press of Colorado.

CHAPTER 7
CARING SCIENCE AND COMPLEXITY SCIENCE GUIDING THE PRACTICE OF HOSPITAL AND NURSING ADMINISTRATIVE PRACTICE

- *Does your organization currently have Schwartz Rounds in place? If yes, how are (or how could you integrate) concepts from caring science and complexity science integrated into the Rounds? If no, what topics would be valuable to discuss?*
- *How would you incorporate performance accountability into your organization?*
- *Think of various examples related to the six Cs of health care economics.*
- *How would you define the caring theorist Sister Roach's six Cs of commitment, compassion, competence, confidence, conscience, and comportment?*
- *In the response, Jim D'Alfonso writes about caring leadership and caring science standards. How would these be integrated into your organization?*

EIGHT

Leadership in Complex Nursing and Health Care Systems

Diana M. Crowell

Three essential elements are needed for effective leadership in both nursing and complex health care systems: personal leadership, team relationship, and leading within a complex system. The key concepts of complexity science relate to leadership by using the framework of the seven core realities of leadership, which clearly exemplify the human–environment mutual process. This perspective develops transformational leadership qualities by fully engaging with others so that leader and follower are raised to higher levels of motivation. Positive deviance (PD) as an exemplar frames the opportunities for transformational leadership in complex organizations. Respondent Shirley addresses the critical role of understanding transformational leadership as a complex system in nursing education and practice. She points out how reflective practice and transformational leadership as a complex systems shifts a nursing educator and practicing nurse from a focus on tasks (a linear view) to a focus on the value of multiple relationships and critical thinking (a nonlinear and creative view). Transformational leadership contributes to building a new leadership philosophy of self-reflection and of developing leadership styles that allow nurses and others to act authentically to cocreate a healthy and caring workforce environment.

INTRODUCTION

A major role for the nurse leader in service or education is to create environments that foster rich relationships. These relationships in turn create the structure and process to transform their complex organizations. Increasing numbers of nurse leaders have been introduced to the concepts of complexity science as they relate to organizations. They are often fascinated intellectually by these concepts and see their value for informing their leadership. It may not be entirely clear how to make these concepts real and workable as they strive to become effective complexity leaders. In addition, leaders must combine their leadership style and actions with their knowledge of complexity to achieve effective complexity leadership. Knowledge alone will not accomplish their aims if leaders continue with directive "top-down" management action.

Leaders who develop and practice their own personal self-reflective activities are more likely to be able to sustain the style and actions of complexity leadership. This personal awareness is the third dimension, knowledge and action being the first two, that support a successful and fulfilling complexity eadership. The seven core realities of complexity leadership serve as a framework to illustrate key concepts of complexity and their application to nursing leadership. They are (1) Who you are is how you lead. (2) You are not in control. (3) There are no observers. (4) All organizations are complex adaptive systems. (5) Chaos and change are the order of the day. (6) Changing the structure does not change the people. (7) People in relationship to other people change the structure. Questions for self-reflection are included to foster personal being and awareness. An overview of complex adaptive systems follows and serves as a foundation for the complexity leadership framework.

Complex Adaptive Systems

An appreciation for organizations as complex adaptive systems derives from a developmental progression of scientific work that spans from early systems theories, the scientific breakthroughs of new physics, to current concepts of complexity. Systems theorists shifted the organizational paradigm from mechanistic and reductionistic foci to one where the study of a phenomenon was not just looking at the sum of its parts, but focusing on the importance of the interaction of those parts (Stacey, 2003).

Organizations can be described as complex adaptive systems. From a leadership perspective, this is more than strategic choice and more than knowledge management where the accepted notion is that leaders can manage change and direction from above or from outside the system. In a complex adaptive system, the leader is actually one agent among all the diverse, independent agents in the organization and not an external observer. Patterns emerge from the relationships of many local interactions. Leaders cannot know the long-term consequences of their choices and actions because new patterns, an emerging coherence, will flow from the self-organization inherent in the interactions. In effect, leadership is distributed as local interactions occur. Diversity is required for a living complex adaptive system. Diversity allows for novelty and new ideas to emerge. Novelty can be expressed as uncertainty and surprise because complex adaptive systems are not stable and predictable (Stacey, 2003). Further explanations and application of these concepts will support and amplify the seven core realties of complexity leadership.

The Seven Core Realities of Complexity Leadership

Who You Are Is How You Lead

The threefold components of complexity leadership are working knowledge of complexity science concepts, action based upon a leadership style that promotes the communication and collaboration necessary to create an environment with rich relationships, and cultivation of a personal being and awareness through self-reflective practices.

This section addresses leadership styles that resonate with a complexity approach. The self-reflective practice of personal being and awareness will be described. Burns' (1978) introduction of transformational leadership was well before the new science of complexity concepts were related to leadership in organizations. The leadership style and actions ascribed to transformational leadership are still contemporary aspects of effective leadership in complex systems. For example, traits such as cultivating relationships; communication, collaboration, and coalition building; engendering motivation; empowering others; taking risks; and being self-reflective all fit the complexity paradigm because they support the value of relationships as important to organizational transformation. One relevant facet of transformational leadership is the phenomenon that when leaders and followers engage in meaningful relationships, they raise one another to higher levels of motivation and morality. Stated in complexity terms, the relationships result in the emergence of higher, more complex levels of motivation and morality.

Transformational Leadership as a Foundation for Complexity Leadership

Contemporary works build upon this foundation. A literature review of leadership styles related to nurse retention conducted by VanOyen Force (2005) identified five themes that have strong implications for effective complexity leadership. They are a dominant transformational leadership style; positive personality traits such as openness, extroversion, and personal power; an organizational structure found in Magnet Hospitals that provides resources and information that support team work, advanced education, and tenure by managers that are perceived to be more effective; and the encouragement of autonomy and empowerment; and recognizing and rewarding staff. Batalden et al. (2003), in their seminal work on clinical microsystems, relate that there is an active process of leading considered helpful in clinical microsystems. This process consists of building knowledge, taking action, and reviewing and reflecting. Building knowledge encompasses practices and processes for improvement of the system, but also seeing the patterns, habits, and traditions in the microsystem. Taking action is more than style and requires active engagement in creating an environment where trust and respect are there for both staff and patients. Creating the space and time for meaningful conversation promotes review and reflection.

Collins (2001) characterizes Level 5 leadership, the leadership found in the most successful companies he studied, as going beyond the ability to inspire high performing standards by setting forth a compelling vision, to becoming an executive who "builds enduring greatness through a paradoxical combination of personal humility plus professional will" (p. 5). These leaders credit others, subjugating their own egos for the larger goals of success of the organization. Collins believes that there seems to be a "seed" inherent in leaders who are capable of moving to Level 5 that may sprout with self-reflection or possibly through a transformative life event.

The transformational traits of focusing on communication and relationships are reinforced by contemporary studies of staff perceptions of effective leaders. Leaders in clinical microsystems employ the relationship-building principles of trust and respect as well. The leader is described as in the middle, embedded in the system and actively engaged on a daily basis with others (Batalden et al., 2003).

The other essential element these effective leaders cultivate is self-reflection (Collins' "seed" to move to Level 5). Personal being and awareness through self-reflective practice is the complement to knowledge (of complexity concepts) and complexity leadership style and action. This requires discovering your own beliefs and values and then acting authentically upon those principles, and is nearly impossible to achieve without a personal self-reflective practice. Turkel and Ray (2004) refer to this as self-renewal. Nurse administrators may neglect their own self-care and renewal as they focus on the bottom line. "To awaken the creative, caring spirit within each of us, it is essential to acknowledge the power of caring for self as being essential to creating a caring harmonious work force environment" (Turkel & Ray, 2004, p. 251).

The complexity leader then can enter into relationship from a centered place where intellectual understanding, effective leadership behavior, and personal being and awareness are congruent. Complexity leadership action flows from a central core of congruence. Living and leading congruently, you are more at ease with yourself and in a position to change and grow along with those whom you lead. Heider's (1986) *Tao of Leadership* contains questions for self-reflection (p. 19).

Questions for Self-Reflection

- Can you mediate emotional issues without taking sides or picking favorites?
- Can you breathe freely and remain relaxed even in the presence of passionate fears and desires?
- Are your own conflicts clarified? Your own house clean?
- Can you be gentle with all factions and lead the group without dominating?
- Can you remain open and receptive, no matter what issues arise?
- Can you know what is emerging, yet keep your peace while others find out for themselves?

You Are Not in Control

Capra (2004) captures the dilemma facing leaders who cling to control.

> The principles of classical management theory have become so deeply ingrained in the ways that we think about organizations that for most managers, the design of formal structures linked by clear lines of communication and coordination and control, had become almost second nature. (p. 103)

As hard as we resist the idea, we really can only be in control of our own actions and ourselves. We want to believe that there is a neat cause and effect pattern: "I do this and you will do that." As Capra reflects, scientific management thinking was based upon that premise.

In complex adaptive systems independent agents in a distributed leadership fashion are the actual actors. Stacey (2003) relates asking managers how a flock of birds fly in formation. They generally responded that the birds follow the leader. But there is no one leader in these types of living systems; birds are acting locally and responding to what is termed simple rules, not to a strategic plan or overall blueprint for the journey. Stacey (2003) describes a computer simulation of "Boids" conducted by Reynolds where the simple rules were to maintain a certain distance from other objects and Boids, match velocities, and move toward the center. An emergent cohesive pattern was produced. Simple rules rather than a detailed complicated list of procedures can help each person to contribute creatively to the whole improvement and change in the system. Independent agents, or systems agents, can be individuals, teams, departments, ideas, factions, or clients, and they are all participants in the organizational process (Olson & Eoyang, 2001).

The scientific concept of attractors has valuable application to understanding organizations and leadership. Attractors are patterns of energy, such as the eddy formed in a rushing river that in time changes the course of the river. Plsek and Wilson (2001) suggest that instead of pushing for change and identifying "resistance," we look for attractors in the system. These patterns of energy are intrinsic motivators. They assert that care must be taken to choose attractors that match the group, not what would be an attractor to the leader. This requires careful attention to what those attractors may be.

Accepting that we are not in control, we can use simple rules and attractors in our leadership environment. Cultivating the quality of detachment can help to accept the lack of control. Detachment isn't a matter of not caring, but of noticing the issue, and taking action, but not overreacting. And think about this: If you are not in control of others, ultimately, no one is in control of you.

Questions for Self-Reflection

- Where can simple rules work better than long, involved, detailed instructions?
- Can I identify attractors in my organization that will motivate or empower others?
- What self-reflective practices will enhance my sense of detachment?
- Can I let go and let others?

There Are No Observers

Physicists now know and tell us that everything in the universe is connected in one self-organizing system. It is impossible to NOT be part of it all. We are, in effect, each other's environments (Rogers, 1970).

F. David Peat, physicist and author, once reflected that even physicists find it hard to believe that no thing or being can be an outside observer, that the minute we look at something, we influence it in some way (Peat, 2003). So it is not difficult to see why we have trouble accepting that we are all part of this "soup" of interacting energies. You can walk into a meeting in a great mood, but soon find yourself feeling gloomy. That is the energy field of those already in the room influencing you. Field theory has evolved from its application to electricity and magnetism to an understanding that fields underlie all matter and are considered regions of influence with characteristic patterns. Attractors can be thought of as these energy patterns. Sheldrake (1995) imagines organizations and organizational space in terms of fields, with employees as waves of energy spreading out in regions of the organization growing in potential. Mindell (1992) goes further to express that fields contain emotional features both positive and destructive. The field is the milieu for relationships. Leaders co-create that field and the quality of those forces and influences. We influence and are influenced by the field, the pattern of energy present. As we develop detachment and can stay centered in a difficult field, the lesser we are influenced by others in negative ways.

Questions for Self-Reflection

- How can I remain centered while the field is full of emotion?
- What part have I played in this difficult situation?

All Organizations Are Complex Adaptive Systems

Emergence and diversity are two essential elements to the complex adaptive systems view. Independent systems' agents, through interactions, produce new ideas, develop new products, and design new ways of being and acting together. These creations are the result of the process of emergence. Emergence is the appearance of new or novel forms that arise from local interactions even without the imposition of top-down direction. The most novel, creative, and effective new structures or processes will come from independent agents with diverse perspectives, mental models, and experiences. Diversity is a prerequisite for emergence of the new.

Leadership in a complex adaptive system is often counterintuitive. The tendency when there is uncertainty and complexity is to be directive to fix the problem, when in reality, that is the time to encourage creative input from people close to the work. The more certainty and agreement about the

Questions for Self-Reflection

- Do I have the tendency to become very directive when the answer to a problem is unclear or there is disagreement?
- Can I trust the diverse or multidisciplinary group to work together to develop ideas and solutions?
- Can I be patient when the situation looks disorganized and uncertain?
- Do I know how to recognize the routine situation that does not require extensive group discussion?
- When assembling a workgroup, do I choose people who think and act much as I do? Or do I seek diversity?

course of action to take, for example, an evidence-based clinical procedure or government regulation requirement, the more directive the approach can be. However, the more diverse the views, ideas, experiences of the participants are, with concomitant uncertainty and disagreement as to what course to follow, the more likely that the interactions of the diverse group will produce the most creative answers (Stacey, 2003).

Chaos and Change Are the Order of the Day

Chaos (disorder and order) and transformative change are just the way this complex adaptive system of a universe operates, however hard we try for stabilization. Writing in *Who Moved My Cheese*, Johnson (1998) declares that stability is a myth and can be dangerous because if we persist in resisting change, we may miss opportunities to avoid disaster or to create better situations. A small event in one part of a system can generate an enormous effect in another part. This sensitivity to initial conditions posits that effects can be disproportional to causes. A small alteration can have wide-ranging consequences. This is known as the butterfly effect based on the weather parable described as the flapping of butterfly wings in China resulting in a hurricane in Florida (Kelly, 1994). In London, Big Ben is calibrated by mechanics on a regular basis to conform to atomic time. The workings are installed in a large space and consist of complicated gears and pulleys. After the calibration, if there is a further slight adjustment to be made, this is accomplished by placing a penny or two on the machinery. This small effort brings Big Ben back into balance. This is a metaphor for understanding when a small action applied thoughtfully can have important effects. In addition, it is a reminder that being careless with small actions can prove disastrous to an organization. The reverse can be true too—a large, top-down strategic initiative may produce very little in lasting results, a consequence that often is puzzling to senior management.

Kevin Kelly wrote about living systems such as termite mounds, beehives, and organizations in his 1995 book *Out of Control, the Rise of the Neobiological Civilization*. He terms the rational logical machine type activities as clockware and the more creative, complex messy events as swarmware (based on bees). He contends we need to use both in our organizations. We can work with others in relationship to set up systems that are tightly organized enough for safety and production but loose enough to be flexible to adapt to change and free enough to create change in order to grow. Two leadership practices based upon Kelly's work are having a good enough vision and encouraging multiple small actions in the organization (Zimmerman, Lindberg & Plsek, 2001). Burns (2001) found in her research that these two practices had the least intuitive support. Managers were uncomfortable with giving minimal specifications and not a detailed plan. They also found it difficult to accept that they did not need to be sure before action was taken. Of particular interest was the finding that, as leaders attempted to lead from a complexity paradigm, they wanted to seek simplicity and control rather than to work to become comfortable with disruption, conflict and disorderly process. Order emerges eventually through information and ethical choice making (Turkel & Ray, 2000). Additional knowledge about complex adaptive systems combined with self-reflective practice can ease the transition from a "plan, organize, control, and direct" paradigm to one that generates comfort with Complexity Leadership.

Questions for Self-Reflection

- Am I holding onto detailed strategic plans that no longer apply?
- How can I use my "pennies" wisely to gain the most results?
- Am I careful not to incite concern or unease in the group by seemingly small remarks?
- How can I become more comfortable with disruption, conflict, and disorderly processes?

Changing the Structure Does Not Change the People

We still somehow believe that if we just change the structure, move the reporting boxes around, build a new building, hand out new titles, and offer bonuses, the people will change. This comes from the old idea that we can do to others and they will do our bidding. "If I do certain things and act in certain ways, I will change others." Complexity Leadership is about "doing with" rather than "doing to" or "imposing upon." It is setting the climate for people in relationships to create the best structure and systems. In a complex adaptive system, the structures emerge from the process, not the other way around.

Questions for Self-Reflection

- Can I refrain from imposing a top-down change in structure to fix a problem?
- What actions can I take to create a relationship-rich climate?

People in Relationship With Others Will Change the Structure

Wheatley (2006) builds upon new science concepts to describe organizations as self-organizing systems. All living systems have the ability to organize themselves into patterns and structures without any externally imposed plan or direction. Things don't fall apart, they are self-organizing. The desired state is not equilibrium but being on the edge of chaos. Organizations are not things but relationships and structures themselves emerge from these relationships rather than being imposed from the top down. Order thus is found in information rather than structure and that information is abundant and open, not closely managed. "Self organization is the ability of all living systems to organize themselves into patterns and structures without externally imposed plan or direction" (Wheatley, p. 13). Rather than highlighting the concept of "organize themselves" in nursing, Ray (1994) and Turkel and Ray (2005) discussed how self-organization (for patients) emerges through caring relationships.

One intervention for promoting change in organizations and communities that exemplifies complexity is Positive Deviance (PD). It requires the role of the leader to shift considerably from a traditional approach to one of complexity.

Positive Deviance (PD) started as a research tool used in nutritional sciences. Jerry Sternin and his wife, Monique, took the basic precepts and developed an action tool that they used worldwide to bring about community improvement. The PD approach is not needs or problem-solving based, but seeks to identify and optimize the solutions already existing in the community. Jerry and Monique Sternin were staffers for Save the Children on assignment in Vietnam to reduce malnutrition in children there. They discovered that whereas most of the children in the villages were malnourished, some were not. Upon inquiry, they determined that the mothers of the well-nourished children were feeding them more often and supplementing the usual rice with shrimps, crabs, and sweet potato leaves. This food was considered unacceptable for small children by most of the villagers. In true PD fashion, the successful mothers were engaged to show and coach the other mothers. In daily meetings they practiced the new behaviors. Over 80% of the children were adequately nourished as a result (Buscell, 2004).

The essence of PD is that it is not the outside experts but the successful insiders who effect the best changes in practice and behavior. PD operates contrary to classic change theories where it is proposed that knowledge itself changes behavior, and the progression is knowledge, attitude, then change of practice. PD turns that around to practice, attitude and then knowledge. The practice changes the attitude, then the knowledge is internalized. Jerry Sternin reflects, "It is easier to act your way into thinking than to think your way into acting" (Buscell, 2004, p. 11).

PD is not for every problem but for those needing social, attitude, and behavior change. It is useful for people who share the same complex system and can adopt the most successful behavior of their peers. There are technical and knowledge problems where clearly an expert is required for changes in practice to occur.

PD is effective in health care situations and has been used with asthma and HIV/AIDS patients to help them keep to arduous medication regimens. One area in which PD can be effective and in which studies are underway, is in hand washing in hospitals. The PD focus on increasing the prevention compliance is a process of a group transforming their culture from within. Staff work together to create solutions that break down the barriers to active surveillance, hand washing, and contact precautions. The closer the staff is to the solution, the more effective and durable these solutions are. In addition to working better and being more self-sustaining, solutions that are created by the staff tend to be simpler and less expensive than those that are mandated or come from outside consultants. Often those imposed from the top are difficult, time consuming, or cumbersome, and compliance falls off (Buscell, 2006).

PD is distributed leadership, with changes in practice that occur at the local level in a self-organizing manner. The characteristics of PD are similar to those of a complex adaptive system, embedded in the culture, generative, building on self, and based on strengths. A leader who facilitates PD is in the role of searcher and inquirer, inviting stories, listening more than talking, allowing the group to bring the solutions and letting go of control because the people are the experts. With PD, leaders are encouraged to consider approaching change management from the PD perspective, taking on a very different role than top down. "Managers overlook the isolated successes under their noses or having appropriated them, re package the discoveries as templates and disseminate them from the top. This seldom generates enthusiasm necessary to create change" (Tanner & Sternin, 2005, p. 1).

Leaders can discover the positive deviants, the innovators who are the key to change. PD puts the leaders in different roles, expert becomes learner, teacher becomes student, and leader becomes follower. The leader in PD is not a pathbreaker, but an inquirer, not problem focused but a discoverer of solutions already imbedded in the system (Tanner & Sternin, 2005). Wheatley and Kellner-Rogers (1996) write in *A Simpler Way* that when we view our organizations as people in relationships able to create systems and solutions, we can nourish them with information and remember that they do self-organize and they can be trusted to do so. Again, this is counterintuitive. However, the lesson from complex adaptive systems is that new structures emerge from relationships, whether gases or chemicals, or people. When the situation is complex and new thinking is needed, two or more people coming together, each with an idea will create something better than any one could accomplish alone.

CONCLUSION

Moving from an intellectual fascination for complexity concepts to actually living Complexity Leadership calls for a congruent combination of knowledge, information style, and personal being and awareness. Some actions to start the journey include the following:

- Enhance your personal being and awareness through self-reflective practices.
- Create opportunities for people to form rich relationships.
- Lessen hierarchy with a free flow of information and communication.
- Accept that we live and work in a complex adaptive system and people do self-organize.
- Celebrate people's self-organizing creativity.
- Be willing to change and grow along with those whom you lead.

Moreover, to continue the journey to complexity leadership in health care organizations, consider the following statements to develop competence in relation to complexity:

1. Discuss how selected concepts of complexity science apply to health care organizations and nursing in particular.
2. Describe how the traits of transformational leaders serve as a foundation for complexity leadership.
3. Apply leadership actions to health care situations involving different disciplines that disagree and where the solution is uncertain.
4. Provide examples of self-reflective practices that support personal being and awareness for complexity leaders.

REFERENCES

Batalden, P., Nelson, E., Godfrey, M., Huber, T., Kosnik, L., & Ashling, K. (2003). Microsystems in health care: Part 5. How leaders are leading. *Joint Commission Journal on Quality and Safety, 29*(6), 297–308.

Burns, J. M. (1978). *Leadership*. New York: Harper & Row.

Burns, J. P. (2001). Complexity science and leadership in healthcare. *The Journal of Nursing Administration, 31*(10), 474–482.

Buscell, P. (2004, August-September-October). The power of positive deviance. *Emerging*, 8–20.

Buscell, P. (2006, Winter). Plexus versus the bacteria. *Emerging*.

Capra, F. (2004). *The hidden connections*. New York: Anchor Books.

Collins, J. (2001). Level 5 leadership—The triumph of humility and fierce resolve. *Harvard Business Review, 79*(1), 5.

Heider, J. (1986). *Tao of leadership—Leadership strategies for a new age*. New York: Bantam.

Johnson, S. (1998). *Who moved my cheese*. New York: Putnam.

Kelly, K. (1994). *Out of control: The rise of neo biological civilization*. Cambridge, MA: Addison Wesley.

Mindell, A. (1992). *The leader as martial artist*. San Francisco: Harper San Francisco.

Mindell, A. (1995). *Sitting in the fire*. Portland, OR: Lao Tse Press.

Olson, E., & Eoyang, G. (2001). *Facilitating organization change*. San Francisco: Jossey-Bass/Pfeiffer.

Peat, F. (2003). *From certainty to uncertainty: The story of science and ideas in the twentieth century*. Washington, DC: Joseph Henry Press.

Plsek, P., & Wilson, T. (2001). Complexity science: Complexity, leadership, and management in healthcare organisations. *British Medical Journal, 323*(7315), 746–749.

Ray, M. (1994). Complex caring dynamics: A unifying model of nursing inquiry. *Theoretical and applied chaos in nursing, 1*(1), 23–32.

Rogers, M. (1970). *An introduction to the theoretical basis of nursing*. Philadelphia: F. A. Davis.

Sheldrake, R. (1995). *Seven experiments that could change the world*. New York: Riverhead Books.

Stacey, R. (2003) *Strategic management and organizational dynamics*. London: Pearson Education.

Tanner, R., & Sternin, J. (2005). Your company's secret change agents. *Harvard Business Review, 83*(5), 1.

Turkel, M. & Ray, M. (2000). Relational caring complexity: A theory of the nurse-patient relationship within an economic context. *Nursing Science Quarterly 13*(4), 306–313.

Turkel, M., & Ray, M. (2004). Creating a caring practice environment through self-renewal. *Nursing Administration Quarterly, 28*(4), 249–254.

VanOyen Force, M. (2005). The relationship between effective nurse managers and nursing retention. *The Journal of Nursing Administration, 35*(7/8), 336–341.

Wheatley, M. (2006) *Leadership and the new science* (3rd ed.). San Francisco: Berrett-Koehler.

Wheatley, M., & Kellner-Rogers, M. (1996). *A simpler way*. San Francisco: Koehler.

Zimmerman, B., Lindberg, C., & Plsek, P. (1998). *EdgeWare*. Irving, TX: VHA.

Response to Chapter 8

Nancy Shirley

REFLECTIVE QUESTIONS

1. What is the relationship between theory and practice (clinical, administrative, or educational) highlighted in this chapter?

Dr. Crowell addresses the critical role of leadership in nursing in relationship to complexity science and complex systems. Health care organizations are, indeed, complex systems, be they acute care facilities or public health agencies. For nursing to be able to provide appropriate leadership, there are a number of skills and tools needed. Leadership content has been part of nursing curricula both undergraduate and graduate for many years. Undergraduate programs have had leadership content for over 50 years although the focus and nature of that content are certainly different from what would be expected in a curriculum preparing for practice in the 21st century.

The components of complexity leadership provide the foundation for leadership development for both new and experienced nurses. The triad of knowledge, action, and self-awareness is critical for effective leadership especially in nursing. Schon proposed the concept of a reflective practitioner in the early 1980s, yet it has been in more recent years that many professions, from teaching to law to nursing, have truly embraced this as foundational for effective leaders. This aspect, although important for all nursing students, becomes extremely important when practicing nurses return to school. In particular, the nurses completing baccalaureate degrees can use this skill of reflection and develop into a reflective practitioner to gain insight into various perspectives of practice. It forces one to go beyond the comfort zones and business as usual to expand one's thinking to fit the complexities of the current and future health care arena. Aspects of nursing leadership as presented here demand a shift from a nursing training perspective focused on tasks to nursing education focused on multiple relationships and critical thinking.

2. How does a nurse or a physician or health care administrator learn about complexity science/s and caring through the presentation of ideas or research?

Nursing education has long identified that nursing is both a science and an art. It is a profession built on relationships—with clients, with other professionals, with other nurses. The various tenets of complexity science reflect these unique characteristics and move nursing and nursing education beyond a linear focus. The intricacies of leadership in nursing education, practice, and administration can be addressed most effectively through a lens such as complexity science, particularly as modeled through the triad in this chapter. The model presented in this chapter focuses on aspects that are critical for effective nursing leadership. The emphasis on action related to environments promoting rich relationships, so important to the practice of nursing in any arena, further explicates the connection between complexity science and nursing. Another major area for learning from this chapter is the recognition of the seven core realities of complexity leadership. A nurse at any level of practice would benefit from an understanding of these realities and how complex adaptive systems function. The multifaceted aspects of the relationships in nursing described above require models such as the one presented here. No longer can simplistic, linear models for leadership be applied with an expectation of effective outcomes. Transformational leadership is also described here and contributes to building a leadership philosophy suited to complex adaptive system functioning.

3. What meanings are illuminated in this chapter for nursing or other health care professions? Direct your answer to education, administration, research, or practice based upon the focus of the chapter.

Throughout this chapter, Dr. Crowell offers questions for self-reflection on the various aspects of complexity leadership and, thus, allows readers to explore their own beliefs, values, and perspectives in relation to the seven core realities of complexity leaders. Readers are exposed to the various exercises that allow them to extract their own meanings from the materials. It becomes very personal and interactive in addition to a presentation of factual materials about complex adaptive systems, complexity science, and complexity leadership. A significant meaning in this chapter is the critical role of self-reflection and the connection to developing a leadership style congruent with one's own value system. Such development of a leadership style provides the foundation for actions described in the model as promoting communication and collaboration. The synergy of these three components mirrors the concepts of complexity science.

4. What is the relevance of this work to the future of the discipline and profession of nursing, health care professions, and health care in general?

The model presented by Dr. Crowell elucidates critical components in the education and preparation of nursing students at all levels. The combination of self-awareness, knowledge, and action provides the foundation for effective leadership. It is clear that the future of nursing is dependent upon shifting paradigms of thinking and developing new skills sets. Nurses in leadership need to be equipped to function in a dynamic, interactive world. The concepts presented in this model are complementary to complexity science and successful leadership in the 21st century. The development of reflective practices that allows the nurse to act authentically upon principles leads to self-renewal and self-care. The questions presented throughout the chapter assist the nurse or other health care providers in the process of self-reflection and understanding complex adaptive systems. Leaders who possess the ability to examine themselves practice from a framework of consistency and congruence with their own personal value system. Such leaders promote healthy, effective, work relationships and create a caring workforce environment. The future of the nursing profession is dependent upon creative leaders who have a healthy self-understanding and emotional intelligence coupled with knowledge of complexity science.

CHAPTER 8
LEADERSHIP IN COMPLEX NURSING AND HEALTH CARE SYSTEMS

- *Crowell speaks to the principles of transformation leadership in her chapter. How do these principles apply to nursing practice or teaching–learning in health care or educational organizations respectively?*
- *What does Positive Deviance mean to you? How would you apply this activity in your organization?*
- *Is there a commitment to the role of the leader in your organization? The clinical nurse leader (CNL) role is developing in many university schools of nursing. How do you see the role being applied in practice?*

NINE

Mathematical Models of the Dynamics of Social Conflict

Larry Liebovitch
Robin Vallacher
Andrzej Nowak
Peter Coleman
Andrea Bartoli
Lan Bui-Wrzosinska

In this chapter, the role that psychological and social mechanisms of human behavior play in the conflicts between individuals, groups of people, and nations is understood by formulating these mechanisms into a mathematical form and using mathematical methods to analyze them and reveal the logical consequences of these mechanisms and alternatives. This approach can shed new light on understanding conflicts and developing ways to resolve them. As researchers, we illustrate this approach by applying it to two general properties of human behavior and to a specific model of the conflict between two groups. The model of conflict between two groups demonstrates how positive or negative feedback between the groups determines the dynamics of conflict. Some of the results of the model's mathematical analysis are consistent with our intuitive understanding of such conflicts, and other results reveal interesting new findings, which may help us better understand and resolve such conflicts. Conboy's response encourages the health care practitioner to consider new ways of managing conflicts, perhaps using variables not previously considered. The overall mathematical formulation of the Liebovitch et al. model corresponds to what has been described as "patterning," and for nursing, Longo's response relates the analysis to the social conflict between nurses. This conflict between nurses is described as "horizontal violence."

INTRODUCTION

Mathematical Models of Human Interactions

Conflicts can be between individual people in a marriage; cultural, political, or racial groups within a nation or between nations. What can a mathematical model tell us about people's behavior in a conflict? Reasoning with words about the psychological and social mechanisms that produce conflicts can lead different people to different conclusions. On the other hand, if we can encode these mechanisms into equations, we can use rigorous mathematics to determine the logical consequences of those mechanisms. The mechanisms that we encode into the equations can be too simple, or even wrong, but at least the mathematics guarantees that the conclusions we draw from those assumptions are logically accurate. We can then test those conclusions against our intuition, experience, observation of conflict situations, and laboratory experiments. Gottman et al. (2002a, 2002b) emphasize that analyzing the properties of a well-defined mathematical model allows us to discover the specific properties of those mechanisms represented in the equations. They eloquently state (Gottman et al., 2002b, pp. 67–68) that

> [t]he goal in formulating the equation or equations of a scientific problem is to write the *dynamics*—that is, the *mechanism* through which the system moves and changes over time. . . . The mathematics forces the development of precise theory about the mechanisms to create movement in the system. . . . We hope to get creative investigators to move away from metaphors of dynamics toward real dynamical equations. . . . Ideally a good model allows us to ask questions that would never have occurred to us had we not written down the model [in mathematical form]. A good model also has surprises; it may organize a wider set of data than we had anticipated or even suggest some discoveries that are entirely new.

In this chapter, we explore how this mathematical approach can be used to help us better understand conflicts and methods of resolving them. First, we see how two rigorous mathematical questions are actually equivalent to two seemingly purely psychological and sociological issues in complex conflicts. This means that the results of rigorous mathematical methods applied to study these mathematical questions may also shed light on these psychological and sociological issues. Second, we formulate a simple nonlinear model of the interactions between two parties (each of which could be individuals, groups of people, or nations) to explore how the dynamics of the conflict between those parties depends on the positive or negative feedback that operates between them.

TWO BASIC MATHEMATICAL QUESTIONS

We now show, perhaps surprisingly, that certain mathematical questions can represent some of the essential features about important issues in human behavior. This means that, in principle, important aspects of these issues can be studied through purely mathematical reasoning. This deductive approach is a fruitful addition to psychological or sociological analysis using inductive reasoning based on observations and experiments.

Consider a system of many units and influence functions that describe how they interact with each other. The "dynamical properties" means how the variables that describe the behavior of each unit evolve in time. These variables' values might approach a limiting value and therefore remain constant, they may cycle around a set of values repeating the same cycle forever, or they may fluctuate in a way that cannot be predicted (Murray, 1989). Our first question is whether the interaction between the units can induce different dynamical behavior that is not originally present in each unit. This is a clear mathematical question. The answer to this mathematical question is unknown. There is a lot of work on whether individual dynamical units in a network synchronize and the dynamics (or lack of it) in their synchronization but surprisingly little work about whether the network of interactions itself changes the dynamical properties of each individual unit. In social terms, this corresponds, for example, to asking whether the violence seen between groups of people arises only from within-group and between-group interactions or whether that propensity to violence must first exist within each person and is only then expressed under certain circumstances.

For our second question, again, consider a system of many units and influence functions that describe how they interact with each other. When can all of the possible types of influence functions that involve three or more units be constructed from influence functions that involve only pairs of units? The answer to this mathematical question is also unknown. There are many examples where adding an additional unit changes the existing interaction between two units. For example, when a third person walks into a room, the existing interaction between the two people already present in the room changes. Proteins, called transcription factors, expressed by some genes bind to the DNA of other genes, which increases or decreases their expression. There are typically several transcription-binding sites in a gene's regulatory region. The regulatory effect of the binding of two transcription factors is different if a third transcription factor is present. This is called "context dependency" (Barash et al., 2003). In both this psychological and biological example, we do not know if a new three-unit influence function is required to model the behavior or if we have just not yet found the right way to combine the pairwise interactions to model the observed behavior.

The psychological and sociological correlates of these two mathematical questions raise very important generic questions about how people behave in conflicts. Both mathematical questions are ambitious enough so that we cannot answer either of them at this time. However, clearly formulating a problem is an important and necessary first step in solving the problem. Sometimes it is the most difficult step. At this time, we shifted our approach from such generic questions to more mechanistic studies of how parties behave in a highly simplified conflict model, specifically trying to understand how the conflict's dynamics depend on the positive or negative feedback between the parties.

A MATHEMATICAL MODEL OF THE CONFLICT BETWEEN TWO PARTIES

In order to understand the role that the behavior of each party and the interaction between them play in a conflict's dynamics, we formulated a simple, nonlinear ordinary differential equation model of the conflict between two parties. We encoded some simple behavior mechanisms and interaction into this mathematical model and used standard analysis methods to determine the dynamical properties of the model. Thus, we determined the necessary logical consequences of the assumed mechanisms and interactions. As described later, even these simple mechanisms and interactions lead to behaviors reminiscent of real conflicts and suggest new insights into understanding and resolving conflicts.

Our approach is based on the work done by Gottman et al. (2002a, 2002b) on the dynamics of the interactions between two people in a marriage. They used their previous observations and experimental studies to formulate the salient mechanisms of how two people interact with each other, translated those mechanisms into mathematical equations, and determined the dynamical behavior of those equations. They then determined the parameters of those equations from scoring each person's behavior recorded on videotapes of interactions between couples in their laboratory and correlated those parameters with whether the couples were later happy and stable, unhappy and stable, or divorced.

Here we describe our model of the conflict between two parties and the results derived from it (Liebovitch et al., 2007). The formulation and analysis of this model is only a necessary first step. We are now facing the challenging (but fun) task of considering how to analyze the data on conflicts to determine the values of the model's parameters and compare the model's behavior to conflict dynamics observed under real-world or laboratory conditions.

The variables in the model, x and y, respectively, are the emotional state of each party. These emotional states can be thought of as each party's internal level of happiness or as operationally defined by an external measure of their behavioral state, such as their facial expressions coded from observations on videotape or physiological measurements such as skin resistance, heart rate, or breathing rate. Here we think of each party as one group of people.

In our mathematical model, we are trying to capture how the interaction between the groups changes in time. Thus, we formulate the model around the use of the derivatives from calculus, which are represented by dx/dt and dy/dt. These derivatives are the change in x given by dx (or the change in y given by dy) divided by a time interval dt, and they represent how fast the values of x (or y) are changing in time. The emotional state's rate of change for each group depends on three factors: (1) their "inertia" to change, (2) the effect of their uninfluenced state, and (3) the influence functions that each group exerts on the other. We represent the inertia to change by the parameters m_1 and m_2. This inertia represents a tendency opposite to any change, so these values will be less than zero in the equations. We represent each group's uninfluenced state, that is, how they would feel if they were alone, by the parameters b_1 and b_2, which would be less than zero if a group is unhappy when it is alone or greater than zero if it is happy when alone.

The form of the influence function that each group exerts on the other must be carefully considered. The effects of such influence functions in escalating or de-escalating conflicts were introduced and studied by Boulding (1962), Richardson (1967), and Pruitt (1969, 2006). We are particularly interested in understanding how the conflict's time evolution, as reflected by $x(t)$ and $y(t)$, depends on the feedback between these two groups. The seminal work of Deutsch (1973, 2006) demonstrated the importance of the role of cooperation or competition in the feedback between two parties. As he describes it, "To put it colloquially, if you're positively linked with another, then you sink or swim together; with negative linkage, if the other sinks, you swim, and if the other swims, you sink" (Deutsch, 2006). In our model, cooperation is modeled as positive feedback between the groups; that is, a positive state of one group increases the positive state of the other group, and a negative state of one group increases the negative state of the other group. Competition is modeled as negative feedback, that is, a positive state of one group increases the negative state of the other group, and a negative state of one group increases the positive state of the other group.

The overall mathematical formulation of this model corresponds to what has been described as "patterning" in nursing. The terms that correspond to how nurses feel about themselves and the nature of nursing practice are represented in the model in the inertia and uninfluenced terms. How nurses interact with clients, each other, and staff is represented by the influence functions. These influence functions represent an interaction along only a single variable. They are meant to bundle together all the different aspects of the nursing interaction into one general interaction. We will see that even this gross oversimplification can lead to quite interesting results. However, it does not of course fully represent the patterning of nursing behavior or do justice to the holistic manner of knowing about the client, the nature of nursing practice, and the nurse as an individual as described by Carper (1978).

What should be the mathematical form of these influence feedback functions? Our intuition was that the influence of one group on another should be proportional at low influences but reach a limiting value at strong influences. That is, these influence functions should look like the letter "S." There are many choices for mathematical functions of this form; we chose to represent them by hyperbolic tangent functions, which are defined by $c_1 \tan h(y) = c_1 (e^{-y} - e^{y})/(e^{-y} + e^{y})$ and $c_2 \tan h(x) = c_2 (e^{-x} - e^{x})/(e^{-x} + e^{x})$. Similar S-shaped functions were used in previous studies on aggression by Puritt (1969, 2006) and Bui-Wrzosinska (2005). Values of c_1 or $c_2 > 0$ correspond to positive feedback (cooperation), and values of c_1 or $c_2 < 0$ correspond to negative feedback (competition) between the groups. These influence functions are illustrated in Figure 9.1.

The model's equations are then given by

$$\frac{dx}{dt} = m_1 x + b_1 + c_1 \tan h(y),$$

$$\frac{dy}{dt} = m_2 y + b_2 + c_2 \tan h(x),$$

where the variables x and y are the emotional state of each group, the derivatives dx/dt and dy/dt are how fast those variables are changing in time, the parameters m_1 and m_2 are the inertia of each group, the parameters b_1 and b_2 are the uninfluenced emotional state of each group, and the parameters c_1 and c_2 determine whether the feedback into each group from the other group is positive feedback (cooperation) or negative feedback (competition).

We are interested in understanding the dynamical behavior of this model, that is, how the behavior of $x(t)$ and $y(t)$ depends on the parameters m_1, m_2, b_1, b_2, c_1, and c_2. In order to do this, we use standard methods to analyze such coupled nonlinear differential equations (Strogatz, 1994; Thompson & Stewart, 1986): (1) We first determine the values of x and y that are the solution to $dx/dt = dy/dt = 0$. These critical points may be either stable points that the values of x and y will reach over long times in the conflict or unstable points from which the values of x and y will diverge over long times. (2) We then use a linear stability analysis, described in the Appendix, to determine whether these critical points are stable or unstable. In nontechnical terms, the stability analysis is equivalent to pushing x and y a little bit off their values at the critical point. If the perturbed values of x and y move

FIGURE 9.1 The Positive (a) and Negative (b) Feedback Influence Functions
The influence of group y on group x is illustrated. The influence of group x on group y would be similar. As shown in (c), when c_1 is greater than zero, group y has positive feedback on group x. When the state of group y is positive (y_1) it produces a positive influence (x_1) on group x, and when the state of group y is negative (y_2) it produces a negative influence (x_2) on group x. As shown in (d), when c_1 is less than zero, group y has negative feedback on group x. When the state of group y is positive (y_3) it produces a negative influence (x_3) on group x, and when the state of group y is negative (y_4) it produces a positive (x_4) influence on group x. (This figure, copyright 2007 by Larry S. Liebovitch, was presented at the International Association for Conflict Management 20th Annual Meeting in Budapest, Hungary in 2007, and is reproduced with permission.)

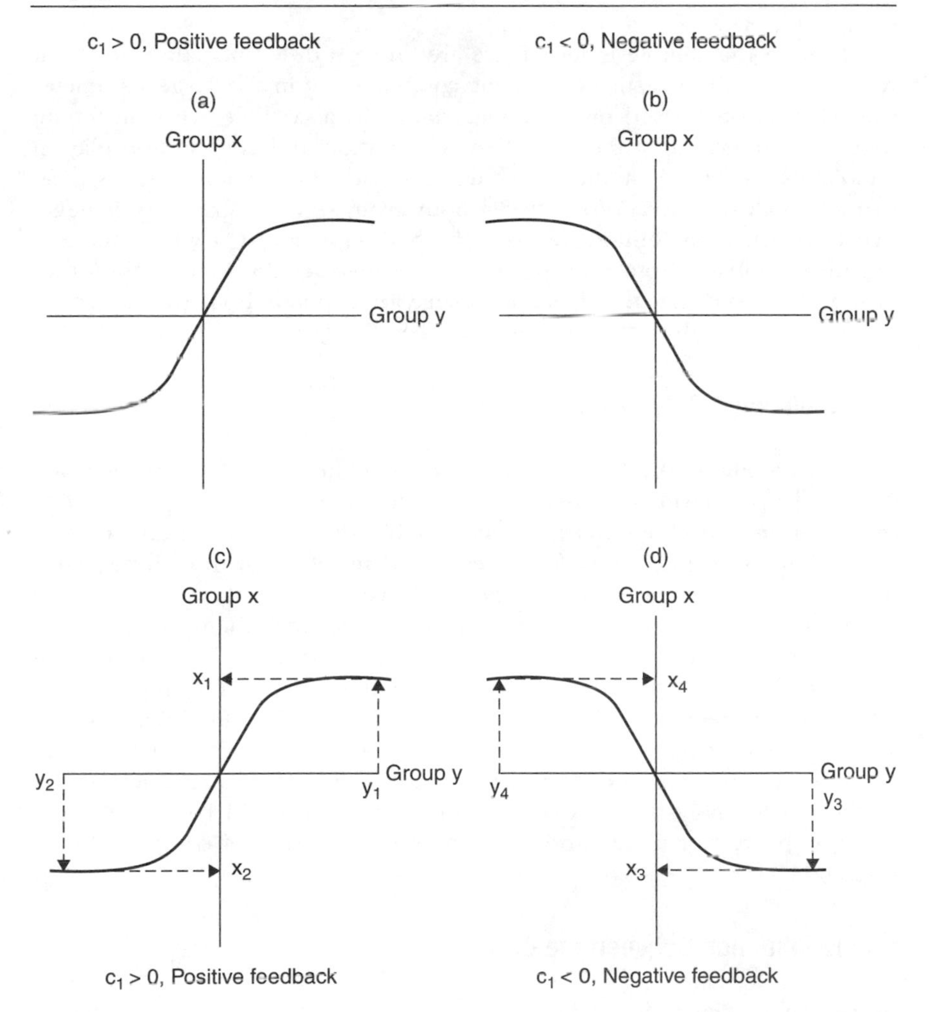

back to their values at the critical point, then the critical point is stable. If on the other hand, the perturbed values of x and y move further away from their values at the critical point, then the critical point is unstable. In technical terms, we determine the eigenvalues at the critical points and whether they are real (positive or negative), imaginary, or complex. We combine the results of the stability analysis performed for different values of the parameters c_1 and c_2 into a bifurcation diagram to understand how the models' dynamical behavior depends on the strength of the feedback and whether it is positive or negative. (3) We also integrated Equation 1 numerically to illustrate the time-dependent behavior of $x(t)$ and $y(t)$ for different values of the parameters.

Even in this simple model, there are enough different parameters that we cannot easily investigate all their combinations in the large parameter space. Therefore, based on their importance in a conflict, we concentrate here on understanding the roles that cooperation and competition play in conflict dynamics. We studied three different cases: (1) where there is positive feedback (cooperation) between both groups, (2) where there is negative feedback (competition) between both groups, and (3) where there is negative feedback from one group and positive feedback from the other group. We found that the dynamical behavior depends both on the nature and on the strength of the feedback between the groups.

Weak Influence Between the Groups

As shown in Figure 9.2, for all three cases, when the strength of the feedback ($|c_1| = |c_1| = 0.5$) is less than the inertia to change ($|m_1| = |m_2| = 0.9$), over long times both groups approach a stable state that is neutral with $x = y = 0$. This is an interesting and unexpected result. Even though the weak influence between the groups is proportional to their state, both groups reach a neutral state as if there was no feedback between them. More surprisingly, as we see next, this behavior changes dramatically when the strength of the feedback between the groups exceeds their inertia to change, that is, when the parameters c_1 and c_2 are larger than the absolute values of m_1 and m_2. The existence of this neutral behavior up to a threshold is a consequence of the simple mechanisms encoded into Equation 1 that is revealed by the mathematical analysis. The existence of such a threshold has been noted in both observational data and other theoretical models of conflicts (Yiu & Cheung, 2005).

Strong Influence Between the Groups

As shown in Figure 9.3, when there is strong positive–positive feedback between the groups, then over long times both groups reach a stable state

FIGURE 9.2 The Numerical Integration of the Model is Used to Compute the Values of *x* (Boxes) and *y* (Plus Signs) as the State of the Two Groups Evolves from Their Initial Values of $x = 1$ and $y = 2$. When the Strength of the Feedback is Less than the Inertia to Change, then All the Models Evolve to the Neutral State $x = 0$ and $y = 0$

(a) Model where there is positive feedback between both groups. (b) Model where there is negative feedback between both groups. (c) Model where the feedback from one group is positive and the feedback from the other group is negative. (This figure, copyright 2007 by Larry S. Liebovitch, was presented at the International Association for Conflict Management 20th Annual Meeting in Budapest, Hungary, in 2007 and is reproduced with permission.)

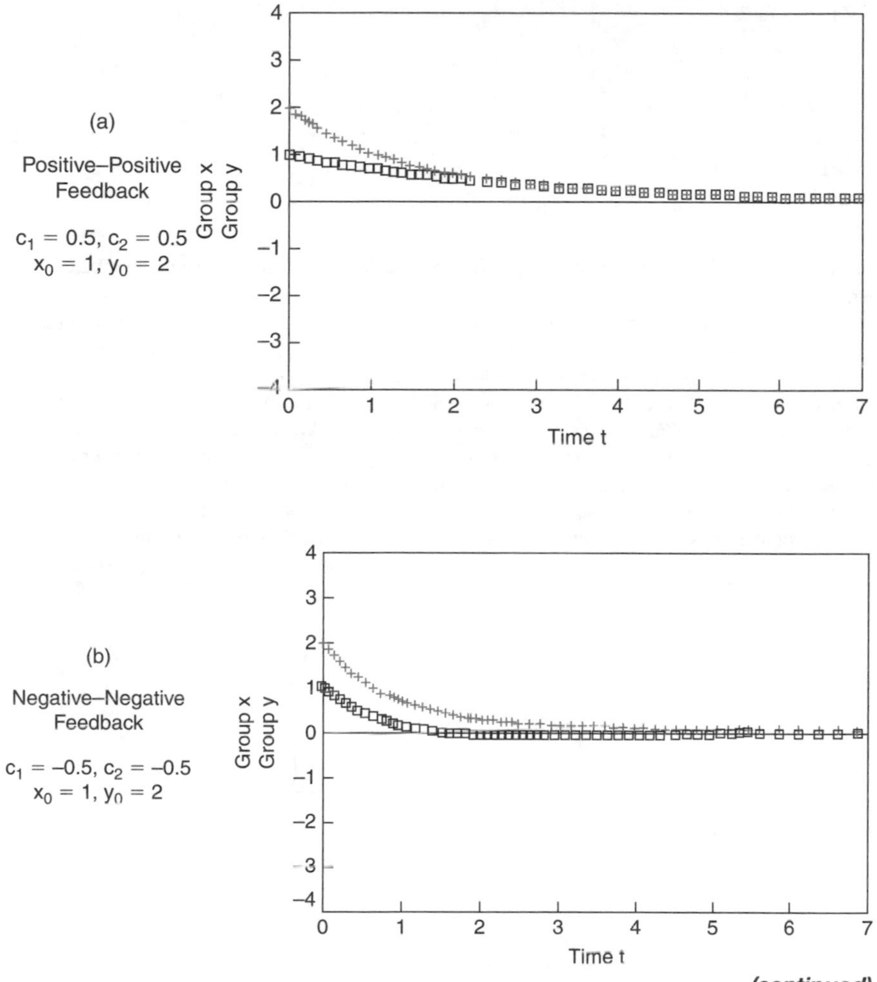

(continued)

FIGURE 9.2 ***(continued)***

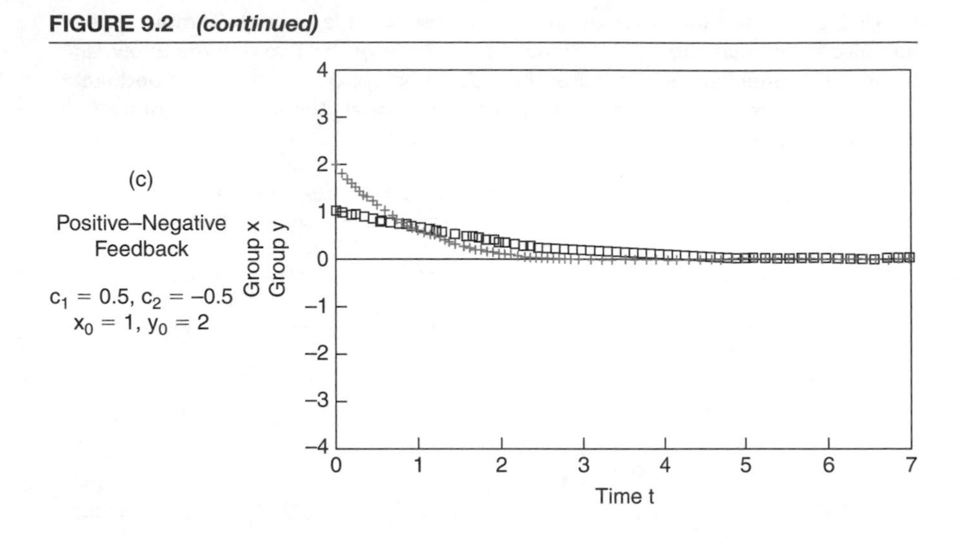

FIGURE 9.3 The Numerical Integration of the Model With Positive Feedback Between Both Groups

The values of x (boxes) and y (plus signs) are shown as the state of the two groups evolves from different initial values of x and y. The values of x and y always evolve toward the stable states where either both are positive or both are negative. (a) When both groups start at the same positive initial value they both evolve to and remain in the same positive state. (b) When both groups start at different positive initial values they both evolve and remain in the same positive state. (c) When group x starts in a positive initial state, but group y starts in a more negative initial state, they both evolve and remain in a negative state. (This figure, copyright 2007 by Larry S. Liebovitch, was presented at the International Association for Conflict Management 20th Annual Meeting in Budapest, Hungary, in 2007 and is reproduced with permission.)

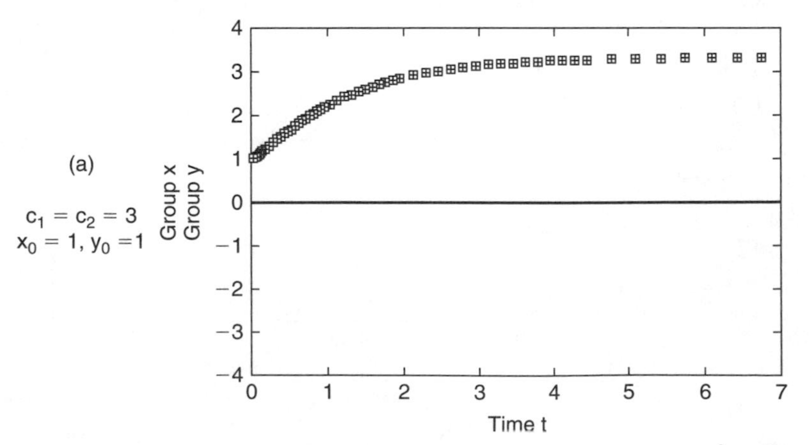

(continued)

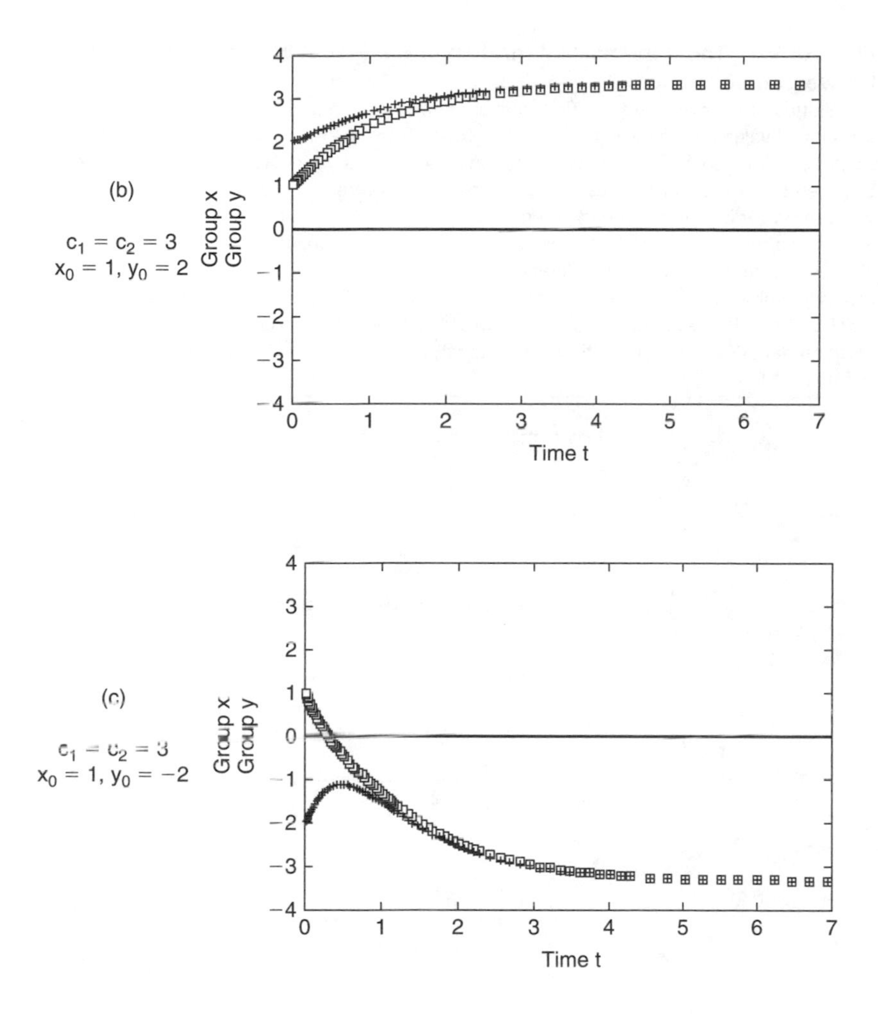

where either both are happy (x and y are both greater than zero) or both are unhappy (x and y are both less than zero). That is, their cooperation reinforces each other's state, so they both evolve to the same state. This matches our intuition about such conflicts. What is new from this mathematical analysis is that this behavior, rather than neutral behavior, emerges only when the strength of the feedback exceeds the inertia to change.

As shown in Figure 9.4, when there is strong negative–negative feedback between the groups, the results are more complex. When both groups have exactly the same values of the model parameters and exactly the same initial values, then over long times both evolve to a stable state that is neutral, where $x = y = 0$. However, if either the initial conditions of each group differ by only 0.1% or if their parameters differ by only 0.1%, then they reach

FIGURE 9.4 The Numerical Integration of the Model With Negative Feedback Between Both Groups

The values of x (boxes) and y (plus signs) are shown as the state of the two groups evolves from initial values of x and y. (a) When the groups have the same parameters and initial conditions they both evolve to the neutral state x = 0 and y = 0. However, small differences between either the initial states or the groups' parameters cause a symmetry breaking and the groups evolve toward very different stable states. (b) When the initial state of group y is 0.1% happier than that of group x, group y evolves to a positive state and group x to a negative state. (c) When the uninfluenced state of group y is 0.1% unhappier than group x, group y evolves to a negative state and group x to a positive state. (This figure, copyright 2007 by Larry S. Liebovitch, was presented at the International Association for Conflict Management 20th Annual Meeting in Budapest, Hungary, in 2007 and is reproduced with permission.)

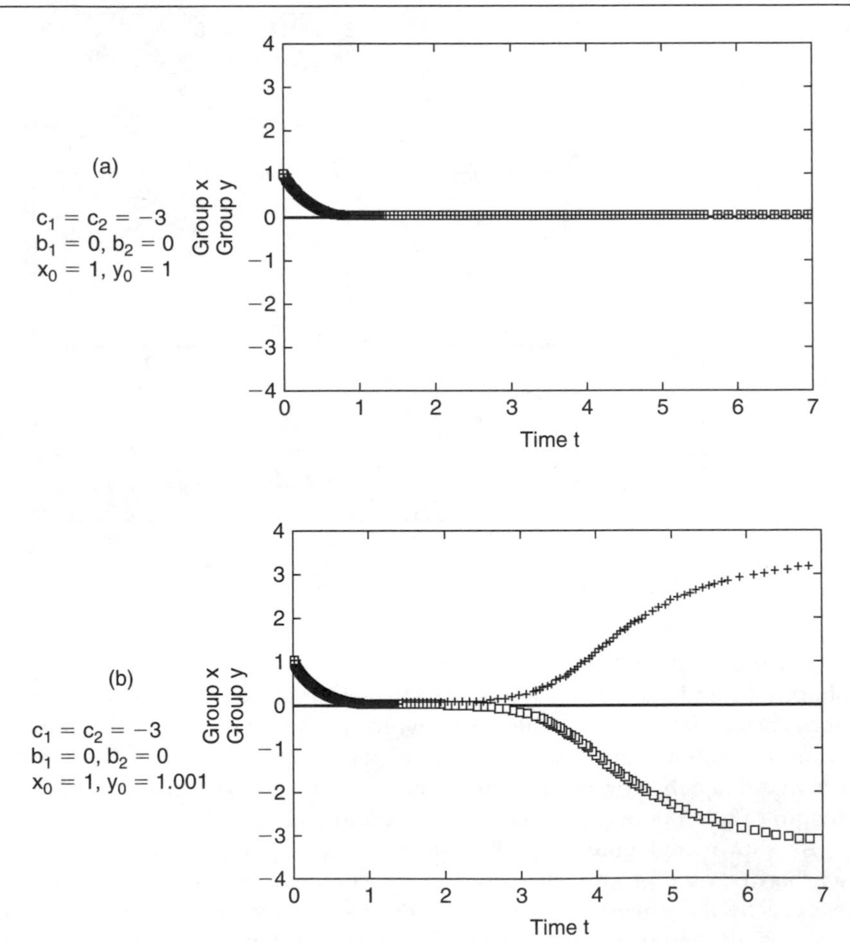

(continued)

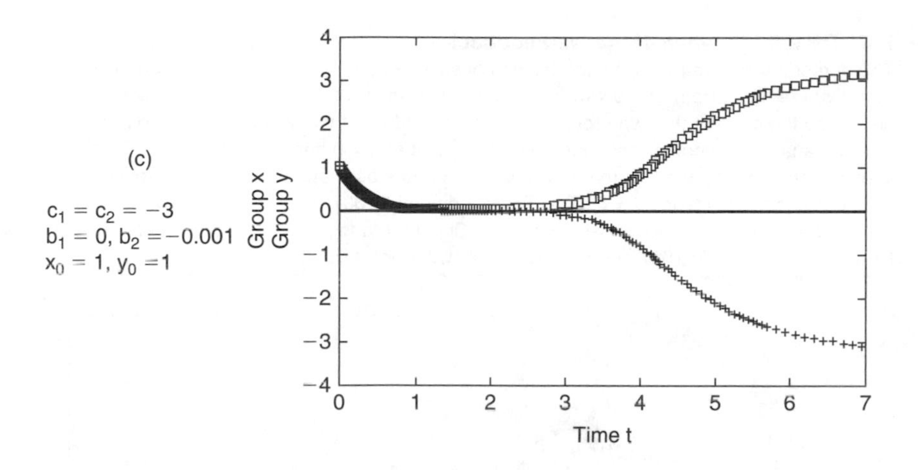

a stable state where one is happy and the other unhappy. It matches our intuition that the competition between the groups would result with one being happy and the other unhappy. However, the mathematical analysis reveals new and unexpected dynamical behavior. When the groups are not identical, the stability analysis reveals that the $x = y = 0$ state is an unstable saddle point; that is, the state of both groups approaches it, but then they diverge toward the stable points where one is happy and the other unhappy. Naive inspection of the time course might suggest that a triggering event at time $t = 3$ generated the escalating divergence in the groups' states. The mathematical analysis instead reveals that both the approach to the neutral state and its divergence from it are the result of the tiny differences between the groups at time $t = 0$. This may give us pause for thought in how we interpret the sudden escalations in conflicts and whether we attribute that escalation to recent events or to the essential response of a negative–negative interaction to long past events.

As shown in Figure 9.5, when positive-negative feedback between the groups, each group's state oscillates for some time before both groups eventually reach a stable state that is neutral with $x = y = 0$. In this case, intuition (at least our intuition) was unclear about what dynamics would be produced by such mixed cooperation–competition feedback between the groups. The mathematical analysis here has revealed two quite interesting and unexpected phenomena. First, the critical point at $x = y = 0$ is a stable spiral, which means that the groups' states must oscillate before they reach their final stable values. Second, both groups eventually reach a stable state that is neutral at $x = y = 0$. This means that the hostile outcome of a negative–negative feedback conflict can be reduced to a neutral state by shifting the influence function of only one party to positive feedback. This matches the experience in some conflicts, where conciliatory actions by one party, without expecting anything in return, can sometimes precipitate action from the other party, which reduces the level of conflict.

FIGURE 9.5 Positive–Negative Feedback

The numerical integration of the model with positive feedback from group y to group x and negative feedback from group x to group y. The values of x (boxes) and y (plus signs) are shown as the state of the two groups evolves from different initial values of x and y. The stability analysis demonstrates that the critical point at x = 0 and y = 0 is a spiral. This is evidenced here by the fact that the values of x and y oscillate as they evolve toward the neutral state x = 0 and y = 0. (This figure, copyright 2007 by Larry S. Liebovitch, was presented at the International Association for Conflict Management 20th Annual Meeting in Budapest, Hungary in 2007 and is reproduced with permission.)

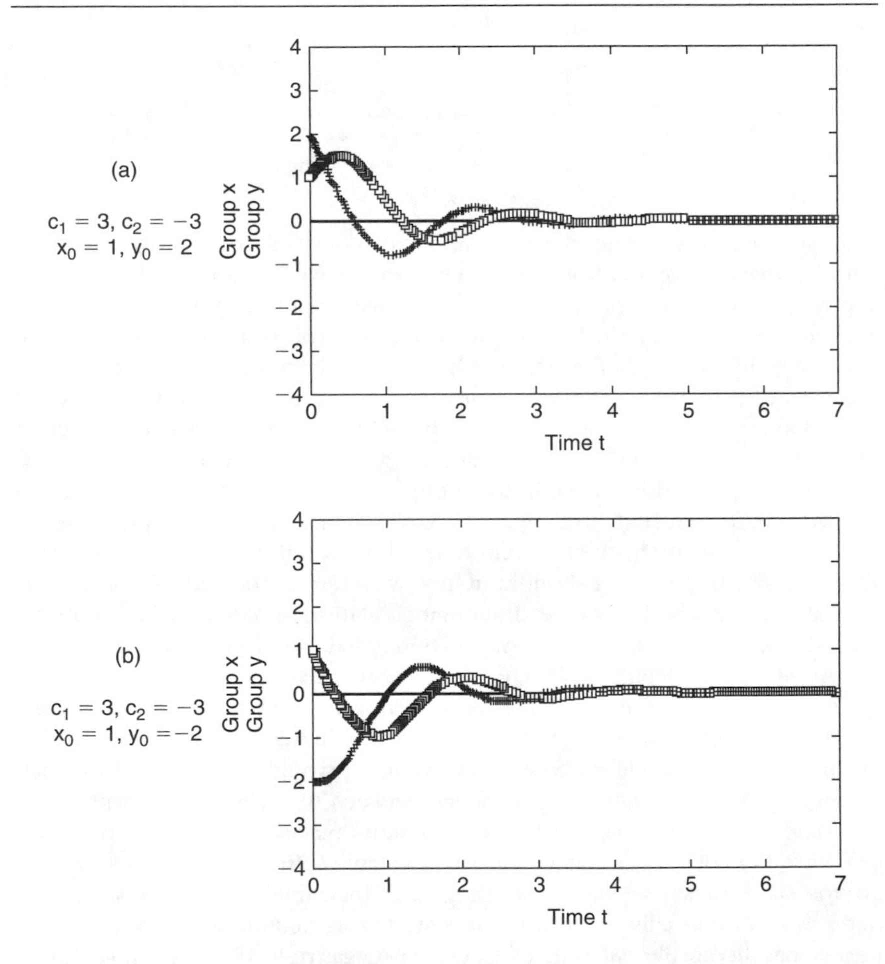

As shown in Figure 9.6, the dynamical behavior of all three cases is summarized by the bifurcation diagram. The rigorous mathematical analysis of these simple influence functions yields results that are consistent with our intuition: (1) that positive–positive feedback (cooperation between both groups mutually reinforcing the same behavior in each other) leads to both groups either swimming or sinking together and (2) that negative–negative

FIGURE 9.6 Dynamical Behavior of All Three Cases Summarized by Bifurcation
The bifurcation diagram summarizes the dependence of the stable states on the strength of the feedback between the groups. (a) When there is positive feedback between both groups the neutral state x = 0 and y = 0 is stable until the strength of the feedback exceeds the inertia to change. When the strength of the positive feedback is larger than that threshold, then the neutral state x = 0 and y = 0 is unstable and there are two stable states with x and y both positive or x and y both negative. (b) When there is negative feedback between both groups the neutral state x = 0 and y = 0 is stable until the strength of the feedback exceeds the inertia to change. When the strength of the negative feedback is larger than that threshold, then the neutral state x = 0 and y = 0 is unstable (if there is any difference in the parameters or the initial states between the groups) and there are two stable states, one with x positive and y negative, and one with x negative and y positive. (c) When there is positive feedback from one group and negative feedback from the other group, then only the neutral state x = 0 and y = 0 is stable. (This figure, copyright 2007 by Larry S. Liebovitch, was presented at the International Association for Conflict Management 20th Annual Meeting in Budapest, Hungary in 2007 and is reproduced with permission.)

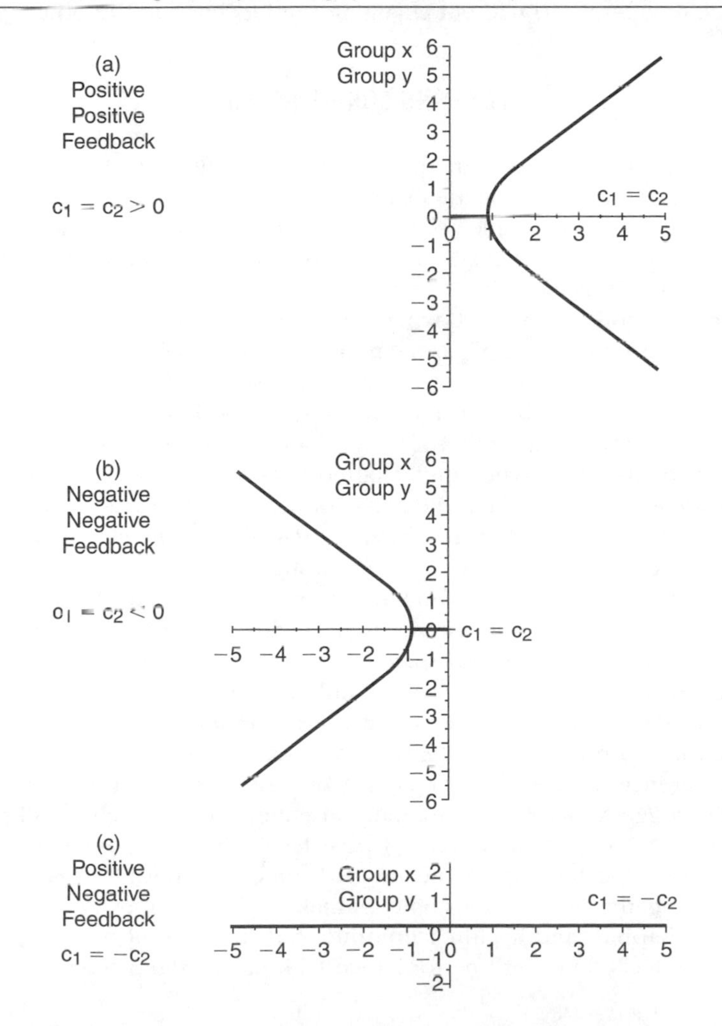

feedback (competition between the groups mutually reinforcing the opposite behavior in each other) leads to one group swimming while the other group sinks. However, the ability to encode these simple feedback influence functions into a mathematical form and then use standard mathematical analysis to determine the logical and necessary consequences also leads to new and interesting results: (1) the existence of a threshold, equal to the inertia to change, that the influence between the groups must exceed before the conflict materializes, (2) that almost unnoticeably small differences between the groups with negative–negative feedback (mutual competition) emerge, over long times, into very large differences between the groups, and (3) that we can switch an intractable conflict between two groups with negative–negative feedback (mutual competition) to at least a much more moderate neutral state by shifting the response of only one of the groups to positive feedback (cooperation) without changing the feedback of the other group.

LESSONS FOR NURSING

We now explore how we can translate these results into the health care environment. Because the mathematical model presented here is far simpler than the full range of complex social interactions, these explorations are not meant as definitive statements, but rather as a starting point to stimulate further thought and discussion.

First, the model shows that when the feedback is below a certain threshold, over long times, each group's emotional state shows no evidence at all of the conflict's existence. However, an increase of the feedback over this threshold can produce dramatically different effects. For example, if the conflicts between nurses and other staff in a small practice, or between needy patients and an overstretched understaffed nursing staff in a hospital, do not show evidence of the strain in the behaviors of those involved, it may not be safe to conclude that small increases in stress will be easily absorbed with no major changes. It may be wise to fully investigate and deal with smaller levels of hostility and stress even though there is no overt behavioral problems, because small changes in those negative feedbacks may well lead to more drastic changes than those anticipated.

Second, which group is happy or unhappy in the model depends sensitively on their initial conditions and their uninfluenced emotional states. We often fail to give proper attention to how small issues can lead to dramatic differences as the social process plays itself out in its highly nonlinear way. This suggests that (1) paying more attention to seemingly small patient needs may establish confidence and trust levels that can have much larger payoffs than expected and (2) meeting the nursing staff's social needs (such as scheduling, training, responsibility in making decisions, etc.), which translate into a happier and healthier uninfluenced state can also have dramatic beneficial effects on overall performance with patients and staff.

Third, the mathematical model shows that an intractable hurting conflict generated by negative–negative feedback can be switched into a neutral state by the actions of one group, alone, changing to positive feedback. That is, even one person, alone, without any change in the other person's attitude, can change the dynamics of their interactions. This means that in patient care and interactions with other hospital staff each person may not have control of other people's attitudes (no matter how much they may want it), but by their own actions they can change the dynamics of the interaction with the other person. Moreover, the mathematical model predicts that in evolving toward this neutral state, both groups will experience ups and downs in their emotional state. We should be careful to recognize that these ups and downs do not necessarily reflect permanent alterations in behaviors.

SUMMARY

It can be hard to formulate our qualitative ideas into the unforgiving clarity of mathematics. However, once we have done so, we can then use the rigorous machinery of mathematics to clearly and fully understand the consequences of the representations that we have encoded in the mathematics. This gives us tremendous power to understand the sometimes surprising consequences of simple psychological or sociological mechanisms.

We showed how developing conflict models raises some basic mathematical questions that cannot yet be answered, but that the answers to those mathematical questions may shed interesting light on the behavior of individuals, groups, and nations.

We also developed a nonlinear model of the interaction between two groups. We showed that the dynamical behavior of this model depends on whether the feedback between the groups is positive, negative, or a mixture of both. It is both surprising and fascinating how the detailed mathematical properties of this simple model can resonate with our experience of complex human behavior as well as suggest new ways to think about conflicts and their resolution.

The development of these mathematical models and their analysis is, of course, only a first step. The next step must be to find ways of measuring these models' parameters from experimental data in field studies or laboratory experiments and then test whether the dynamical behavior predicted by these models matches, or does not match, the observed dynamics of those conflicts.

ACKNOWLEDGMENT

This work was supported in part by a grant from the James S. McDonnell Foundation, Intractable Conflict as a Dynamical System, Peter T. Coleman (P.I.).

APPENDIX: STABILITY ANALYSIS OF THE MODEL OF CONFLICT BETWEEN TWO GROUPS

We now present the stability analysis of the model of conflict between two groups.

Let x and y be the state of each group at time t, $c_{1,2} > 0$ corresponds to positive feedback, and $c_{1,2} < 0$ corresponds to negative feedback. The equations of the model are:

$$\frac{dx}{dt} = m_1 x + b_1 + c_1 \tan h(y),$$

$$\frac{dy}{dt} = m_2 y + b_2 + c_2 \tan h(x). \tag{A1}$$

We analyze the linear stability around a critical point $x = x_s$ and $y = y_s$. The eigenvalues of Equation A1 are

$$\lambda = \left(\frac{1}{2}\right)\left\{(m_1 + m_2) \pm \left[(m_1 - m_2)^2 + 4c_1c_2 \sec h^2(x_S) \sec h^2(y_S)\right]^{\frac{1}{2}}\right\}. \tag{A2}$$

For the models studied here, $m_1 = m_2 = m < 0$ and thus Equation A2 simplifies to

$$\lambda = m \pm \left[c_1c_2 \sec h^2(x_S) \sec h^2(y_S)\right]^{\frac{1}{2}} \tag{A3}$$

When $|c| < |m|$, the only critical points are at $x_S = y_S = 0$ and therefore Equation (A3) becomes

$$\lambda = m \pm [c_1c_2]^{\frac{1}{2}}, \tag{A4}$$

For either positive–positive feedback ($c_1 = c_2 > 0$) or negative–negative feedback ($c_1 = c_2 < 0$), then $c_1c_2 > 0$. Because $m < 0$ and $|c| < |m|$, both eigenvalues $\lambda < 0$ and so $x_S = y_S = 0$ is stable. When there is positive–negative feedback, then $c_1c_2 < 0$ and so the eigenvalues are complex, but with real part $m < 0$. Thus, this critical point is a stable spiral.

When $|c| > |m|$, the critical point at $x_S = y_S = 0$, as before, λ is given by eq. (A4). For either positive feedback between both groups ($c_1 = c_2 > 0$) or negative feedback between both groups ($c_1 = c_2 < 0$) because $m < 0$ and $|c| > |m|$, there is one negative and one positive eigenvalue and thus this critical point is an unstable saddle. When there is positive–negative feedback, both the eigenvalues are complex, but because the real part $m < 0$, this is a stable spiral. The two other critical points for either positive–positive or negative–negative feedback occur at $x_s = -c_1/m_1$, and $y_s = -c_2/m_2$, and the eigenvalues from Equation A3 become

$$\lambda = m \pm \varepsilon \tag{A5}$$

where $\varepsilon < 1$. Because $m < 0$, both these eigenvalues are less than zero and both critical points are stable.

REFERENCES

Barash, Y., Elidan, G., Friedman, N., & Kaplan, T. (2003). Modeling dependencies in protein-DNA binding sites. In M. Vingron, S. Istrail, P. Pevzner, & M. Waterman (Eds.), *Proceedings of the seventh annual international conference on research in computational molecular biology* (pp. 28–37). Berlin, Germany: ACM (Association for Computing Machinery). Retrieved from http://portal.acm.org/citation.cfm?id=640079

Boulding, K. E. (1962). *Conflict and defense.* New York: Harper.

Carper, B. (1978). Fundamental patterns of knowing in nursing. *Advances in Nursing Science, 11*, 13–23.

Deutsch, M. (1973). *The resolution of conflict: Constructive and destructive processes.* New Haven: Yale University Press.

Deutsch, M. (2006). Cooperation and competition. In M. Deutsch, P. T. Coleman, & E. C. Marcus (Eds.), *The handbook of conflict resolution: Theory and practice* (pp. 23–42). San Francisco: Wiley.

Gottman, J., Swanson, C., & Swanson, K. (2002a). A general systems theory of marriage: Nonlinear difference equation modeling of marital interaction. *Personality and Social Psychology Review, 4*, 326–340.

Gottman, J. M., Murray, J. D., Swanson, C. C., Tyson, R., & Swanson, K. R. (2002b). *The mathematics of marriage.* Cambridge: MIT Press.

Liebovitch, L. S., Vallacher, R., Nowak, A., Bui-Wrzosinska, L., & Coleman, P. (2007). *Dynamics of Two-Actor Cooperation-Competition Conflict Models.* International Association for Conflict Management, 20th Annual Conference, Budapest, Hungary, July 1–4, 2007. Retrieved from http://www.ccs.fau.edu/~liebovitch/IACM2007.pdf

Murray, J. D. (1989). *Mathematical biology.* New York: Springer-Verlag.

Pruitt, D. G. (1969). Stability and sudden change in interpersonal and international affairs. *Journal of Conflict Resolution, 13*, 18–38.

Pruitt, D. G. (2006). *A graphical interpretation of escalation and de-escalation.* Presented at Dynamics and Complexity of Intractable Conflicts. Kamimierz, Poland, Oct. 19–22, 2006.

Richardson, L. F. (1967). *Arms and insecurity.* Chicago: Quadrangle.

Strogatz, S. H. (1994). *Nonlinear dynamics and chaos.* Reading, MA: Addison-Wesley.

Thompson, J. M. T., & Stewart, H. B. (1986). *Nonlinear dynamics and chaos.* New York: Wiley.

Yiu, K. T. W., & Cheung, S. O. (2005). A catastrophe model of construction conflict behavior. *Building and Environment, 41*, 438–447.

Response to Chapter 9

Lisa Conboy

CONCEPTS PRESENTED IN THIS CHAPTER THAT ARE RELEVANT TO THE COMPLEX SYSTEMS SEEN IN NURSING, MEDICINE, AND/OR OTHER EXAMPLES OF HEALTH CARE PRACTICE

Conflict dynamics	Interaction
Competition	Uninfluenced state
Cooperation	Weak influence
Feedback	Strong influence

HOW INFORMATION CAN BE COLLECTED ON THESE CONCEPTS. INCLUDE INFORMATION SCALING FROM MICRO THROUGH MESO TO MACRO

SCALE	CONCEPT	RELATED TO	SEEN IN ASSESSMENT DATA
Micro	Conflict dynamics	Nurse–patient interaction	Clarity and completeness of patient intake
	Uninfluenced state	Nurse–patient interaction	Nurse's success in predicting effectiveness of an interaction style
	Weak/strong influence	Patient treatment decision making	Strong patient feelings concerning selected treatments are related to conflict
	Weak/strong influence	Health education	Weak/vague patient feelings concerning potential outcomes of health behaviors may not lead to system (behavioral) changes
Meso	Conflict dynamics	Nursing student and professor interactions	Emotional state (e.g., satisfaction) with hospital division of labor (student/nurse duties)
	Uninfluenced state	Interactions between patients' families	The initial emotional state of a patient's family is influential in occurrence and outcome of negative interactions with other patients' families

(continued)

SCALE	CONCEPT	RELATED TO	SEEN IN ASSESSMENT DATA
Macro	Conflict dynamics	Interaction between health service institutions	Effectiveness of resource sharing between institutions
	Feedback	Interaction between health service institutions	Patterns of resource sharing

EXAMPLES OF HOW THESE CONCEPTS AND/OR METHODS CAN BE APPLIED IN PRACTICE

Conflicts that have low levels of interchange or feedback show little change over time. Yet, beyond a certain threshold, small increases in negative or positive feedback can have a dramatic influence on outcomes. Thus, addressing problem areas early is warranted.

If conflict exists already, and both groups are offering negative feedback, the conflict can be made more neutral more easily than might be expected. Only one member of the dyad in conflict needs to decrease the negativity of the feedback for the dynamics of the conflict to move to neutral.

Conflicts can to some extent be predicted if we know the emotional state of an individual or group when not in interaction with another potentially conflicting individual or group. Much like the first point above, making small consolations to maintain a positive emotional state can help to avoid larger conflicts.

CONCLUSION: EVALUATING THE USEFULNESS OF THE CHAPTER CONCEPTS AND METHODS FOR NURSING AND HEALTH CARE PRACTICE

The concepts outlined in this chapter are relevant and applicable to health care practice. The authors have found evidence suggesting predictability to the dynamics of conflicts. The results of this deductive investigation suggest ways to predict, avoid, and ameliorate conflicts between individuals or groups.

This chapter will be helpful in understanding methodologies for nursing and health care practice. The results are relevant across scales of inquiry, allowing for applicability in interpersonal as well as group and institutional interactions. Similarly, the study encourages the health care practitioner to consider new ways of managing conflicts, perhaps using variables not previously considered.

Last, the models are relatively simple, the writing is clear, and the conclusions are immediately applicable. The study offers some results that overlap with current interaction theory, as well as a few new ideas to explore further.

Response to Chapter 9

Joy Longo

Understanding relationships between individuals and between groups is instrumental for nurses and foundational to the practice of nursing. In relationships, caring can be acknowledged. Boykin and Schoenhofer (2001) state "The nature of relationships is transformed through caring . . . A relationship experienced through caring holds at its heart the importance of person-as-person" (p. 4). Despite the best attempts at fostering caring relationships, conflicts can arise between the patient and the nurse and can also extend beyond the nurse–patient dyad and emerge as a separate entity among health care providers. Though nurses recognize the need for a humanistic approach to health and relationships, health care is a commodity, and it is expected that its members provide cost-effective, quality care. In substantiating these outcomes, health care facilities are called upon to provide information regarding success and failure *rates* of proposed indicators. In the chapter of Liebovitch, Vallecher, Nowak, Coleman, and Bartoli, they propose a model with this now-familiar language of mathematical formulas and numbers to examine social conflicts. This model provides a lens in which to view the conflicts encountered by nurses, and those which will be discussed further in this paper.

The most important relationship in which the nurse engages in is that between the nurse and the patient. With this relationship comes the recognition of the patient as an individual with unique needs. A conflict may be fueled by the often complex world of health care delivery resulting in the failure of the nurse to truly come to know the patient. Factors such as short staffing, high patient acuities, and inadequate resources play a role in nurses' ability to provide individualized care, which is sought by the patient possibly resulting in discord. This can be equated by what is illustrated in the Liebovitch et al. model of social conflict as mutual competition, or negative–negative feedback. According to Liebovitch et al., if both groups involved in a conflict start at the same point, over time a steady state will be reached. If the groups are not identical, rather than reach a stable state they will move away from each other to where one group is happy, and one group is unhappy. In nursing this can be demonstrated when two parties start with different values. A patient presents his or her own value set, which influences their health care behaviors and decisions, and nurses have a personal value set in addition to holding the values of the profession. An incongruency in values could result in disdain. Nursing relationships can also be represented in the model when positive–negative feedback occurs between groups. In these circumstances, Liebovitch et al. found that there is some oscillation before a steady state is reached and postulate that this leads to the understanding that a negative–negative conflict can be neutralized if only one party shifts to positive feedback. This is exemplified by the nurse who approaches a situation fully prepared to engage. An example of this would be the patient who is labeled as "a problem patient" and "complainer." The label is passed on from nurse to nurse without resolve, thus perpetuating conflict between staff and patient. When a nurse enters into this situation from a caring perspective, time is taken to listen and to come to understand the patient's story. As pointed out by Liebovitch et al., attention to the little things, such as listening, can mean a lot to the patient and enforce the importance of a humanistic approach to nursing. The positive feedback provided by the nurse represents a shift that could counterbalance the conflict.

Interest in the social conflict that occurs among health care workers is rising due to new Joint Commission standards and the recognition of the impact of conflict on patient safety. The Joint Commission (2008), an accrediting body for health care facilities, recently issued a sentinel alert concerning the potential impact of disruptive behaviors among health care workers on patient safety and implemented a leadership standard requiring that health care facilities assume responsibility for such behaviors. For nurses, relationships wrought with a type of disruptive behavior known as horizontal violence represent an example of a social conflict between nurses. Horizontal violence is defined as subtle or overt acts of aggression that are acted out between colleagues (Longo & Sherman, 2007). The behaviors are attributed to the theory of oppressed group behavior (Roberts, 1983). Due to oppressive circumstances in the environment, nurses may be prevented from fully acting upon their potential, thus creating a feeling of low self-esteem and leading to the belief that circumstances cannot be changed. The resulting frustration can be manifested in passive–aggressive behaviors displayed toward coworkers. The oppressive forces are similar to the factors that ignite nurse–patient conflict and include perceived deficiency of resources and lack of recognition. The outward display of aggressive behaviors can be viewed as disruptive because they can impact communication and collaboration in an environment where teamwork is a strong concern. In the model, Liebovitch et al. describe cooperation as being associated with positive–positive feedback that leads to a state where both groups are happy or both groups are unhappy. In some circumstances, despite positive feedback from all groups, an unintended outcome such as a poor patient prognosis or irresolvable work issues may result. In these circumstances, though a joint effort is attempted, neither party is satisfied with the outcome.

When nurses encounter disruptive behaviors, a common method of dealing with the conflict is avoidance (Hogh & Dotradottir, 2001). Liebovitch et al.'s description of a neutralized situation concerning negative–negative feedback when one party shifts to positive feedback can be applied here.

In terms of staff, if one person attempts to bridge the gap in a conflict, it could change the dynamics of the conflict. The concern for nurses is that they are primed to take this step. Proper coaching and appropriate communication skills can help with this often daunting task (Longo, 2010).

As suggested by Liebovitch et al., an important message for nurses to take from the model is that underlying emotions can impact conflict. For nurses, underlying emotional issues may result from internal tension and frustration caused by the often oppressive issues in the environment. It is imperative that nurse managers do not assume that because things appear status quo on the surface, there are no problems. Nurse managers need to take a proactive approach by evaluating the environment for any issues that can potentiate stress beyond what is inherent in health care. One way to accomplish this is by "rounding" with professional staff to determine better outcomes (Studer, 2008) where, on a daily basis, the manager spends time with the nurses to evaluate and to learn their concerns and to understand ways conflicts can arise between staff and patients or between staff members themselves. This will allow hidden issues to surface and be dealt with in an appropriate manner.

In the Liebovitch et al. chapter, the researchers outline a mathematical model to help explain social conflict between groups. The health care environment is inherently complex and has the potential to grow more complex with coming health care reform. The relevance of this work to the future of the nursing discipline is to come to understand and to address conflict at a time when it is being seen as a threat to patient safety. This responsibility lies with nurses.

REFERENCES

Boykin A., & Schoenhofer, S. O. (2001). *Nursing as caring: A model for transforming practice.* Boston: Jones and Bartlett.

Hogh, A., & Dotradottir, A. (2001). Coping with bullying in the workplace. *European Journal of Work and Organizational Psychology, 10,* 485–495.

Joint Commission. (2008). *Behaviors that undermine a culture of safety.* Retrieved July 8, 2008, from http://www.jointcommission.org/SentinelEvents/Sentineleventalert/sea_40.htm

Longo, J. (2010, January 31). Combating disruptive behaviors: Strategies to promote a healthy work environment. *OJIN: The Online Journal of Issues in Nursing, 15*(1), Manuscript 5.

Longo, J., & Sherman, R. (2007) Leveling horizontal violence. *Nursing Management, 38*(3), 34–37, 50–51.

Roberts, S. J. (1983). Oppressed group behavior: Implications for nursing. *Advances in Nursing Science, 7,* 21–30.

Studer, Q. (2008). *Results that last.* Hoboken, NJ: John Wiley & Sons.

CHAPTER 9
MATHEMATICAL MODELS OF THE DYNAMICS OF SOCIAL CONFLICT

- *Social conflict is prevalent wherever there are human relations. Explore how complexity theorists have defined social conflict. How did Conboy address the issues in her template of the social conflict chapter?*
- *How is social conflict defined in nursing? Is there a difference?*
- *Is it possible to predict conflict from the dynamics of relationships in groups? How do you manage conflict in your work life?*
- *Longo speaks about "horizontal violence" in nursing. Why do you think this has become so prevalent in the health care culture? How can we deal with social conflict in unit management and patient care nursing situations, or in the college or university environment?*

TEN

Modeling the Complexity of Story Theory for Nursing Practice

Patricia Liehr
Mary Jane Smith

Visual models reflect the interaction of dynamic concepts, and accentuate the intricate complex relationship between concepts and how the interaction is limited by the page on which it is printed. In the model of story, three ellipses map a vortex that depicts a continuing, dynamic, evolving process consistent with the unitary transformative paradigm of Newman and an energy-laden integrated whole. Pattern mapping of a phenomenon involves paying attention to the links between essential elements to bring to life the energy-laden nature of the phenomenon. The context in which a phenomenon is situated infuses it with meaning and guides pattern mapping. To connect with self-in-relation, people see themselves as existing and growing in a context of other people and times with sensitivity to bodily expression and a sense of history and future in the present moment. Remembering disjointed story moments and embracing hidden meanings facilitate flow and create a sense of ease.

Respondent Hain brings to life the complexity of "story" in terms of what it means to be a patient with chronic kidney disease. Hain's reflection shows that listening to the story of a patient reveals more than the patterns of physical disease and treatment regimes, and gives a more holistic revelation of the fears, pain, and hopes that a patient experiences when faced with a devastating diagnosis. Hain shows how "story theory and method" has importance in nursing practice and how story theory will now be integrated into the new Doctor of Nursing Practice (DNP) program.

INTRODUCTION

This is a story about model making. Like any story, it has a beginning, middle, and end, with complicating, developmental, and resolving dimensions. The complicating dimension is the challenge of picturing in two-dimensional space a dynamic complex process. The development of the model shows the shifting look of the model picture as scholars strive to push two-dimensional space to its limits. This chapter represents a resolving dimension, as the editors of this book who are knowledgeable about complexity science invited this story of modeling complexity. This invitation named a process, modeling complexity, which has been lived by the authors, but which, until now, was an unnamed experience. The story begins with an understanding of the complexity of modeling complexity, placing emphasis on configuration of relationships and context for configuration.

Pattern mapping of a holistic phenomenon reveals the meaning of the relationships among the phenomenon's essential elements (Ray, 1994). When mapping such a phenomenon, it is important to configure the relationships as dynamic processes and pay attention to links between the essential elements, because depiction of these links can bring to life the energy-laden nature of the phenomenon being mapped. The context in which a phenomenon is situated infuses it with meaning and guides pattern mapping.

Modeling is a sort of pattern mapping, and story theory is situated within the context of the unitary transformative paradigm (Newman, Sime, & Corcoran-Perry, 1991) for mapping the theory. Concepts derive meaning from theory niches that emerge (Paley, 1996). In addition, a theory derives meaning from the paradigm niche, and this niche also informs mapping. In the unitary transformative paradigm (Newman et al., 1991), humans are appreciated as irreducible beings uniquely emerging over time with ever-increasing complexity in relation with the environment.

STORY THEORY

Story theory proposes that story is a narrative happening of connecting with self-in-relation through nurse–person intentional dialogue to create ease (Liehr & Smith, 2008). Ease emerges as a person acquiesces in the whole story, even for a moment. Nurses recognize ease in the midst of story gathering when the person sharing the story has an "aha" experience with a "click of connection" as story moments come together creating "just recognized" meaning. Nursing encounters are inherently complex. The patient's story is intricately interwoven with that of family, friends, nurses, doctors, and an array of other health care providers participating in the current health experience. In gathering the patient's story, the nurse and patient are like weavers of an intricate tapestry.

The nurse stays attentive to the thread of what "matters most" to the patient while being sensitive to the warp and weft of contributing story threads that necessarily contribute to the emerging tapestry pattern. The nurse abandons preexisting assumptions, respects the storyteller as the expert, and queries vague story directions, thereby intentionally engaging the other to enable connecting with self-in-relation and ease.

The theory is based on three assumptions underpinning the conceptual structure. The assumptions state that people (1) change as they interrelate with their world in a vast array of flowing connected dimensions; (2) live in an expanded present moment where past and future events are transformed in the here and now; and, (3) experience meaning as a resonating awareness in the creative unfolding of human potential (Smith & Liehr, 2003). These assumptions are consistent with a unitary-transformative "view of the world," an inherently complex view; they establish a value structure that infuses the theory concepts with meaning. The three concepts of the theory are intentional dialogue, connecting with self-in-relation and creating ease.

Intentional dialogue is the central activity between nurse and person, which brings story to life; it is querying the emergence of a health challenge story in true presence (Smith & Liehr, 1999). This way of being with the other is imbued with energy that has implications for model making. It is a dynamic rather than linear process, demanding depiction beyond simplistic connecting lines and arrows.

Connecting with self in-relation occurs as reflective awareness on personal history, past and future all alive in the expanded present moment. It is an active process occurring during intentional dialogue when a person sees self in relationship with others who are contributing to a personal developing story plot (Smith & Liehr, 1999). As the story is shared, connection to bodily sensations, other people and times, and one's history and future surface into awareness. This sense of the whole challenges model making. Simultaneously and dynamically intentional dialogue, connecting with self-in-relation, and creating ease happen as an integrated unity.

Creating ease is remembering disjointed story moments to experience flow in the midst of anchoring (Smith & Liehr, 1999). As a person anchors even for a moment, comprehending the complex whole all at once, easiness with self ensues, expressed as flow and lived as movement toward resolving a health challenge.

Story theory comes to life in research and practice through complicating, developmental, and resolving processes, essential elements of all stories. "When gathering health story data, the complicating process focuses on a health challenge that arises when there is a change in the person's life; the developmental process is composed of the story plot that links to the health challenge and suffuses it with meaning; and the resolving process is a shift in view that enables progressing with new understanding" (Liehr & Smith, 2008, p. 213).

EVOLUTION OF MAPPING THE MODEL OF STORY

In a classic paper about how scientists really think, Root-Bernstein (1989) identifies model making as one of the "tools of invention" along with pattern recognition, aesthetics, and abstracting.

> *... The attraction of abstraction*
> *Is its simplifying feat.*
> *He who coddles newborn models*
> *Helps raise theories as he ought;*
> *Play and punning, just for funning,*
> *Can yield most surprising thought.* (p. 487)

Root-Bernstein (1989) describes modeling in conjunction with abstracting, where abstracting is the elimination of descriptors specific to individuals, leaving only essential elements, whereas modeling arranges essential elements in relation to one another. "Models in science are almost always simplifications of a complex situation that cannot be analyzed completely" (p. 481).

The challenge of mapping story theory crystallized around the issue of highlighting dynamic connection reflecting the wholeness and evolving transformational processes enveloping the central concepts of connecting with self-in-relation, intentional dialogue, and creating ease. Although these theory concepts have remained consistent over the years, the approach for mapping the connection between the concepts has changed.

The first model was introduced in 1999 (Smith & Liehr, 1999) when the theory was named "Attentively Embracing Story." We were certain that we wanted to highlight the connection of nurse–person intentional dialogue, conveying the energy-infused quality of the engagement. The nurse–person connection of the first model attempted to capture the energy of intentional dialogue through an "energy wave process" connecting the nurse and client. The particular encircling arrows (Figure 10.1) were selected because their frayed edges suggested movement. In this inaugural attempt to map story theory, we included information that was critical to the birth of a theory; each of the subconcepts associated with the three main concepts was included in the model so that the reader had a sense of the qualities composing connecting with self-in-relation (personal history and reflective awareness), intentional dialogue (true presence and querying emergence), and creating ease (remembering disjointed story moments and flow in the midst of anchoring). In addition, the intentional dialogue connection was depicted as occurring between nurse and client, documenting the reality that the theory origins were strongly associated with nursing practice; that the theory was born in practice. However, we had not yet begun to systematically consider the methodology associated with taking the theory back to practice or to research. Rather,

FIGURE 10.1 Attentively Embracing Story

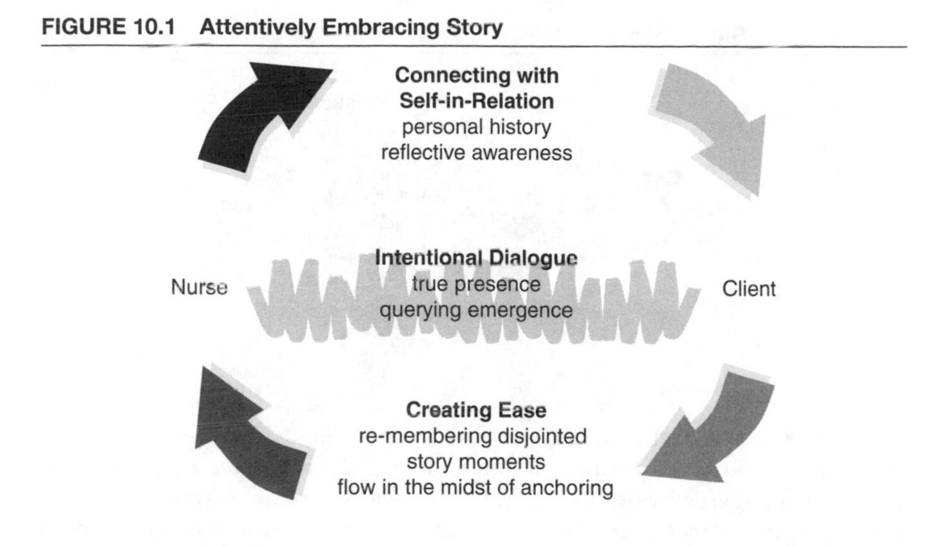

our focus was on ontology, or the essential meaning of story theory, and the epistemology, coming to know by understanding the concepts and the relationships between the concepts.

In spite of attempts to reflect the dynamic nature of story, our students and colleagues critiqued the model as "too linear" . . . like a circle that one cannot escape. This critique suggested that the model depicted a process that was antithetical to the meaning of the theory. It was necessary to take a close look at how the links between the concepts were mapped and consider other approaches for depicting the model. Pink (2005) addresses design as one of the crucial six senses to move from the information to the conceptual age. He defines design as "utility enhanced by significance" (p. 70), emphasizing the merging of practical and aesthetic qualities. In retrospect, it seems that we moved forward to design the model so that it aesthetically reflected the complex process in a way that it could be readily used in research and practice.

The current model (Figure 10.2), which was presented in 2003, spreads the "energy wave process" connecting nurse and person so that it infuses all of the concepts in the theory while highlighting the central activity of nurse–person intentional dialogue. This central activity is further emphasized with a heavy dotted ellipse between nurse and person. Overall, there are three ellipses in the design of this model, mapping a vortex intended to convey a dynamic, all-at-once, holistic process, encompassing the three theory concepts. This design shifts the model from one with an essential linear nature to one with an essential dynamic nature consistent with the unitary transformative paradigm, the theory contextual niche. The theory is mapped to depict an energy-laden integrated whole.

FIGURE 10.2 Story Theory

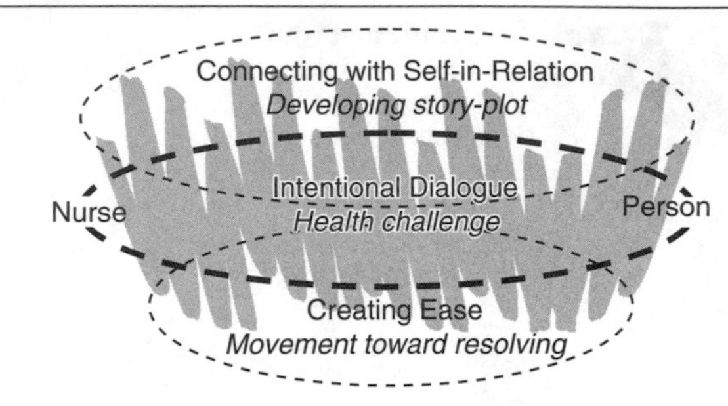

In this design of the model, information about methodology is associated with each of the theory concepts. Intentional dialogue about a complicating health challenge describes the central nurse–person activity when using the theory to guide practice or research. Connecting with self-in-relation happens in tandem with the developing story plot; as the story is shared and story moments are attentively integrated and strung together, the plot emerges and the storyteller connects with personal thoughts and feelings as they occur in relation to others. Creating ease is sometimes a burst, sometimes an edging movement, sometimes a slow steady movement toward a unique miniscule sensation of relief; an extended moment of comfort; or an "ah-ha" experience that permeates the story plot and enables a shift to a new way of being as the storyteller moves toward resolving a health challenge. At this stage of theory development and model design, ontology, epistemology, and methodology intertwine in a meaning-making endeavor characterized by both aesthetics and practicality, essential qualities of design in a conceptual age (Pink, 2005).

The reader will notice that a change in theory name occurred since it was first introduced in 1999 (Smith & Liehr, 1999). The name change was first introduced in 2006 (Liehr & Smith, 2006). Although the process of "attentive embracing" is still incorporated into the theory's meaning, the language was removed from the theory name, resulting in the name story theory. This shift in name suggests that the theory ideas could be used any time a nurse sits with a person to intentionally dialogue about what matters most. We had always envisioned the theory as a guiding structure in nursing situations where the nurse was coming to understand what matters most to another, but the original name suggested a direction for the intentional dialogue, where the storyteller resolved a health challenge by embracing it. This sometimes happens in practice, but it is not a given even when a caring nurse attentively queries a health challenge story.

In our original conversation about story more than a decade ago, well before the theory was created, we talked about the people each of us worked with, pregnant teens and people with heart disease (Smith & Liehr, 2003).

We recognized that people who embraced their story of pregnancy or broken heart were likely to be the people who accepted their "new normal" and got on with meaningful day-to-day living. The "Attentively Embracing Story" theory name was born there, in nursing practice. As we began to use the theory to guide practice and research, it became clear that "embracing story" was not an "all-or-nothing" phenomenon but rather an incremental activity, sometimes inch by inch and sometimes leaps and bounds. The original name confined the process in a way that limited the expression of the complexity that is naturally inherent in the emergent human story. Though the name "story" theory is simpler than the first, it more accurately reflects the complex process.

Finally, this story is still evolving. All along the theory development path, we have listened closely to students and colleagues who critique the work and made changes to enhance practicality while staying sensitive to aesthetics. As the theory is being used more frequently in research (Gobble, 2009; Hain, 2008; Ito, Takahashi, & Liehr, 2007; Liehr, Meininger, Vogler, Chan, Frazier, Smalling, & Fuentes, 2006; Ramsey, 2008; Williams, 2003, 2007), there will continue to be changes. Recently, we have described a story inquiry method (Liehr & Smith, 2007, 2008) that will influence ongoing use and development of the theory. As long as we must rely on two-dimensional space to depict the theory, the current model stands as is, representing its unitary transformative contextual niche and configuring the inherent complexity of story theory through lively connection of the theory concepts. Just like the characters in *Flatland* published in 1952 by Edwin A. Abbott, what we see is limited by our view. Though our mind's eye recognizes the dynamic complex process that is story, the snapshot of it that occurs on a page reduces it to "static" and we are challenged to make static "dynamic." That is what we have done to the best of our ability, and this is our story of modeling complexity.

REFERENCES

Abbott, E. A. (1952). *Flatland*. New York: Dover.

Gobble, C. D. (2009). The value of story theory in providing culturally sensitive advanced practice nursing in rural Appalachia. *Online Journal of Rural Nursing and Health Care, 9*(1), 94–105.

Hain, D. (2008). Cognitive function and adherence of older adults undergoing hemodialysis. *Nephrology Nursing Journal, 35*(1), 23–29.

Ito, M., Takahashi, R., & Liehr, P. (2007). Heeding the behavioral message of elders with dementia in day care. *Holistic Nursing Practice, 21*(1), 12–18.

Liehr, P., Meininger, J. C.,Vogler, R., Chan, W., Frazier, L., Smalling, S., et al. (2006). Adding story-centered care to standard lifestyle intervention for people with Stage 1 hypertension. *Applied Nursing Research, 19*, 16–21.

Liehr, P., & Smith, M. J. (2006). Middle range theories. In J. J. Fitzpatrick & M. Wallace (Eds.), *Encyclopedia of nursing research* (2nd ed., pp. 341–353). New York: Springer.

Liehr, P., & Smith, M. J. (2007). Story inquiry: A method for research. *Archives of Psychiatric Nursing, 21*, 120–121.

Liehr, P., & Smith, M. J. (2008). Story theory. In M. J. Smith & P. Liehr, *Middle range theory for nursing* (2nd ed., pp. 205–225). New York: Springer.

Newman, M., Sime, A. M., & Corcoran-Perry, S. A. (1991). The focus of the discipline of nursing. *Advances in Nursing Science, 14*, 1–6.

Paley, J. (1996). How not to clarify concepts in nursing. *Journal of Advanced Nursing, 24*, 572–578.

Pink, D. H. (2005). *A whole new mind: Moving from the information age to the conceptual age*. New York: Penguin Group.

Ramsey, A. R. (2008). The concept of yearning to be recognized: Caring for women who suffer from migraine. *International Journal for Human Caring, 12*, 26–30.

Ray, M. A. (1994). Complex caring dynamics: A unifying model of nursing inquiry. *Theoretic and Applied Chaos in Nursing, 1*, 23–32.

Ray, M. A. (1998). Complexity and nursing science. *Nursing Science Quarterly, 11*, 91–93.

Root-Bernstein, R. S. (1989). How scientists really think. *Perspectives in Biology and Medicine, 32*, 472–488.

Smith, M. J., & Liehr, P. (1999). Attentively embracing story: A middle range theory with practice and research implications. *Scholarly Inquiry for Nursing Practice: An International Journal, 13*, 187–204.

Smith, M. J., & Liehr, P. (2003). The theory of attentively embracing story. In M. J. Smith & P. Liehr, *Middle range theory for nursing* (pp. 167–188). New York: Springer.

Williams, L. A. (2003). Informal caregiving dynamics with a case study in blood and marrow transplantation. *Oncology Nursing Forum, 30*, 679–688.

Williams, L. A. (2007). Whatever it takes: Informal caregiving dynamics in blood and marrow transplantation. *Oncology Nursing Forum, 34*, 379–387.

Response to Chapter 10

Debra Hain

REFLECTIVE QUESTIONS

1. What is the relationship between theory and practice (clinical, administrative, or educational) highlighted in this chapter?

Stories are an integral part of nursing practice; nurses often gather stories to assist with decision making. The story-gathering method, guided by story theory (Liehr & Smith, 2008), is a way to discover a health challenge and to identify approaches used to resolve the health challenge as the nurses care enough to value and honor what "matters most" to the storyteller. Story has been used in clinical practice to guide an educational intervention for individuals experiencing chronic kidney disease (CKD).

James, a 62-year-old male who has been diagnosed with CKD secondary to diabetes, has been informed that he would need renal replacement therapy (hemodialysis, peritoneal dialysis, or transplantation) within the next 2 to 3 months.

James' story provides an exemplar of the *dynamic* nature of sharing story as a way to individualize an educational intervention in a nephrology office setting. When James presented for CKD education, which included discussion about impending renal replacement therapy, he experienced a perplexing moment as he was immobilized by his fear, fear of the unknown. The nurse knew that this was a complex situation and intentionally abandoned preconceived assumptions and respected James as expert about "self" as she cared enough to be truly present. Sharing story is a dynamic complicating process that focuses on a health challenge that changes a person's life, requiring a new understanding of self. As the nurse and patient engage in dialogue, they connect with the "self-in-relation," the mutual human-to-human process where the story unfolds in the present moment—where the past and future are transformed into the here and now. In this story of the "self-in-relation," James experienced a reflective awareness that dialysis was separate and distinct from who he is and that he would not be defined by it; he was able to move toward resolving the health challenge, enabling him to make a decision about which renal replacement therapy was best for him.

2. How does a nurse or a physician or health care administrator learn about complexity science/s and caring through the presentation of ideas or research?

A recent qualitative study used story theory, a middle-range nursing theory, to guide story gathering and story analysis to identify health challenges of making lifestyle changes in older adults undergoing hemodialysis and to identify approaches for resolving the health challenges (Hain, Wand, & Liehr, 2010). The findings of this study were presented at a nursing conference where practicing nurses and nursing students were able to learn more about how story could be used to gain an insight into the challenges older adults face and some possible strategies professionals may use to help the person undergoing hemodialysis resolve his or her particular health challenges. Presentation of the results allowed for open dialogue between the presenter and the audience. Nurses disclosed personal knowing about themselves as they related the findings to their own practice settings.

Nurses obtain complex information when caring for this population, which requires them to consider individual variables while understanding how mutually dependent the nurse and the patient are. For example, time, an important concept, has many meanings to people undergoing hemodialysis. This research provided insight into the paradox of time. The time-consuming nature of dialysis has distinct implications depending on the context in which it is viewed. The nonlinearity of the concept of time can be seen as time is needed to get to and from dialysis, time spent undergoing hemodialysis and the time it takes to recover from the treatment once the person goes home; yet what is striking is that the person is given more time (*life-sustaining time*) because of dialysis. Even though time was important to the study participants, what was mutually discovered was the fact that many other complex health challenges had to be confronted. Three central themes emerged from the data: (1) living a restriction-driven existence; (2) balancing independence/dependence; and (3) struggling with those providing care. Although these are not discussed at length in this response, it is important to recognize that this population of patients faces multifaceted issues.

Complexity science provides a model for nurses to examine the human–environment process by connecting what they know about the intricate needs of this population of patients with aspects of nursing research that offer evidence of how story theory can be used in research and applied to practice. As first author, I espouse the use of story theory to practicing nurses and other health care professionals. In fact, recent findings from this study supported the use of story theory in practice. Story theory has been identified as a resource for best practice to improve quality of life and reduce mortality in the first year of dialysis (Performance, Excellence, & Accountability in Kidney Care, 2010). The best practice guideline recommending story theory as a way to obtain important information states "enhance and maintain quality of life by incorporating an individual and holistic educational approach about the physical and psychosocial impact of dialysis on patients' lives." There are several suggested interventions for achieving this best practice guideline; however, it is recommended that story theory be used as an assessment tool to identify challenges people face: what is important to them right now; what have they done in the past to "get through" difficult times; and what are their future hopes and dreams.

3. **What meanings are illuminated in this chapter for nursing or other heath care professions?**

Nurses are essential members of interprofessional health care teams who care for people with a wide range of health challenges across diverse health care settings. The Liehr and Smith chapter illuminates how story theory can be incorporated into both research and practice to address complex issues of caring for people as they face health challenges. It shows the importance of paying attention to essential elements that may appear as individual when in fact there are interrelated factors; when considered in relation to the other, individual issues transform to the whole. In particular, Liehr and Smith's chapter provides insight into the dynamic nature of the nurse–person relationship and the emerging story as a way to assist an individual to resolve the health challenge. In day-to-day nursing practice, nurses can get caught up in the task of providing education as they follow physician's orders. They often miss the opportunity to truly *listen* to the person, and therefore, they are unaware of how the human story can give direction to an education session. When a nurse cares enough to attentively listen as the storyteller dialogues about a health challenge, the nurse–patient relationship intensifies, and the situation transforms from a static to a dynamic process in which the person's preferences and wishes are considered as they mutually set goals.

4. **What is the relevance of this work to the future of the discipline and profession of nursing, health care professionals, and health care in general?**

As health care reform evolves, nurses will be called upon to establish innovative methods aimed at reducing health care costs while maintaining quality of care. As the discipline of nursing moves toward educating more doctoral prepared nurses, it is important to acknowledge the power of stories for scholarship—education, research, and practice. The American Association of Colleges of Nursing has proposed a change in education that is required for nurses who will be practicing at the most advanced level of nursing. By 2015, the Doctor of Nursing Practice (DNP) will be essential for entry into advanced nursing practice (American Association of Colleges of Nursing, 2004, 2006). The need for competent nurses to provide care in an increasingly complex health care environment has been one of the driving forces behind this initiative. Story theory, a middle-range nursing theory, provides a systematic approach for scholarly inquiry that focuses on listening and communication strategies to improve the health and well-being of the recipients of nurse caring. By coming to know "person through story," the nurse can identify what is most important and develop goals that the person is most likely to embrace, which may result in better health outcomes.

REFERENCES

American Association of Colleges of Nursing. (2004). *AACN position statement on the practice doctorate in nursing*. Retrieved March 11, 2010, from http://www.aacn.nche.edu/dnp/dnppositionstatement.htm

American Association of Colleges of Nursing. (2006). *AACN essentials of doctoral education for advanced nursing practice*. Retrieved March 11, 2010, from http://www.aacn.nche.edu/dnp/pdf/essentials.pdf

Hain, D. J., Wands, L. M., & Liehr, P. (2010). Approaches to resolve health challenges in a population of older adults undergoing hemodialysis. Online advanced release. *Research in Gerontological Nursing*, doi:10.3928/19404921-20100330-01

Liehr, P., & Smith, M. J. (2008). Story theory. In M. J. Smith & P. R. Liehr (Eds.) *Middle range theory for nursing* (2nd ed., pp. 205–224). New York: Springer.

Performance, Excellence, and Accountability in Kidney Care. (2010). *PEAK Patient tools and resources best practice #2*. Retrieved March 11, 2010, from www.kidneycarequality.com

CHAPTER 10
MODELING THE COMPLEXITY OF STORY THEORY FOR NURSING PRACTICE

- *Take a few moments to reflect on a patient experience where you came to know and understand the patient as a person. Could you relate to the theory Attentively Embracing Story?*
- *How were the concepts of connecting with self-in-relation, intentional dialogue, and creating ease part of the story?*
- *Write the story or share the story with a colleague.*
- *In the response, Dr. Hain wrote about story with respect to a patient with chronic kidney disease. What did her story reveal? How does that have relevance for the care of patients with chronic diseases?*

ELEVEN

Providing Nursing Care in a Complex Health Care Environment

Michael Bleich

The properties of complex adaptive systems are manifest in complex health care environments, the nurse–patient encounter, including diversity, nonlinearity, embeddedness, distributed control, self-organization, and interconnectedness. Caring for patients with chronic, complex, and/or multisystem disorders in complex health care organization contexts requires new nonlinear ways in relation to old, traditional, or linear ways of understanding assessed patterns, administering care, and evaluating results. Gambino (2008), a colleague of Bleich, discusses the chapter within a template that highlights complexity science concepts and presents an information scaling model outlining the micro scale in terms of cocreation and emergence as patient-centered care and complex responsive processes, the meso scale in terms of cocreation and emergence as shared governance in complex health care systems, and the macro scale as distributed control in professional practice for improved patient outcomes and professional relationships. Respondent Gambino presents nonlinear exemplars with ideas such as positive deviance (PD), debriefing sessions for nonlinear education, "afternoon tea" where care providers and patients/families meet for sharing or cocreating treatment plans, and so forth. Gambino also presents two models to assist in the teaching–learning process of nonlinear dynamics and collaborative organizational understanding for transformation.

INTRODUCTION

Health care environments may well represent the most complex settings within global institutional structures, marked as they are by overlapping and hierarchical power structures with confusion over where boundaries—and even clear delineation about who the consumer is—exist (Christensen, Bohmer, & Kenagy, 2000; Leape, 1994; Reiser, 1994). Arguably, the patient is at the center of attention, yet employers, insurers, and physicians, acting in their own role as health care resource consumers are often linked in a struggle as to who can access, purchase, and orchestrate what the patient receives in services. Nurses provide steady care in these varying settings, only occasionally distressing about their complexity and without a full appreciation that non–health care institutions are less fraught with ambiguity (Bergeson & Dean, 2006; Price, 2006).

An academic health science center, for example, does not have one mission, as do most organizations; rather, it has several that overlap and compete for resources. For example, physicians often function under their own autonomous organizational structure, which self-regulates out of hospital administrators' control, with regulators intervening to ensure that a complementary relationship exists. Some internal organizations in the health science center have labor affiliations, which introduce additional parties into work design and function. A myriad of supplier–vendor relationships also exist with a range of accountability for organizational risk. These partnerships directly influence service delivery in the orchestrated set of clinical functions that comprise a care delivery model or system. The partnerships with pharmaceutical, information system, and other equipment suppliers are examples of vendor relationships that shape how, where, and when nurses allocate their time and energy in serving their patients/clients. This is just a partial description of the confounding care environment where many nurses practice (Tan, Wen, & Awad, 2005).

The notions aligned with complexity science should not be confused with these complicated or "complex" care environments described earlier. Too often, the concepts associated with complexity science are substituted for the "complicated" patient, or for "confusing" delivery systems, leading health care providers to believe that they understand or are expert in complexity science concepts. Further, the oft-substituted term of "chaos theory" resonates with nurses who try to bring order to care delivery organizational dynamics. This chapter differentiates complexity science concepts from the conundrum of clinical care experiences and discusses decision making and new ways for nurses to conceptually engage in providing care within complex work settings. It will distinguish how complexity science is a way of thinking, behaving, and approaching care. The chapter also posits the enormous influence that nurses can have when using complexity science approaches to influence clinical outcomes from a patient-centered approach (Holden, 2005; Westley, Zimmerman, & Patton, 2006).

The Dominant Paradigm of Reductionism

In examining traditional approaches to health care, the medical model has the longest history, with its centuries-old focus on the naming, treatment, and evaluation of clinical problems. Simplistically, this naming is reflected in a disease taxonomy, where a set of symptoms exists, the disease is named, treatments and cures are tested, and disease resolution comes about as a result of scientific treatments. Science that follows this ideology has traditionally been based on a dominant paradigm of problem solving through reductionism, namely, taking apart the various pieces in order to understand the pathology at the cellular, organ, or body systems level (Engel, 1977; Rosenberg & Golden, 1992). Today, the ideology extends further, incorporating new knowledge about the human genome and the intricacies of precursors to pre-cellular development. In recent years, physician leaders have come to acknowledge that reductionism as a single worldview falls short of meeting patient needs, yet reimbursement structures are built on and perpetuate the specialization of services that arise from reductionism. However far it extends, this paradigm has worked in so many cases that most clinicians—regardless of the discipline—view the world through this lens. Outside of health care, but parallel with developing technologies, the Industrial Revolution gave further credence to reductionism, as inventions were created to meet societal needs and organizations used the assembly line to connect the pieces in rapid-fire processing to mass-produce transportation, telecommunication devices, and other utilitarian products. The model was marked by low cost and high volume production, creating profits for business owners. The model infiltrated health care.

Attempts were also made to standardize nursing and clinical care delivery using the reductionism model. In the history of asylums, the practice of following highly structured routines and procedures was part of the therapeutic milieu, where a high volume of "social deviants" were to be restored to health and societal functioning (Katz, 1986; Rothman, 1977). In spite of poor success, these beliefs permeated institutional structures, and nurses were taught to strictly adhere to the structured regimens that came to define nursing work. Nurses followed orders and ensured routines were met in a non-questioning culture.

To be clear, there was and is a value to reductionism, particularly when causality has been determined through the knowledge that this science has generated. For instance, we know much about the causal intricacies aligned with procedures such as chemotherapy or blood administration. The structure of administering these caustic but curative agents within the human body requires careful clinical attention, cautiously titrating dosing, and specified observations and protocol-driven responses to the patient to ensure desired outcomes. So, as linear (stepwise) as these procedures are, it cannot be denied that *complexity* as used in the common vernacular aptly describes the knowledge needed to carry out specific procedures of this sort, yet as noted previously, these complicated and prescriptive regimens are different than complexity science.

There is no question that linearity in clinical performance built on logic, analytic rules, and causal predictability has value. However, the question that nurses and other clinicians must address is this: "Are there other models or useful theories that can inform nursing work, particularly when faced with new problems such as when a disease emerges, or where a syndrome for which no identified treatment exists using the rational science model?" And, "Are there other ways of approaching the clinical relationship other than the dissection of the relational aspects of care?" The answer is yes. There are alternative ways of addressing human realities that are based on non-reductionist approaches. These approaches frame the clinical problem in terms of what I define as the *science of the interactive, interdynamic whole* or, in other terms, a *complexity science* approach to clinical care.

Complexity Science Approaches to Clinical Decision Making

Increasingly, we are aware of the frailty of the planet, of our ecosystems, and of human existence within the global world. This awareness comes from an expansion of consciousness made possible by physicists, organizational scientists, emerging biologic sciences, and technology. The instant exchange of information—the ability to experience the world as a whole—and heightened awareness of influencing dynamics between and among species and cultures—all lead to an appreciation of how embedded and interconnected we are to each other and the environment. In health care, this embedded connection links health and disease, holistic approaches to disease, environmental health determinants, and cultural and community connectedness to the lived human experience. Likewise, economic systems are linked to patient and family life, employment, and access to care. Decision making based on the concept of how systems are embedded within other decisions can reflect heightened maturity and a more expansive understanding of clinical consequences.

Clinical decision making is an important part of nursing and the clinical practice of other health disciplines (Pesut, 2008; Sepucha, Fowler, & Mulley, 2004; Tanner & Chesla, 2009). Through adept decision making, the result of critical thinking, comes critical action taken to resolve illness or otherwise positively impact patients. There has been in nursing a true desire to build science and to take a rightful place in the scientific world. As such, we have used the scientific approach, this reductionist view of science, to be seated at the table with others who have been recognized and rewarded for this approach to clinical practice.

Clinical decision making in nursing is, however, not just about the process of diagnosing and treating disease, medicine's predominant domain.

In fact, although nursing complements medicine and nurses can acquire diagnostic skills in disease identification and treatment quite readily, our fundamental focus on holistic care has never been fully actualized by this approach. As a discipline, nursing has a history of focusing on the context of care, that is, how the patient comes to require clinical care embedded in family and community, and how the patient desires wholeness—as when the biophysical end of life is near, that psychosocial and spiritual dimensions may thrive to create peace at the time of death. Nursing and caring are linked together and can never be fully understood in terms of causal dynamics, even though we have created evidence that links human touch to changed biophysical measures. Spiritual connectedness—aiding the patient to a deeper appreciation of healing, health, forgiveness, and meaning and purpose in life—is adrift if left to purely reductionist descriptions.

Nursing has gained the public's trust by acknowledging the intuitive and emotional perspective of the human experience. Even in highly bureaucratic and structured environments, we have persisted in being holistic, using lateral thinking to connect what could not otherwise be accounted for in terms of clinical algorithms, care pathways, checklists, and prescriptive methods. Instead, nursing uses *both* linear and nonlinear approaches to care delivery, realizing that approaches to care may need to surpass either/or decisions, but rather to create both/and dynamics that will deliver the most effective care (Gambino, 2008). To summarize this point, nurses may follow a clinical protocol for diabetes management, yet concurrently engage the family, focus on anxiety, and deal with environmental variables, which, when combined, creates behavioral and physiologic changes in the patient and family's best interest.

The professional nurse considers the nature of the decision needed (its predictability), the complexity of the decision (the risk level), the nature of the environment (stability factors), and the maturity of the relationships as he or she engages in the decision-making process, with its intensity and temporal nature, in order to determine whether and how to merge reductionist/linear approaches to care with complexity/nonlinear, whole systems approaches (Kinnaman & Bleich, 2004).

Complexity Concepts in Clinical Practice

In clinical practice, nurses who contextualize care bring significant value to the patient, family, and community. For instance, in the intensive care setting, the professional nurse realizes that besides the moment-by-moment physiologic care management that is linked with survival, there is the need to "see beyond" life saving. She realizes that attention to areas like skin management, family role functions, and end-of-life dynamics provides *value* to the human experience: that is, avoiding debilitating skin breakdown that could render the patient vulnerable to infection, subject the patient to contractures that would permanently reduce capacity for performing activities of daily living, or misusing family and societal resources that would permanently alter quality of life. The complexity principle that "small things can have a large impact," "the butterfly effect," comes to play in this scenario, so attending to movement and skin integrity may appear small in context with a physiologic crisis, but have a huge impact on life (Goldstein, 2008). Similarly, the large investment of time and resources on physiologic stability may have a low impact if the patient is already in an end-of-life state, such that the patient could be better served through palliative or hospice care that addresses comfort and spiritual needs.

In complexity science, the principle of abiding by and acknowledging less prescriptive, simple rules is another concept that has clinical merit (Kluger, 2008; Maeda, 2006). The idea that the patient is a social being and has a need for human connections and relationships is a simple rule that we understand as a profession, and adhere to in terms of valuing the person's dignity. In an organization where this simple rule is honored, the nurse has the flexibility to negotiate with the patient, family, and significant others the ways and means to be present in such a way that honors the patient's needs concurrently with those who love and care for the patient. In a rule-based organization, the opportunity for interconnection is overcome by strict adherence to visiting hours and other rule-derived practices. In recent times, families have been allowed to be present during cardiac-arrest situations; this would have defied norms that nurses would have, in the past, never allowed. With the concept of simple rules, principles and values surface as more important than rules that may be counterproductive to the experience of illness as part of the human experience. As suggested earlier, the nurse engages *with* the patient and others to determine the nature of the event, the complexity associated with it, the environment, and the maturity of relationships to determine how to embrace the simple rule rather than function with a mental model of absolutism.

The clinical practice environment can be enhanced with another principle, that of emergence (Stange, 2009). In this situation, the nurse does not bring the answers to the decision-making table; instead, he or she brings the knowledge, skill, experience, and perspective to cocreate and shape the answers. What might at first glance be seen as the surrender of authority and responsibility is misguided, for in this situation, self-awareness and the capacity to move fluidly as a contributor to the energy surrounding clinical events becomes a paramount contribution to care. Imagine a family conference where a determination is being made about long-term care placement. When approaching the family with a linear and predictive approach, the nurse has the "answers" to what is right and intervenes by directing the patient and family in a set direction. "What will be best for your mother is for her to have the safety of a nursing home, realizing that you cannot care for her special feeding and medical needs." Although no nurse would be that direct, the dominant logic would seep through more subtle communication. Using a complexity science approach of emergence, the nurse approaches the patient and family as part of a dialogue, observing and encouraging the family's energy and willingness to secure alternatives to care possibilities and providing guidance about ways to simplify and alter care by working with the family unit to reach an acceptable alternative solution. Knowledge of resources, regulations, ethics, and other experiences allows the nurse's professionalism to emerge as part of a less-directive process. Although in the end the mother may end up in a nursing home, the approach is fundamentally and energetically different in the approach to care, resulting in higher satisfaction with the decision.

The same clinical example also models the complexity science principle of distributed control. Distributive control is where a clinical objective is set but where all the parties engage in self-regulated behavior to achieve that goal, adapting and creating solutions to ensure balance. In the previously noted clinical scenario, the mother could have been placed in the nursing home without a sense of control or satisfaction. When the patient, her family, and clinicians work toward a common goal, in this case, the mother's safety and quality of life, then all energy is transformed to meet that goal. In a distributed control model, alternatives to long-term care could be considered, but in the event that nursing home placement is the result, the post-hospital discharge to the nursing home would ensure continued energy to moderate the experience to the patient's satisfaction. For nurses, distributed control cocreates satisfying clinical solutions that are acceptable to the patient.

In these examples, it should be apparent that complexity science is based on a systems approach to caregiving. Systems thinking is often touted in the nursing literature as a key element of nursing practice, but often it is shortchanged by an overdependence on linear rules. In systems thinking that is based on the notion of *adaptable* systems, the focus is not aligned with the idea of *linear* systems, which sees a system as a sequence of pieces and parts that function as a whole. An adaptable system has elements that change in a dynamic manner. This is an important distinction and one that is not often referenced in systems work. Rather than the Donabedian (1988) idea of structure—process—and outcome, where a defined set of resources are aligned in a sequential manner to produce a prespecified outcome and the components together are considered systemic in nature, in complexity science, an adaptable system recognizes that a predictable outcome may not exist; that there are covariables embedded within systems that create new patterns from which meaning can be derived. If we examine the disaster response from the events that followed the World Trade Towers disaster in New York City, we know that hospitals and care systems had disaster plans in place, with structures, processes, and outcomes preestablished. But the very nature of the disaster and its magnitude quickly created a scenario wherein the linear system could not prevail by itself. Instead, other elements of other systems interacted in such a manner that the health care workers' traditional roles were expanded, other systems were cocreated on the spot to facilitate communications and care, and players were drawn into the system that had not even been considered, for example, the psychoemotional dynamics of *all* providers, community members, and survivors and their families. The cohesive principles that were derived from that event demonstrated the magnitude of adaptable and interdependent systems, not with *predictive* capacity, but with *adaptive* capacity. Nursing needs *both* kinds of systems knowledge in order to produce effective professional practitioners (Trochim, Cabrera, Milstein, Gallagher, & Leischow, 2006; Wilson & Holt, 2001).

The Journey for Practice Excellence Continues

In a reformed health care environment, nursing is at the nexus of opportunity. We will be called to action in terms of our ability to create new models for care delivery and to reframe existing models to better serve a growing and changing demographic that needs—and will demand—health care in a different format than that which is currently being delivered.

It seems clear that those current delivery models, which are based on reductionist care models and often treat the patient as a moving part through a disjointed system, will implode in the near future. Whether we can reform care delivery into one with an adaptive systems capacity will define nursing's contribution to the new and emerging approach to health care, not unlike how Swiss watchmakers had to redefine their linear work at the onset of digital technology. An understanding of the two system types—linear and adaptive—is needed as an approach to personalized care *and* as a platform for creating care-delivery structures. New structures are already emerging in schools, workplaces, and other settings, such as malls. The signs around us show that consumers are reaching out to alternative sources for service, in part, because of access and the capacity of alternative sources to adapt their service delivery to perceived consumer need.

Decision making is an important part of nursing's professional portfolio. It is supported by critical thinking, but thinking and decision making must lead to critical action. A model of action, that embraces *both* linear and nonlinear science and adaptive capacity to respond to new syndromes, to human responses to aging, to the environment, and to resource limitations, and that brings technology and social networking capacity into the fold, is needed. Nurses have been hardwired to think and respond in these terms, but we have failed to fully acknowledge how our adaptive capacity and use of a single model for systems thinking has limited our potential. A full recognition of complexity science dynamics will position the discipline to embrace adaptive systems as part of our future legacy.

REFERENCES

Bergeson, S. C., & Dean, J. D. (2006). A systems approach to patient-centered care. *The Journal of the American Medical Association, 296*(23), 2848–2850.

Christensen, C. M., Bohmer, R., & Kenagy, J. (2000). Will disruptive innovations cure health care? *Harvard Business Review, 78*(5), 102–112.

Donabedian, A. (1988). The quality of care: How can it be assessed? *Journal of the American Medical Association, 260*(12), 1743–1748.

Engel, G. L. (1977). The need for a new medical model: A challenge for bio-medicine. *Science, 196*(4286), 129–136.

Gambino, M. L. (2008). Complexity and nursing theory: A seismic shift. In C. Lindberg, S. Nash, & C. Lindberg (Eds.), *On the edge: Nursing in the age of complexity* (pp. 49–71). Bordertown, NJ: Plexus Press.

Goldstein, J. (2008). Resource guide and glossary for nonlinear/complex systems terms. In C. Lindberg, S. Nash, & C. Lindberg (Eds.), *On the edge: Nursing in the age of complexity* (pp. 263–290). Bordertown, NJ: Plexus Press.

Holden, L. M. (2005). Complex adaptive systems: Concept analysis. *Journal of Advanced Nursing, 52*(6), 651–657.

Katz, M. B. (1986). *In the shadow of the poorhouse: A social history of welfare in America*. New York: Basic Books.

Kinnaman, M. L., & Bleich, M. (2004). Collaboration: Aligning resources to create and sustain partnerships. *Journal of Professional Nursing, 20*(5), 310–322.

Kluger, J. (2008). *Simplexity: Why simple things become complex (and how complex things can be made simple)*. New York: Hyperion.

Leape, L. L. (1994). Error in medicine. *Journal of the American Medical Association, 272*(23), 1851–1857.

Maeda, J. (2006). *The laws of simplicity*. Cambridge, MA: MIT Press.

Pesut, D. J. (2008). Thoughts on thinking with complexity in mind. In C. Lindberg, S. Nash, & C. Lindberg (Eds.), *On the edge: Nursing in the age of complexity* (pp. 211–238). Bordertown, NJ: Plexus Press.

Price, B. (2006). Exploring person-centered care. *Nursing Standard, 20*(50), 49–56.

Reiser, S. J. (1994). The ethical life of health care organizations. *The Hastings Center Report, 24*(6), 28–35.

Rosenberg, C. E., & Golden, J. L. (Eds.). (1992). *Framing disease: Studies in cultural history*. New Brunswick, NJ: Rutgers University Press.

Rothman, D. J. (1971). *The discovery of the asylum: Social order and disorder in the new republic*. Boston-Toronto: Little, Brown.

Sepucha, K. R., Fowler, F. J., Jr., & Mulley, A. G., Jr. (2004). Policy support for patient-centered care: The need for measurable improvements in decision quality. *Health Affairs, 7*, 54–62.

Stange, K. C. (2009). A science of connectedness. *Annals of Family Medicine, 7*(5), 387–395.

Tan, J., Wen, J., & Awad, N. (2005). Health care and services delivery as complex adaptive systems. *Communications of the ACM, 48*(5), 36–44.

Tanner, C. A., & Chesla, C. A. (2009). *Expertise in nursing practice: Caring, clinical judgment and ethics*. New York: Springer.

Trochim, W. M., Cabrera, D. A., Milstein, B., Gallagher, R. S., & Leischow, S. J. (2006). Practical challenges of systems thinking and modeling in public health. *American Journal of Public Health, 96*(3), 538–546.

Westley, F., Zimmerman, B., & Patton, M. Q. (2006). *Getting to maybe: How the world is changed.* Mississauga, ON: Random House Canada.

Wilson, T., & Holt, T. (2001). Complexity and clinical care. *British Medical Journal, 323*, 685–688.

Response to Chapter 11

Mary Gambino

CONCEPTS PRESENTED IN THIS CHAPTER THAT ARE RELEVANT TO THE COMPLEX SYSTEMS SEEN IN NURSING, MEDICINE, AND/OR OTHER EXAMPLES OF HEALTH CARE PRACTICE

Cocreate	Simple rules
Emergence	Dialogue
Adaptable/adaptability	Distributed control
Nonlinear	Small changes can have a large impact
Embedded	Interconnected/network

HOW INFORMATION CAN BE COLLECTED ON THESE CONCEPTS. INCLUDE INFORMATION SCALING FROM MICRO THROUGH MESO TO MACRO

SCALE	CONCEPT	RELATED TO	SEEN IN ASSESSMENT DATA
Micro	Cocreate	Patient-centered care	Listening to and including the patient in care plan development
	Emergence	Complex responsive processes	Dialog/collaboration among health care professionals and the patient
			Interdisciplinary patient rounds with all voices heard/respected
Meso	Cocreate & emergence	Shared governance	Meeting minutes reflecting decisions about professional practice
Macro	Distributed control	Professional practice	Use of simple rules and provision of boundaries to foster professional practice
			Patient satisfaction
			Health care professional satisfaction
			Improved outcomes of care
			Improved relationships among health care professionals

MODELS FOR NURSING AND HEALTH CARE PRACTICE

FIGURE 11.1 Nonlinear and Unpredictable Self-Organizing Dynamics

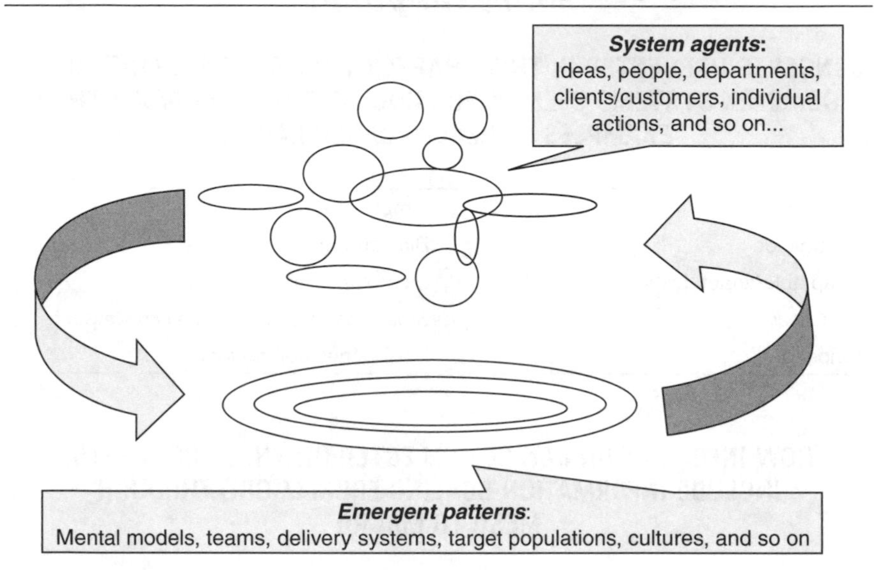

FIGURE 11.2 Organizational Context

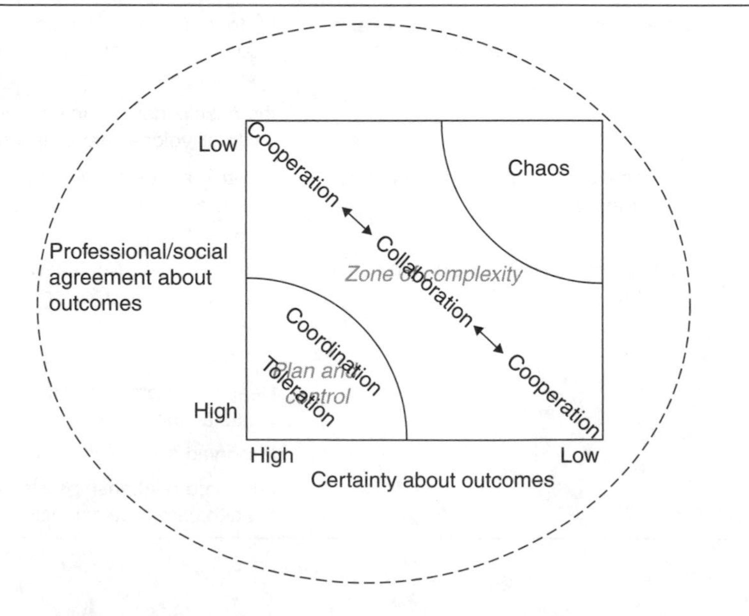

EXAMPLES OF HOW THESE CONCEPTS AND/OR METHODS CAN BE APPLIED IN PRACTICE

1. Picard, C., & Henneman, E. A. (2007). Theory-guided evidence-based reflective practice: An orientation to education for quality care. *Nursing Science Quarterly, 20*(1), 39–42.
   a. Provides an excellent and poignant example of Mike's last sentence on page 6.
      i. It is 8 AM in the intensive care unit. I (Beth) am a critical care nurse with 27 years' experience. I work a few days a month in this unit. I am standing at the bedside of a 40-year-old woman (Mary) who is dying of respiratory failure. This patient has cystic fibrosis and it is nothing short of a miracle and her mother's care that she is still alive today. The goal for the day is to transfer Mary onto a bed that continually rotates in order to improve her oxygenation levels. Although I am aware that there is some research supporting the use of this type of bed, I am dismayed that it is being used in this case. The thought of transferring this fragile, dying woman is just too much—it is ridiculous, actually. I wonder if anyone has spoken to the family and discussed end of life issues with them. I am fairly confident that they have not. I spend a moment shaking my head to myself and move on to administer the 20 medications that my dying patient is ordered to receive at 8 AM. (Picard & Henneman, 2007, p. 40)
2. Implementation of a shared governance model is an example of cocreation and emergence.
3. Appreciative inquiry is an example of emergence and cocreation.
4. Use of simulation with debriefing is an example of how to teach in a nonlinear way, as well as prepare a nursing workforce that is adaptable.
5. According to Sue Nash, PhD, RN (personal communication), M.D. Anderson in Houston has implemented "afternoon tea," where care providers and families meet in a comfortable setting to share tea and discuss and formulate treatment plans—cocreation.
6. Transforming Care at the Bedside (TCAB) initiatives, as well as a number of Methicillin-resistant Staphylococcus aureus (MRSA) initiatives, have used distributed control, cocreation, and emergence to create solutions to clinical problems that fit the context.
7. Positive deviance is an example of dialogue, cocreation, emergence, and distributed control for solving intractable problems like MRSA.
8. Curley, C., McEachern, J. E., & Speroff, T. (1998). A firm trial of interdisciplinary rounds on the inpatient medical wards. *Medical Care, 36*(Suppl.), AS4–A12.
   a. These authors demonstrated improved outcomes by implementing dialogue and distributed control rather than clinical paths.

9. Adaptability is demonstrated in TCAB initiatives using group huddles to redistribute workload on nursing units as needed.

CONCLUSION: EVALUATION OF THE USEFULNESS OF THE CHAPTER CONCEPTS AND METHODS FOR NURSING AND HEALTH CARE PRACTICE

This chapter provides an outstanding explanation and examples of the linear/nonlinear dynamic that is so needed in nursing care delivery. It also challenges the reader to adapt by incorporating this model with the linear model as a way to achieve the nursing legacy that is possible. These ideas are applicable to care at the individual, unit, system, and community/population level as well as to wellness and illness.

CHAPTER 11
PROVIDING NURSING CARE IN A COMPLEX HEALTH CARE ENVIRONMENT

- *What concepts can you identify from this chapter on complexity science and health care that have relevance to you either in education or nursing and health care practice?*
- *Reflect on and determine the micro, meso, and macro scaling variances in your health care or educational environments.*
- *What is the meaning of embeddedness in an organization?*
- *Can you share an exemplar about complexity theory and evidence-based practice from your experience?*
- *Gambino concludes that this chapter provides an outstanding explanation of linear and nonlinear dynamical models in nursing care delivery. What do those ideas or concepts mean to you in nursing and health care or educational systems?*

REFERENCES

Curley, C., McEachern, J., & Speroff, T. (1998). A firm trial of interdisciplinary rounds on the inpatient medical wards. *Medical Care, 36*(Suppl.), AS4–12.

Kinnaman, M. L., & Bleich, M. R. (2004). Collaboration: Aligning resources to create and sustain partnerships. *Journal of Professional Nursing, 20*, 310–322.

Olson, E. E., & Eoyang, G. H. (2001). *Facilitating organization change: Lessons from complexity science*. San Francisco, CA: Jossey-Bass/Pfeiffer.

Picard, C., & Henneman, E. A. (2007). Theory-guided evidence-based reflective practice: An orientation to education for quality care. *Nursing Science Quarterly, 20*(1), 39–42.

TWELVE

The Complexity of Diabetes and the Caring Role of the Nurse Practitioner

Jane Faith Kapustin

Patients with chronic disease such as diabetes mellitus face problems associated with self-management, and most experience difficulty with reaching optimal glucose control. Navigating our convoluted health care system is difficult, and the patient can become overwhelmed, further increasing the risks for complications. Nurse practitioners (NPs) can achieve good health outcomes by partnering with patients to reach realistic goals and adherence patterns. Because over 50% of complications associated with chronic diseases can be prevented, NPs need to emphasize basic health promotion measures. Applying the caring theory, NPs can provide care coordination in complex systems and guide patients with diabetes along the continuum toward health. Respondent Coffman discusses the movement of how the complexity of diabetes impacts a patient from the cellular level to the individual and family and the community levels. The complexity of diabetes begins with glycemic and dietary control for patients, but the lifestyle, habits of thought, opinions, knowledge, and experience impact not only the individual but also the family and community. Patients, family members, and community public health officials, and especially nurses, need to stay open to the constant flow-through of energy and information for creative emergence toward healing and health of individuals with diabetes and other chronic illness, and facilitate the transformation of health care policy.

INTRODUCTION

Complexity and Scope

Over 90 million people in the United States live with chronic disease such as heart disease, diabetes, cancer, lung disease, and stroke. Approximately 70% of all deaths are related to chronic disease, and many of them are preventable. As our population ages, the risk of developing chronic disease escalates as well (Centers for Disease Control and Prevention [CDC], 2007; Wilper, 2008). If current trends continue, our health care capacity could be depleted in the not too distant future. Up to 70% of the nation's $2 trillion spent annually for medical care is attributed to management of chronic disease (McGinnis, Williams-Russo, & Knickman, 2002).

Type 2 diabetes mellitus (T2DM) is a prime example of a complex chronic disease that can be attributed to the effects of poor lifestyle choices and is preventable in the majority of cases. Consider the epidemiology of T2DM: 23.6 million people in the Unites States have DM, representing about 8% of the population. Of the 23.6 million cases, approximately 17.9 million are diagnosed, 5.7 million undiagnosed, and another 57 million have "prediabetes." The trends indicate that more people are being diagnosed, and the sad fact is that most of the cases could be prevented or at least managed with conservative lifestyle interventions. T2DM is associated with risk factors such as having a family history of T2DM, being overweight or obese, belonging to certain ethnic groups such as African American or Hispanic, having gestational diabetes, delivering a baby over 9 pounds, and leading a sedentary lifestyle (American Diabetes Association [ADA], 2010).

Having prediabetes is a strong risk factor associated with the eventual development of T2DM. Because of the insidious nature of the disorder, at the time of diagnosis, many patients with T2DM have already endured hyperglycemia and its related health consequences for several years. Macrovascular and microvascular complications that damage major organs and the microvasculature are prevalent among patients with T2DM, and they escalate with poor overall glycemic control. Insulin resistance and progressively declining beta-cell function that occurs over time for most patients with diabetes further decrease optimal management (ADA, 2010; National Institute of Diabetes and Digestive and Kidney Diseases [NIDDK], 2008).

Complex Treatment Options

Treatment for T2DM has become increasingly complex because there are numerous comorbidities associated with T2DM, and there are many treatments available to manage both. Research has demonstrated conclusively that tight control of blood glucose (BG) significantly reduces the incidence and cost of complications of diabetes (Montori & Fernandez-Balsells, 2009).

As a result, almost half of Americans diagnosed with diabetes are prescribed glucose meters to monitor their BG levels at home. The value of this approach is well known to be limited by an extraordinarily high rate of patient noncompliance with glucose testing, where patients test less often than prescribed (ADA, 2010).

The introduction of lifestyle changes is the mainstay of therapy to initially manage diabetes. Research demonstrates that patients benefit when they are actively involved in their treatment plan, and they need to be taught self-management techniques such as measuring glucose levels, choosing appropriate food, exercising daily, and maintaining other health parameters such as blood pressure or cholesterol levels (ADA, 2010; Donohue-Porter, 2009). Many patients will need to consider implementing major lifestyle modifications to manage T2DM effectively, such as losing weight and making better food choices. Limiting intake of high carbohydrates, fats, excess calories, alcohol, and other substances will need to be adopted, and patients need to be counseled to stop smoking (ADA, 2010).

Patients need to learn how to perform self-monitoring of glucose. Most patients will need to check glucose levels two or three times a day, and some patients' routines require more frequent monitoring. Patients may need to dose their mealtime insulin based on their glucose readings, and they may also need to test more often if they are using an insulin pump. Because patients with T2DM have problems with controlling postprandial glucose excursions, they will often test 2 hr after a meal for a complete picture of their glucose control (Danne, 2009).

Pharmacological management of T2DM is equally complex. Because of the differing target sites for antidiabetic medications, patients with T2DM can take single oral medication, multiple types of oral medications, insulin, other injected medications, or combinations of oral medications with insulin. Often, the patient will be prescribed an expensive, complex, pharmacologic regimen that could involve multiple oral medications taken several times a day along with as many as two to four injections of glucagon-like peptide-1 (GLP-1) agonists and/or insulin. GLP-1 agents mimic the effects of incretins that are produced in the gut and are glucose dependent to lower glucose levels. Because patients with T2DM frequently have other comorbidities such as hypertension and hyperlipidemia, this pharmacologic routine will most likely include many other agents to control them (ADA, 2010; Joslin Diabetes Center, 2009).

Because diabetes is a complex and progressive disease, most patients will require multiple pharmacologic agents to reach glycemic control. Patients should anticipate that treatment will be augmented periodically in spite of close compliance with the treatment plan. This fact can be very frustrating for patients as they watch their glucose readings deteriorate, so it is important to remind them periodically (ADA, 2010).

To illustrate this phenomenon, consider a typical medication regimen for a newly diagnosed patient, Robert, whose glucose readings were ranging from 150 to 225 mg/dL fasting in the morning (normal glucose 70–100 mg/dL). Robert's glycosylated hemoglobin A1C (Hb A1C), a measure of long-term glucose control, is 8.2% (normal is less than 7%). Because these readings are elevated but not exceedingly high, oral medication can be started before insulin is tried. Robert is a 52-year-old obese male (body mass index of 33—obese range) with no prior medical history of significance.

In addition to recommending lifestyle modifications such as weight loss, exercise, carbohydrate counting, and smoking cessation, the NP can choose from numerous medications for monotherapy such as either a sulfonylurea (to increase secretion of insulin) or biguinide (metformin, to decrease gluconeogenesis) (see Table 12.1).

The NP chooses metformin because Robert presents with other signs and symptoms of insulin resistance or metabolic syndrome: waist circumference is 48 inches, BP is 156/88 mmHg, triglycerides are 250 mg/dL, HDL is 30 mg/dL, and his fasting glucose is elevated (National Cholesterol Education Program [NCEP], 2002) (Exhibit 12.1). Because it has associated gastrointestinal side effects, metformin is titrated slowly at 500 mg per day for 3 days, 500 mg twice daily for a week, and then eventually increased to 2,000 mg daily to achieve glucose goals.

Even after adhering to the newly developed treatment plan, Robert begins to experience glucose level elevations within a year. A second medication that targets a different site can be added such as a thiazolidinedione (TZD; used to enhance insulin sensitivity), a GLP-1, or alpha-glucosidase inhibitor (to decrease carbohydrate absorption in the intestines). Because Roberts Hb A1C is still below 8%, oral medications can be continued before insulin is added as long as the goal of 7% can still be achieved after new medications are added. At some point, Robert may need insulin to keep his glucose within goal.

To add another layer to the complexity of T2DM management, highly motivated patients can also be successfully managed with insulin pump therapy. If patients use an insulin pump, they could be testing glucose more than six times daily and bolusing insulin to correct or anticipate glucose excursions based on carbohydrate counts. Many patients are now being placed on continuous glucose monitoring insulin pump devices that require more dexterity and skill. Both forms of delivery are very complex and involve extensive patient commitment as well as extensive teaching prior to use (Danne, 2009).

After considering the many adjustments in lifestyle, the adoption of close monitoring of glucose levels, and participating in multitiered treatments, it is obvious that chronic disease management is quite complicated and demands major lifestyle changes to cope. It is not surprising that many patients become overwhelmed with their new situation (Solowiejczyk, 2010). In fact, the rate of noncompliance with self-management (Egede, Ellis, & Grubaugh, 2009) and the prevalence of depression among patients with diabetes are quite high. The rate of depression among patients with T2DM ranges from 21% to 27% (Anderson, Freedland, Clouse, & Lustman, 2001; Schwartz et al., 2009;

TABLE 12.1 Oral Medications to Treat Type 2 Diabetes Mellitus

CLASS	GENERIC NAME	BRAND NAME	ACTION	COMMENTS
Sulfonylurea	Glyburide second generation	Micronase Diabeta Glynase	Promotes more insulin	Take one to two times a day. May cause lows
	Glipizide second generation	Glucotrol Glucotrol XL	Promotes more insulin	Take two times a day or once with XL. May cause lows
	Glimepiride third-generation	Amaryl	Promotes more insulin	Take once a day. May cause lows
Biguanides	Metformin	Glucophage	Reduces glucose production in liver and promotes insulin sensitivity	May reduce effectiveness of birth control pills. Not used with renal problems Check creatinine clearance if over 65 years of age
Alpha-glucosidase inhibitors	Acarbose Miglitol	Precose Glyset	Reduces glucose absorption in GI tract	May have side effects in the GI tract
Thiazolidinediones	Rosiglitazone Pioglitazone	Avandia Actos	Increases insulin sensitivity	Not used with heart failure. Check liver enzymes as directed
Meglitinides	Repaglinide	Prandin	Promotes more insulin	May cause lows
Glucagon-like peptide-1 (GLP-1) or DPP-4 inhibitor	Exenatide Sitagliptin	Byetta Januvia	Reduces mealtime glucose by mimicking action of incretin	May help with weight loss

Exhibit 12.1

Criteria for Clinical Diagnosis of Metabolic Syndrome

1. Abdominal obesity: waist circumference > 35 inches in women or 40 inches in men
2. Triglycerides > 150 mg/dL
3. HDL cholesterol < 50 mg/dL in women or < 40 mg/dL in men
4. Blood pressure ≥ 130/85 mm Hg
5. Fasting plasma glucose > 110 mg/dL

National Cholesterol Education Program. (2002). Adult Treatment Panel III final report. *Circulation, 106*, 3143–3421.

Wild, Roglic, Green, Sicree, & King, 2004), and depressive symptoms have been linked with worsened BG levels and diabetes complications such as heart disease (deGroot et al., 2010).

Cost Considerations

To further add to the complexity and other problems associated with management of T2DM, the cost of treating this chronic disease is staggering. Results from the National Diabetes Economic and Barometer Study estimates costs of over $217 billion annually for undiagnosed prediabetes and diabetes due to direct medical costs as well as lost productivity (Dall et al., 2010). As the diabetes epidemic continues to expand, so does the cost of diabetes associated with diagnosis and treatment of the disease and its associated complications. Medical expenses are about 2.3 times higher for people with diabetes than for those without, and approximately 10% of health care dollars are directed to diabetes and its complications.

Huang, Basu, O'Grady, and Capretta (2009) predict that costs to treat diabetes will triple by 2034 and could reach $336 billion as prevalence of obesity and T2DM continue to rise. The increasing diabetes epidemic places more burden on our health care system already stretched by costs associated with our aging baby boomer generation, and significant changes in public and private strategies are needed in order to help address the rise in prevalence of diabetes.

The ever-increasing numbers of children and adolescents as well as young adults being diagnosed with T2DM is a critical indicator of the future crisis facing our nation. Considering that long-term complications associated with diabetes increase as the patient ages, the consequences of younger patients now dealing with the disease will compound the current health care system burden.

An Ounce of Prevention Is Worth a Pound of Cure

The complexity of this preventable chronic disease needs to be dissected to consider the many factors that influence health. The five determinants of population health are highly interrelated and have major influence on overall health status. However, they are not restricted to genetics or access to medical care. The other determinants, namely social circumstances, environmental exposure, and behavioral/lifestyle choices, have far more potential for influencing health and require more attention if the trajectory of chronic disease will be altered (CDC, 2007; McGinnis et al., 2002; National Association of Chronic Disease Directors, 2007). For example, genetic conditions are often underlying for some chronic diseases, yet they are not evident until social or environmental factors exert their influences. As an example of obesity, many people possess the genetic code that predisposes them to being overweight or obese, yet not all of them become overweight or obese. Social or environmental influences that favor obesity, such as learning unhealthy eating patterns or leading a sedentary lifestyle, are often the triggers that lead to weight gain.

Chronic diseases such as diabetes, heart disease, and cancer have genetic components that are often expressed later in life. These chronic disorders are multifactorial in nature such that numerous interactive genes are implicated for their expressions. In the case of obesity, up to two-thirds of the risk may be genetic in nature, but the controllable lifestyle factors signal the expression of obesity (Bamshad, 2005). Promising research in genomics will lead to better understanding of this powerful interaction so that specific interventions can be developed to alter an inevitable sequence of disease expression. The Institute of Medicine of the National Academies (2009) has proposed that genomics is one of the key areas for public health education, and the Centers for Disease Control and Prevention (CDC) Office of Genomics and Disease Prevention (2009) maintains that the integration of genetics and public health holds the future for managing population health (Byck, Lemke, Lea, Brennan & Beckett, 2005; CDC, 2009; Collins, Marks & Koplan, 2009).

So what can be done to interrupt this trajectory for economic as well as human tragedy? The greatest intervention to improve our health is to change personal behavior to prevent disease development. Considering that obesity and physical inactivity are the main drivers of causes of premature death, efforts need to focus on the unhealthy behavior (Schroeder, 2007). Obesity is now considered a major preventable cause of death and disability in the United States, yet sustained solutions remain elusive.

Daily choices of food intake, exercise, gender, substance use, and maintenance of weight influence our development of chronic disease. The evidence base supports the association between being overweight/obese and leading a sedentary lifestyle with a higher risk of T2DM and heart disease (ADA, 2010). Tobacco use is still considered the leading single factor associated with death and disability, and taken all together, behavior choices account for close to one million deaths annually in the United States. These are all regarded as "early" deaths that are associated with significant disease burden. This health determinant is also the most controllable, responsible for the highest number of preventable diseases (McGinnis et al., 2002; National Association of Chronic Disease Directors, 2007).

So how can the NP respond? It is necessary to consider that our health care system is designed for disease management, not health promotion or disease prevention. Technology and other innovations are produced to support eradication of disease, and there are few resources left over to support prevention programs. Adequate disease prevention systems would mandate major shifts in funding that support policies to promote health (Forrest & Riley, 2004; McGinnis et al., 2002). Because primary prevention targets children, emphasis needs to be placed on interventions that are not designed to be delivered in the typical examination room. The great majority of children will benefit from preventive efforts that focus on environmental forces, school interventions, and social policies that are designed to improve the health of children. These interventions are complex and have multifactorial root causes. This makes them difficult to deliver using one-dimensional modalities (Forrest & Riley, 2004).

Using childhood obesity as an example, policies that affect school meals and snacks, physical education programs, and health and nutrition courses to guide healthy choices require complex coordination of efforts. Because they are not intended to affect the health of a child in isolation, the necessary coordination of community and home interventions to provide safe places to play outdoors and to help limit the amount of television viewing and video games must also be orchestrated carefully. Supporting the family and providing social learning theory concepts that suggest the way to affect the child's behavior is by first influencing the parent's behavior need to be included. Multiple systems must be activated to ensure a thorough approach to the problem (Davis et al., 2007; Forrest & Riley, 2004).

Taking an active role in policy and legislative initiatives are paramount for NPs, and it is important to point out that many prevention initiatives are outside of the traditional health care field. Exerting legislative influence on promoting use of community centers for exercise can be accomplished through forming coalitions with the public health sector and other interest groups (McGinnis et al., 2002; National Association of Chronic Disease Directors, 2007).

Prevention is quite complex and mandates multiple upstream efforts to limit chronic disease. This is contrasted with medical treatments that often focus on single symptoms or presentations (McGinnis et al., 2004). In the case of T2DM, treatment is focused on managing blood sugar with multiple treatments; however, the *prevention* of diabetes involves multiple upstream efforts such as working with schools to promote healthy eating, physical activity, and limiting unhealthy snacks and sodas. Other diabetes prevention solutions require coordinated community efforts to ensure adequate outdoor space for exercise, funding of public service announcements, and diet and nutrition counseling. Naturally, these efforts require major and sometimes very different funding streams, so the costs associated with these interventions are hard to calculate and can be quite high (McGinnis et al., 2004; Teutsch, 2006; Tuomilehto, Lindstrom, Eriksson, Valle, Hamalainen, et al., 2001).

Prevention interventions often involve changes to behavior and value systems. Another major barrier to providing adequate prevention programs involves lack of financial support. The traditional fee-for-service system of our health care infrastructure is not designed to support prevention or health promotion. Consider the interventions proposed for diabetes prevention. There are no billable procedures for school, home, or community interventions that address prevention of childhood obesity. And our current public health system is not adequate to absorb these initiatives without restructuring funding sources to support them (Davis et al., 2007; McGinnis et al., 2004).

New technologies and innovative approaches to health care that are specifically designed for disease prevention will have a prominent role in our future health care system. The quest for new medications that treat diabetes more effectively or prevent the onset of type 1 diabetes continues, and major gains have been made (Goldman, Baoping, Bhattacharya, Garber, Hurd, et al., 2005). Obesity management technology should improve to offer more treatment options. The focus on obesity prevention needs to be augmented and directed to our children. It is far better to prevent obesity than to treat it and its associated complications such as diabetes and heart disease.

In addition to developing comprehensive public health strategies such as public service announcements, school policies, community policies, and health department interventions, employers must share a role in promoting healthy choices. In addition, the employer has an economic incentive to reduce the burden of obesity, because health care costs for obese individuals due to weight-related problems are significantly higher than for nonobese individuals (Blackburn, 2008). Financial incentives and disincentives through health insurance premiums, deductibles, bonus payments, and other benefits will center on intensive efforts to encourage healthy lifestyles, weight management, and exercise (Hodge, Garcia, & Shah, 2008; Watts & Segal, 2009).

The fact is that maintaining a healthy lifestyle works to prevent disease. Consider the findings from the Diabetes Prevention Program Study (Knowler, Barrett-Connor, Fowler, et al., 2002) that established the significance of lifestyle modification in reducing risk for developing T2DM. In this large clinical trial, individuals with glucose intolerance were randomized to three groups. The study group received intensive lifestyle counseling, another group was given metformin, and the control group received a placebo. The lifestyle modification group reduced their risk of developing T2DM by 58%, significantly more than the metformin group (31% reduction). This trial demonstrated that lifestyle interventions alone will significantly reduce the risk of chronic disease development.

The cost savings associated with this intervention cannot be understated. However, as the researchers pointed out, the costs associated with the rigorous intervention of dietitian visits, exercise coaches, exercise equipment, counseling sessions, and health club memberships made this somewhat impractical for general population use. Research for more practical ways to implement lifestyle modifications along with ensuring sustainability should be the next step. But the implications for health care providers are tremendous and further support health promotion efforts.

Health Promotion and the Theory of Primary Caring

The original role of the NP was created years ago to provide the best of nursing: direct patient care for patients in need of primary care. Pioneer NPs realized that providing holistic care as the aspects of primary care are addressed was a clear advantage that nursing added. The role of the NP has been founded on the theory of caring, and that philosophy still guides NP interactions with patients, families, and communities (Hagedorn & Quinn, 2004).

Hagedorn's Theory of Primary Caring (Hagedorn & Quinn, 2004) includes five domains. *Connection* is the relationship-based caring engagement that NPs form with patients, family, and community. The effectiveness of NPs is accomplished with authentic listening, serving with respect, and applying compassion. *Consistency* refers to the importance of providing competent care within a health care environment that is constant and dependable. The NP is bound by ethical care that is delivered in confidence with respect and compassion to demonstrate *commitment* to the patient. *Community* is the context within the NP provides care that meets the patient's needs. Access to care and culturally competent care are the main principles under which the NP operates to be sure patients obtain the services necessary to achieve optimal health outcomes. Finally, *change* explains the introduction of innovation and how the NP strives to facilitate meeting patient needs. The NP is expected to assume the role of change agent to support the patient and community health initiatives.

Consider the following scenario to illustrate the Theory of Primary Caring (Hagedorn & Quinn, 2004) from the NP perspective—complexity in the health care arena: Tonya is a 47-year-old female diagnosed with T2DM 12 years ago. Her medical history includes hypertension, hyperlipidemia, neuropathy, and past history of heroin abuse (quit 5 years ago, managed on a methadone program). She has demonstrated poor glycemic control for at least 2 years, often presenting with home glucose levels in the 300 to 400 mg/dL range and Hb A1C levels ranging from 9.9% to 12%. She insists that she takes her insulin every day and seems highly distressed when her levels remain consistently elevated. Today, her finger-stick is 424 mg/dL, HbA1c is 10.2%, and urine ketones are positive (ketones are formed when glucose levels are markedly elevated and inadequate insulin is available).

The NP engages in an in-depth conversation with Tonya about her care routines. Tonya checks her finger-sticks at least daily but admits that she does not always check during the day because she has trouble affording the test strips. She is somewhat evasive about her responses and insists that she is managing financially. She takes her Humalog Mix insulin (75/25% insulin) as instructed and is still taking seven other oral medications to manage her diabetes and comorbidities.

The NP orders 10 units of Humalog, a rapid-acting analog insulin product, given to Tonya now in the clinic, offers Tonya a bottle of water to drink, and rechecks her glucose later. The NP is using the insulin as a test dose to make sure the patient still responds readily, and Tonya's follow-up finger-stick drops to 232 mg/dL about 90 minutes later. After more discussion, the NP gently informs Tonya that her levels would not be as high if she was indeed taking her insulin as ordered and prompts her to discuss in further detail. Tearfully, Tonya admits that she shares her insulin with her elderly mother because they cannot always afford full doses of insulin, especially at the end of the month before their next Social Security check arrives. She only takes her full doses of insulin after she is sure there is enough for her mother as well. Even though they are both covered by insurance for their medications, the family does not receive adequate amounts to last a full month. Admitting that she is embarrassed, Tonya has a low health literacy level and has trouble navigating the health care system. Tonya was unsuccessful with placing phone calls to her insurance to correct the situation, so the NP refers Tonya to the center's diabetes educator who assists her with contacting her insurance to obtain adequate amounts of insulin and to arrange for home delivery of her medications. On the next return visit 3 months later, Tonya achieves much improved glycemic control and has engaged in more consistent glucose monitoring. She is ready for the next steps of adopting more health promotion interventions such as focusing on carbohydrate counting, weight loss, and exercise.

In this case, the NP presents in a nonchallenging way to become a trusted health care provider before the patient confides in her. She confronts the patient gently and with respect, so that the patient will open up and tell her what the issues are. By displaying a caring attitude, the NP and Tonya can partner to meet her health care needs and achieve more optimal health outcomes. The medical, social, and environmental factors are highly complex in this case as the patient becomes overwhelmed with managing her own care (Donohue-Porter, 2009). Providing culturally competent and compassionate care is the goal of the NP as holistic care is offered. Forming a close, trusting relationship is the hallmark of the NP–patient relationship. Through the Theory of Caring, the NP can attend to much more than the medical needs of the patient (Hagedorn & Quinn, 2004).

Implications for NPs

It is important that we strive to achieve parity for preventive services. We can approach this mission by supporting the Department of Health and Human Services' Healthy People 2020 health objectives (Collins et al., 2009) and the strategies outlined in the newly minted health care reform bill passed in March 2010. Medical care and treatment alone is inadequate: Health outcomes can best be achieved by combining forces with preventive systems. Policies on education, community health, and other social determinants help shape the health of our nation, and cannot be underestimated. There is no better time than now for NPs to get involved in local, state, and federal initiatives to fund prevention programs. With the focus on health care reform, legislators and policy makers are looking for solutions to our health care dilemmas, and NPs have been targeted for assuming roles in the health promotion arena. The role of the NP is a perfect fit because health promotion and disease prevention is a basic underpinning of education. Therefore, NPs are well positioned to assume the role of the medical home leader for many populations (Kapustin, 2010a). NPs will need to continue to strive for independent practice status in all states and for credentialing parity for third-party insurers so all patients can receive adequate care. The Health Care Reform Act will help make that a reality.

NPs should spend the greatest effort in this area to make a substantive difference in the rates of chronic disease prevalence. Opportunities for health promotion are unlimited (Kapustin, 2010a). The NP role was created over 40 years ago to help fill the primary care gap in rural areas, and health promotion and disease prevention were emphasized. Today's NP curriculum still reflects those same approaches. The American Academy of Nurse Practitioners' (2009) Web site indicates that NPs perform a variety of services, but they "focus on health promotion, disease prevention, health education, and counseling." This role is also stressed by the American Colleges of Nurse Practitioners (2009), and "disease prevention" is specifically emphasized.

Acknowledging that prevention does not exist solely in the clinical arena, NPs need to get accustomed to coordinating health promotion interventions in the community, schools, homes, and the workplace. Too many resources are directed to medical care, and NPs are masters at providing disease prevention strategies. Every health care encounter needs to include elements of advising patients to modify unhealthy lifestyles and to engage in more healthy practices, and if disconnects are discovered, NPs can use their expertise to investigate and resolve barriers. NPs can exert influence at the local, state, and national levels to stop the cycle of chronic disease, especially as the measures in the Health Care Reform Bill are operationalized (Kapustin, 2010a, 2010b). Through the application of the Theory of Primary Caring, NPs can continue their mission to provide excellent primary care.

REFERENCES

American Academy of Nurse Practitioners. (2009). *FAQs: about NPs*. Retrieved January 22, 2010, from http://www.aanp.org/NR/rdonlyres/67BE3A60-6E44-42DF-9008-DF7C1F0955F7/0/09FAQsWhatIsAnNP.pdf

American College of Nurse Practitioners. (2009). *What is a nurse practitioner?* Retrieved January 22, 2010, from http://www.acnpweb.org/i4a/pages/index.cfm?pageid=3479

American Diabetes Association. (2010). Clinical practice recommendation. *Diabetes Care, 33*(Suppl. 1), S1–S96.

Anderson, R. J., Freedland, K., Clouse, R., & Lustman, P. J. (2001). The prevalence of comorbid depression in adults with diabetes: A meta-analysis. *Diabetes Care, 24,* 106–110.

Bamshad, M. (2005). Genetic influences on health. Does race matter? *Journal of the American Medical Association, 294*(8), 937–946.

Blackburn, D. (2008). Making obesity everybody's business: What is the employer's role. *Obesity Management, 8,* 169–175.

Byck, G., Lemke, A. A., Lea, D., Brennan, M. B., & Beckett, D. (2005). *Public health genomics: Assessing the state's role*. Academy Health Meeting, Boston, MA, p. 22, abstract 3739.

Centers for Disease Control and Prevention. (2007). Department of Health and Human Services. Retrieved January 22, 2010, from http://www.cdc.gov/nccdphp/overview.htm#2

Centers for Disease Control and Prevention. (2009). *Public health genomics*. Retrieved January 22, 2010, from http://www.cdc.gov/genomics/

Collins, J. L., Marks, J. S., & Koplan, J. P. (2009). Chronic disease prevention and control: Coming of age at the Centers for Disease and Control and Prevention. *Preventing Chronic Disease. Public Health Research Practice and Policy, 6*(3), 1–6.

Dall, T. M., Zhang, Y., Chen, Y. J., Quick, W. W., Yang, W. G., & Fogli, J. (2010). The economic burden of diabetes. *Health Affairs, 29*(2), 297–303.

Danne, T. (2009). Self-monitoring of blood glucose (SMBG): From theory to clinical practice. *Medscape CME Diabetes & Endocrinology*. Retrieved January 22, 2010, from http://cme.medscape.com/viewarticle/709187

Davis, M. M., Gance-Cleveland, B., Hassink, S., Johnson, R., Paradis, G., & Resnicow, K. (2007). Recommendations for prevention of childhood obesity. *Pediatrics, 120,* S229–S253.

deGroot, M., Kushnick, M., Doyle, T., Merrill, J., McGlynn, M., Shubrook, J., et al. (2010). Depression among adults with diabetes: Prevalence, impact, and treatment options. *Diabetes Spectrum, 23*(1), 15–18.

Donohue-Porter, P. (2009). Diabetes education nurses handle complexity with care. *Nursing 2009, 39*(1, Suppl.), 14–15.

Egede, L. Ellis, C., & Grubaugh, A. (2009). The effect of depression on self-care behaviors and quality of care in a national sample of adults with diabetes. *General Hospital Psychiatry, 31,* 422–427.

Forrest, C. B., & Riley, A. W. (2004). Childhood origins of adult health: A basis for life-course health policy. *Health Affairs, 23*(5), 155–164.

Goldman, D. P., Baoping, S., Bhattacharya, J., et al. (2005). Consequences of health trends and medical innovation for the future elderly. *Health Affairs—Web Exclusive, W5,* R5–R17.

Hagedorn, S., & Quinn, A. A. (2004). Theory-based nurse practitioner practice: Caring in action. *Medscape: Topics in Advanced Practice Nursing eJournal*. Retrieved January 22, 2010, from http://www.medscape.com/viewarticle/496718

Herman, W. H., Brandle, M., Zhang, P., et al. (2003). Costs associated with the primary prevention of type 2 diabetes mellitus in the diabetes prevention program. *Diabetes Care, 6*(1), 36–47.

Hodge, J. G., Garcia, A. M., & Shah, S. (2008). Legal themes concerning obesity regulation in the United States: Theory and practice. *Australia and New Zealand Health Policy, 5,* 5–14.

Huang, E. S., Basu, A., O'Grady, M., & Capretta, J. C. (2009). Projecting the future diabetes population size and related costs for the US. *Diabetes Care, 32*(12), 2225–2229.

Institute of Medicine of the National Academies. (2009). *Roundtable on translating genomic-based research for health.* Retrieved January 22, 2010, from http://www.iom.edu/CMS/3740/44443.aspx

Joslin Diabetes Center. (2009). *Guidelines for pharmacological management of type 2 diabetes.* Retrieved January 24, 2010, from http://www.joslin.org/docs/Clinical-Guidelines-for-Pharmacological-Management-of-Type2-Diabetes.pdf

Kapustin, J. (2010a). Chronic disease across the lifespan. *Journal for Nurse Practitioners, 6*(1), 16–26.

Kapustin, J. (2010b). Guest Editorial: Chronic disease prevention. *Journal for Nurse Practitioners, 6*(1), 6.

Knowler, W. C., Barrett-Connor, E., Fowler, S. E., Hamman, R. F., Lachin, J. M., et al. (2002).Reduction in the incidence of type 2 diabetes with lifestyle intervention or metformin. Diabetes Prevention Program Research Group. *New England Journal of Medicine, 346*(6), 393–403.

McGinnis, J. M., Williams-Russo, P., & Knickman, J. R. (2002). The case for more active policy attention to health promotion. *Health Affairs, 21*(2), 78–93.

Montori, V. M., & Fernandez-Balsells, M. (2009). Glycemic control in type 2 diabetes: Time for an evidence-based about-face? *Annals of Internal Medicine*, 150, 803–808.

National Cholesterol Education Program. (2002). Adult Treatment Panel III final report. *Circulation, 106,* 3143–3421.

National Institute of Diabetes and Digestive and Kidney Diseases, National Institutes of Health. (2008). *Insulin resistance and pre-diabetes* (NIH Publication No. 09-4893). Retrieved from http://diabetes.niddk.nih.gov/DM/pubs/insulinresistance/

Schroeder, S. A. (2007). We can do better—improving the health of the American people. *New England Journal of Medicine, 357,* 1221–1228.

Schwartz, F., Ruhil, A., Denham, S., Shubrook, J., Simpson, C., & Boyd, S. L. (2009). High self-reported prevalence of diabetes mellitus, heart disease, and stroke in eleven counties in rural Appalachian Ohio. *Journal of Rural Health, 25*, 225–229.

Solowiejczyk, J. (2010). Diabetes and depression: Some thoughts to think about. *Diabetes Spectrum, 23*(1), 11–15.

Teutsch, S. (2006). Cost-effectiveness of prevention. *Medscape Public Health and Prevention, 4*(2). Retrieved from http://www.medscape.com/viewarticle/540199

Tuomilehto, J., Lindstrom, J., Eriksson, J. G., et al. (2001). Prevention of type 2 diabetes mellitus by changes in lifestyle among subjects with impaired glucose tolerance. *The New England Journal of Medicine, 344*(18), 1343–1350.

Watts, J. M., & Segal, L. (2009). Market failure, policy failure and other distortions in chronic disease markets. *BMC Health Service Research, 9,* 102. Retrieved January 24, 2010, from http://www.pubmedcentral.nih.gov/articlerender.fcgi?artid=2704185

Wild, S., Roglic, G., Green, A., Sicree, R., & King, H. (2004). Global prevalence of diabetes: Estimates for the year 2000 and projections for 2030. *Diabetes Care, 27,* 1047–1053.

Wilper, A. P., Woolhandler, S., Lasser, K. E., McCormick, D., Bor, D. H., & Himmelstein, D. U. (2008). A national study of chronic disease prevalence and access to care in uninsured U.S. adults. *Annals of Internal Medicine, 149,* 170–176.

Response to Chapter 12

Sherrilyn Coffman

In their prologue, Davidson, Ray, and Turkel suggest that examination of parts and relationships can provide information that guides action. Furthermore, examination at different scales—from very small (cells) to larger (individuals and families), to very large (communities)—reveals repeating patterns that best reflect the larger essence of the whole. Kapustin's review on the complexity of diabetes covers all aspects and implications of diabetes from the cellular to the community level. This reflection on Kapustin's work is organized according to these levels.

DIABETES: THE CELLULAR LEVEL

Adaptability is one of the defining characteristics of health, indicating the functional responsiveness of body systems. Normal glucose control by the healthy pancreas creates rhythmic behavior and variability, indicating physiologic complexity. In chaos theory, this behavior is labeled "chaotic dynamics" (West, 2006). West argues that disease indicates the *loss of complexity* at a physiological level. Illness occurs when the body's regulatory systems work overtime, indicating the breakdown of one of the many regulatory mechanisms. In diabetes, the regulation of glucose metabolism is affected, either from a lack of insulin (type 1) or a resistance to insulin (type 2).

It is the treatment of diabetes that meets the definition of complexity. Kapustin describes the pharmacological management for type 2 diabetes, which is becoming increasingly more complex as clinicians attempt to mimic normal physiology. Specifically, the goal of therapy is to mimic the normal variability in glucose levels seen in healthy individuals. Insulin may be delivered by an insulin pump, with variable doses administered in response to carbohydrate intake or changes in body physiology at different times of day. GLP-1 agents mimic the effects of incretins produced in the intestine, which lower glucose levels and affect weight gain. Most type 2 patients are prescribed multiple pharmacologic agents to reach glycemic control, based on the varying effects of different medications.

A dilemma for "diabetes care provider" is the determination of how to measure "glycemic control." Average BG or hemoglobin A1C is one indicator. However, clinicians are beginning to look more at the variability of BG levels over a 24 hour period (and not just average glucose levels) as a measure of short-term control. Use of continuous BG monitors is becoming more routine in type 1 diabetes. This practice is directly in line with chaos theory. West (2006) argues that fluctuations in physiologic phenomena (e.g., BG levels) contain valuable information that can be used to diagnose and treat patients with chronic illness.

DIABETES: THE INDIVIDUAL AND FAMILY LEVEL

Complexity scientists maintain that living life requires more than logic and linear reasoning. "It requires an aesthetic sense—a feeling for what fits, what is in harmony" (Briggs & Peat, 1999, p. 8). When a family member is diagnosed with diabetes, this aesthetic sense is lost, chaos ensues, and the entire family's lifestyle requires reorganization. Self-organization is the goal. Chaos theory suggests that a family system that self-organizes out of chaos must stay open to a constant flow-through of energy and material to survive. Habits of thought, opinions, and experiences are often required to change, to move individuals out of negative feedback loops that are preventing change. Only by letting go of old habits, can a creative self-organization become possible (Briggs & Peat, 1999).

Self-organizing systems possess the ability to reorganize themselves to deal with new information. They are adaptive and resilient rather than rigid and stable (Wheatley, 1999). Successful control of a chronic condition like diabetes requires creativity and self-organization. Fear of failure or of making mistakes, an obsession with control, and a refusal to move out of one's comfort zone can block the action of creativity in one's life. This requires a shift in psychological perspective (Briggs & Peat, 1999). Changing eating and exercise habits, learning new tasks such as insulin injections, and learning when to ask for help are activities in diabetes self-management that require a new motivation and sense of purpose. The individual and family must be open to the possibility that they can create a new lifestyle and accomplish the new tasks asked of them. This requires opening oneself to uncertainty and discovering new ways to live one's life.

Complexity science focuses on wholeness, providing additional insight into the management of a chronic disorder such as diabetes. Wheatley notes that "we can discover the whole by going further into its parts" (1999, p. 141), but attention must be held at both levels simultaneously. The dynamic interplay between diet, exercise, and insulin or diabetes medications requires this perspective. The relationship between these factors is more important than any one factor by itself, and can change glycemic control. Furthermore, the same input at different times can create different effects. Self-management of diabetes is a complex process, a dynamic state at the edge of chaos, requiring self-organization to create order or stability through constant change. Thus, management of diabetes is homeodynamic and irreversible.

Diabetes management is a labor-intensive process that involves the patient, family, and diabetes care provider. Registered nurses, certified diabetes educators, and NPs assist patients and families in learning diabetes self-management. Nurses who function from a caring perspective can be especially effective. "Nursing as caring" is a philosophy of care that echoes *emergence*, a concept at the heart of complexity science (Coffman, 2010). Emergence suggests that the universe and people in it are ceaselessly creative. With creative agency come values, meaning, and doing, all of which are as real in the universe as particles in motion (Kauffman, 2008). Because of this ceaseless creativity, the nurse, patient, and family member approach each situation with openness, not knowing exactly what the outcome will be. In addition to relaying knowledge, the nurse attempts to instill courage, faith, and hope. As Kierkegaard says, "we live our lives forward." Kauffman further describes this approach to the patient and family in the statement "We live our lives forward into mystery, and do so with faith and courage" (2008, p. xi). With trust in the nurse as teacher and coach, patients and families are more open to learning and more motivated to make substantial life changes.

DIABETES: THE COMMUNITY LEVEL

Kapustin describes a variety of social issues and legislative challenges that impact the health of society, including health promotion, disease prevention, and care of groups of patients with diabetes. Complexity science teaches us that emergent organizations are more likely to be successful in creating solutions to broad problems, for example, the new health care reform act signed into law in 2010 in the United States. An emergent system is truly greater than the sum of its parts (Marion, 1999). Individuals within a system who understand the "whole" of the organization or community, including the legislative community have a grasp of a larger picture than is known to the parts. The emergent system is far more functional than its constituents and can carry and use a great deal of information. As a system grows, it becomes stronger, and can maintain its integrity in the face of adversity. A system always communicates back to its individual members and maintains an influence on their behavior, thus further ensuring its vitality and transformation.

As nurses become more involved in social and legislative initiatives, they can examine the systems (organizations and groups) in which they participate. The new worldview of complexity has demolished the concept that individuals are separate and unconnected. Wheatley (1999) recommends that effective team members learn how to facilitate process and become savvy about how to foster relationships. Members need to become better at listening, conversing, and respecting one another's uniqueness, because these abilities are important ways to strengthen relationships. Organizational power is relational, and when power is shared through participative management and self-managed teams, positive creative power emerges.

REFERENCES

Briggs, J., & Peat, D. (1999). *Seven life lessons of chaos: Spiritual wisdom from the science of change*. New York: Harper Perennial.

Coffman, S. (2010). Marilyn Anne Ray: Theory of bureaucratic caring (Chapter 8). In M. Alligood & A. Marriner-Tomey (Eds.), *Nursing theorists and their work* (7th ed., pp. 113–136). St. Louis, MO: Mosby.

Kauffman, S. A. (2008). *Reinventing the sacred: A new view of science, reason and religion*. New York: Perseus Books Group.

Marion, R. (1999). *The edge of organization: Chaos and complexity theories of formal social systems*. Thousand Oaks, CA: Sage.

West, B. (2006). *Where medicine went wrong: Rediscovering the path to complexity*. Hackensack, NJ: World Scientific.

Wheatley, M. (1999). *Leadership and the new science: Discovering order in a chaotic world*. San Francisco: Berrett-Koehler.

CHAPTER 12
THE COMPLEXITY OF DIABETES AND THE CARING ROLE OF THE NURSE PRACTITIONER

- *How does the information discussed relevant to the complexity of diabetes relate to your area of practice?*
- *Is there another chronic disease that can be viewed in relation to complexity science?*
- *How can you apply the Theory of Primary Caring Nurse Practitioners to your role as a health care provider?*
- *In the response, Dr. Coffman discussed the dilemma related to glycemic control. Is this an issue for you as a practitioner or for the patient population you serve?*

THIRTEEN

Lessons in Complexity Science: Preparing Student Nurses for Practice in Complex Health Care Systems

Claire E. Lindberg

Novice nurses usually begin their careers in practice situations dominated by large, complex, highly technological, and error-prone health care systems. Within these systems, nurses, including novices, are expected to care simultaneously for multiple patients who suffer from chronic and/or complex conditions, while coping with demands and stressors inherent in the system. In addition, the nurse's role has become increasingly complex, and now includes multidisciplinary team coordination and information management. Although these conditions are challenging for all nurses, the high attrition rate among nurses in the first years of practice suggests that new nurses are not well prepared for practice in today's complex health care systems. Integration of complexity science theoretical content and classroom management strategies into the nursing curriculum will provide future nurses with a better foundation for entering today's practice world. This chapter begins by outlining the forces in today's health care system that make it necessary for nursing students to graduate with a strong foundation in complexity science. Suggested complexity science content and placement in the curriculum is described. The principles of complexity science, particularly those related to complex adaptive systems that are at work in the classroom setting, are explained, such as the coexistence of order and disorder, adaptability, distributed control, and embeddedness. Classroom methods that increase both experiential, learning and student participation in the learning experience, and that are consistent with complexity science principles are described, along with illustrative examples. An analysis of barriers to implementation is provided in order to assist educators in preparing to transition their curricula. The Institute of Medicine (IOM) has proposed a health care system that is based on principles derived from complexity science. This chapter assists nursing educators and administrators to envision new ways of preparing nursing students to become future leaders in such a system. Respondent Dittman focuses her remarks on the nature of complexity and complexity sciences in nursing education. She discusses how complexity science is used in the academic setting to bridge the disparity between complexity science and practice. Dittman supports Lindberg in the way in which the nurse educator must shift her focus from decontextualized knowledge to teaching that is understood within nursing and health care situations where clinical reasoning, multiple ways of thinking, and cocreative emergence are included. Transformational change by application of complexity science principles minimizes the education–practice gap.

INTRODUCTION

Challenges to Caring in Complex Health Care Systems

The complexity of the 21st century American health care system is well recognized. Factors contributing to this complexity include increasing numbers of elderly patients and patients with chronic and/or complex multisystem disorders, cultural diversity among both patients and providers, and the number and variety of different providers involved in the care of each patient. In addition to these human factors, the highly technological nature of health care services, the increased use of complex pharmacological agents and other treatment routines, and the increasing demands of and oversight by payers and regulatory agencies contribute to this complexity. Although many of these things have led to positive changes, there have also been costs and negative outcomes. At the turn of this century, the IOM pointed out that the growing complexity of the health care system contributes to medical errors and other adverse patient events (IOM, 2000). According to the IOM, although errors may appear to be the fault of an individual, they are more likely to be related to system properties. Complex systems are highly prone to accidents and errors because "one component of the system can interact with multiple other components, sometimes in unexpected or invisible ways" (IOM, 2000, p. 58). Not only does the IOM describe the complexity of the health care system, it explicitly advocates using a complex systems framework to create new systems focused on patient safety and quality improvement (2001).

As the health care system has become more complex so has the role of the nurse. In addition to increasing in complexity, the nursing role has widened in scope. Benner, Sutphen, Leonard, and Day (2010) note that many responsibilities formerly assumed by physicians have, over the last half century or so, moved into the domain of nursing. This shift has required acquisition of knowledge and skills beyond those that nursing care traditionally required.

In addition, because of their direct care role, nurses are held responsible for actions of other individuals who are not present at the bedside. For example, although a medication error may originate with the prescribing physician or the dispensing pharmacist, the nurse remains ultimately responsible for a medication that she administers.

Nurses, now identified as key agents of patient care quality and safety, (IOM, 2004) are said to function at the "sharp end" of health care. The term "sharp end" refers to the level of an organization where the effects of action are directly experienced and therefore, where errors may have critical consequences. For example, many people are involved in producing, packaging, prescribing, and delivering a medication for pain to the hospital pharmacy. These actions take place at the "blunt end," that is, they are removed from direct interaction with the patient. The sharp end, wherein the nurse administers the medication, is where the immediate effect on the patient occurs.

Not only does health care now include increasingly more complicated and sophisticated treatment routines, the patients themselves present with more difficult conditions. Advances in knowledge, nutrition, and care technology have allowed Americans to live longer with more chronic conditions. It is not uncommon for a single patient to be elderly and to also have multiple coexisting chronic conditions such as diabetes, hypertension, cardiovascular disease, autoimmune disease, and lung disease. Social, economic, and family situations may further complicate the course of the patient's illness. Such highly complex patients are the norm in most care settings today. In most settings, a nurse cares for several patients simultaneously, each with a different set of medical conditions and unique needs, adding multiple layers of complexity to her work. The challenges to the nurse's clinical thinking and judgment, prioritizing, and time management grow exponentially as more patients are added to her caseload.

Providing safe nursing care involves continuously assessing, monitoring, and diagnosing patient conditions, setting and resetting priorities throughout the shift, managing highly technological monitoring and treatment processes, and juggling competing tasks, all the while coping with multiple interruptions and changing conditions (Ebright, Patterson, Chalko, & Render, 2003). As the providers at the hub of patient care, nurses receive information from multiple sources and utilize that information in direct patient care. In addition, as part of their coordinator role, nurses filter and distribute information to appropriate others, including other professionals, the patient, and family members. Nurses must also be skilled at helping patients and family members understand information, events, and actions, and at assisting others to achieve consensus. Thus, among the most important skills a nurse must have are those related to managing interpersonal relationships.

In addition to meeting the growing challenges of patient care and interpersonal relationships, nurses are increasingly required to meet system-related demands such as documentation, learning new protocols and procedures, adapting to changes in unit staff, and to management and administrative changes on the unit and in other departments. Krichbaum et al. (2007) call attention to the fact that 40% of nurses' time is taken up by system demands that have been superimposed on patient care duties. Thus, in a phenomenon that these authors call "complexity compression," nurses are expected to provide care to increasingly complex patients, in an increasingly complex system, in a shortened time period. Because it is complex and error fraught, the current health care system is a difficult and stressful place in which to work.

It is clear that these workplace factors are critical causes of nursing workforce instability and nursing shortages. Attrition rates among nurses are unacceptably high and are highest within the first 2 years after graduation. Studies show attrition rates ranging from 13% to 30% in the first year, to 57% within the first 2 years (Bowles & Candela, 2005; Kovner et al., 2007). Another study showed that 37% of new nurses intended to look for a new position within 1 year and 41.5% stated they would leave their job if they had the freedom to do so (Kovner et al., 2007). These statistics may reflect, at least in part, inadequate preparation for the responsibilities and stressors of the role (PricewaterhouseCoopers' Health Research Institute, 2007).

It is also obvious that novice nurses are facing mighty challenges in adjusting to their new role. Novice nurses state they feel frustrated and pressured, have unreasonable workloads, and have difficulty in getting their work done, all of which lead to job-related stress, burnout, feelings of inadequacy, and intention to change jobs or leave nursing (Kovner et al., 2007; Pellico, Brewer, & Kovner, 2009). Many of the factors that cause dissatisfaction and lead to attrition among new nurses appear to be related to the complex conditions in the workplace, for which these novices are not adequately prepared.

Recognizing nursing's challenges in the age of complexity, Benner et al. (2010) make the point that nurses must graduate with multiple skill sets that allow them to function safely and efficiently in a demanding workplace. In their Carnegie Foundation study, these experts conclude that newly graduated nurses are not adequately prepared for the demands of today's practice or for today's complex practice settings, citing "significant gaps" (p. 4) between practice demands and the preparation nursing students receive. The IOM cites the lack of adequate preparation for practice in a complex environment among newly licensed nurses as one factor in the patient safety crisis (IOM, 2004). Accordingly, considering the current health care milieu in which novice nurses will practice, redesigning nursing education is urgent.

Educational System Redesign: A Call for Action

There are many issues to be addressed in the redesign of nursing education. Benner et al. (2010) call for a curriculum that "integrates knowledge, skilled know-how, and ethical comportment" along with the attitudes and skills required for lifelong learning (p. 10). Their 2010 book, *Educating Nurses: A Call for Radical Transformation*, advocates experiential learning including progressive, guided clinical experiences in simulation labs and later with patients in clinical settings. However, the primary focus of that book is on preparing students to care for individual patients. It does not discuss how to adequately prepare nurses to function within the complexity of our current health care system. Thus, it does not address the 40% of nurses' time that is spent dealing with system issues and problems. This highlights the fact that it is necessary to convey to nursing students an understanding of complexity science theory and its applications in order to better prepare them to adapt to and function safely within the health care system of today and of the future.

The American Association of Colleges of Nursing (AACN), the professional organization for nursing schools and nursing faculty, recognizes the necessity of including content on complexity science in the nursing curriculum at both the undergraduate and graduate levels. Early in the 21st century, AACN recognized the importance of complexity science knowledge for nurses by including it as a key curricular element in the clinical nurse leader curriculum (AACN, 2003). In the most recent set of guidelines for baccalaureate nursing education, AACN lists "Basic Organizational and Systems Leadership for Quality Care and Patient Safety" as the second of nine essential elements to be included in the undergraduate nursing curriculum. Within these essential elements, it is specifically listed that baccalaureate nurses should graduate with an understanding of "complex organizational systems" and that program content should include complexity science (AACN, 2008). Similarly, *Draft: The Essentials of Masters Education in Nursing* (AACN, 2009) places this content in Essential II as required leadership content.

Despite these important recommendations, inclusion of complexity science in the nursing curriculum is in a nascent stage, perhaps because this content is not well understood by nurse educators. Most nurse educators were schooled in traditional nursing theory and have not been exposed to complexity science content. It is this author's experience that there also is resistance to inclusion of complexity science content, as well. When speaking about complexity science and nursing curriculum, I have been challenged by faculty who heatedly contested this content's inclusion in the curriculum. Others expressed concern that inclusion of complexity science content would necessitate displacing other curricular content, most specifically traditional nursing theories. However, as the above discussion suggests, nurses must understand complexity science in order to provide safe and high quality health care in our complex health care system and to survive as care providers within that system. This requires inclusion in the nursing school curriculum.

Spinning the Thread: Integrating Complexity Science in the Curriculum

Spider webs are an often used analogy in complexity science. Spider webs are delicate structures, carefully woven with attention to detail in the creation and placement of each thread. Thus, the spider creates a complex, interconnected structure from discrete threads that allows for communication between parts and supports the integrity of the mission of the whole. Inclusion of complexity science content into the nursing curriculum similarly requires understanding of mission, careful planning, and attention to detail to ensure that content is woven into the curriculum in a clear, developmental manner that supports development and preparation of these future practitioners.

Teaching complexity science to nursing students requires attention to both content and process. A complete content integration discussion is beyond the scope of this chapter, but some major points are presented below. It is suggested that complexity science be woven as a thread throughout the curriculum; however, certain topics lend themselves to certain locations in the typical nursing undergraduate or graduate curriculum. Although each educational program has a slightly different curricular structure, there are many commonalities, a fact that is both grounded in tradition and necessitated by the NCLEX-RN and other exams required for licensure or advanced practice certification. Despite differences, general principles for introducing complexity science into the curriculum can be presented. Table 13.1 provides a brief summary of suggested complexity science content, along with a few key references for each topic and suggestions for content placement in a generic nursing curriculum. Educators should be able to extrapolate from this table ideas on content placement in their own undergraduate or graduate nursing education programs.

Teaching the basics of complexity science theory early in the curriculum allows students to build upon and utilize this knowledge as they progress throughout the curriculum. Integrating content in both classroom and clinical courses encourages the development of a sense of salience regarding use of complexity science in clinical and interpersonal interactions. Having a "sense of salience" implies the ability to readily apply information and insights from complexity science to various clinical and classroom situations. This ability will only develop with repetition, opportunities for application to practice and discussion of clinical experiences. For example, understanding physiology through a complexity science lens provides a basis for understanding the normal physiologic fluctuations in beat-to-beat variability of the heart and thus leads to an understanding the implications of heart rate variability loss in common disease states encountered in both pathophysiology and clinical courses. In the latter courses, students should learn to understand the implications of heart rate variability in such conditions as congestive heart failure, which manifests decreased heart rate variability and ventricular fibrillation, in which heart rate variability is greatly increased (West, 2006). On the other side of life, students studying maternal–fetal physiology and/or care of the laboring woman will learn that loss of fetal heart variability is an indicator of impending fetal distress, a further illustration of physiologic variability principles. Likewise, gait variability loss is a manifestation of certain neurological disorders, such as Parkinson's disease (West, 2006) and loss of hormone secretion variability is related to endocrine disorders

TABLE 13.1 Suggested Complexity Science Content, Placement in the Typical Curriculum, and Resources

CONTENT	CURRICULAR PLACEMENT	RESOURCES*
Complexity Science Theory	Nursing Theory Course Introductory Nursing Course Role Course	1. Lindberg, Nash, and Lindberg (2008). 2. Zimmerman et al. (2001).
Applications of Complexity Science to Human Health and Disease	Biology Courses Physiology/ Pathophysiology Courses Clinical Courses	1. West (2006). 2. Lindberg et al. (2008).
Research Compatible With a Complexity Science Perspective	Research or Research Methods Course Evidence-Based Practice Course	1. Lindberg et al. (2008).
Health Care System as Complex Adaptive System Applications of Complexity Science in Leadership Patient Safety and Quality Nursing Care in the Complex Health Care System	Nursing Leadership Course Health Care Issues Course Role Course	1. Zimmerman et al. (2001). 2. Lindberg et al. (2008). 3. Ebright et al. (2003).
Complexity Science Perspective on Human Interactions	Human Development Course Communications Course Family Course Role Course	1. Lindberg et al. (2008). 2. Doane & Varcoe (2005). 3. Stacey (2003).

* See Reference List for full citations.

such as diabetes. Students caring for patients with the above conditions can be asked to reflect back on what complexity science tells us about these and other conditions and then apply that knowledge to the patient at hand. This process allows students to apply their knowledge of complexity science to sharpen their assessment, critical thinking, and planning skills.

Benner et al. state that formal education about relational practices is neglected in favor of technical training, cataloguing, and classification activities in current nursing education programs (2010), despite the fact that relational aspects of practice have increased in importance as the nursing role has evolved to emphasize multidisciplinary team coordination and patient advocacy. Communication with patients and families has also become more complex as patients enter care "sicker and leave sicker" than in the past. Thus understanding and applying the principles behind complex human relationships is essential. The Theory of Complex Responsive Processes is a newer complexity science-related theory, which attempts to describe and explain human interactions (Stacey, 2003). This theory explains human relationships as evolving social processes wherein the brain simultaneously shapes, and is shaped by, iterative and nonlinear interactions between individuals and groups. Thus, repetitive communication and interaction patterns are formed over time. These patterns are lasting but may be transformed by a change initiated by any one of the participants. Examining and working with this aspect of complexity science helps students to understand constructive and destructive patterns in communication and their effect on relationships and behavior. Initially students should learn to identify, assess, and monitor these iterative patterns in their own communications and among other individuals in their immediate social circle. Reflection on their own experiences with communication and interaction is an important early experience for students to engage in. Experimenting with changing their own communication patterns in positive ways and observing how this effects dyadic and group interactions can help students prepare to participate in more complex interactions in the clinical setting. As students progress multiple opportunities will present themselves for pattern analysis. With experience, students can learn to influence interactions in the clinical setting by controlling their own actions and reactions, thus becoming adept facilitators of communication among individuals and groups. Students should progress to applying the skills they develop in this area to client and family situations and to group and interdisciplinary team interactions. The ability to understand and manage dyadic and group communication patterns is a critical skill in intercultural communications, dealing with difficult personalities, and in conflict situations.

A third essential content area relates to complexity-inspired leadership principles. Most nursing curricula have a course that aims to impart the basics of leadership theory, and most undergraduate programs provide some type of leadership experience, such as acting as "team leader" for a group of students on a clinical unit or possibly shadowing a nursing manager or administrator. Begun (2008) posits a broader definition of nursing leadership, wherein nurses in any and all positions have the potential to be leaders in the service of advancing the profession. Both of these leadership interpretations are important. Introducing nursing students to the basic concepts of complexity science theory up front in the curriculum provides students with the needed material for observation and analysis of leadership systems later in their program. In the leadership segment of their program, students can be introduced to specific applications of complexity science as it is applied to management, including the nine "emerging and connected organizational and leadership principles" described by Zimmerman, Lindberg, and Plsek (2001, p. 23). Undergraduate students can be encouraged to analyze and critique the leadership styles of managers they observe and to observe the complex power dynamics that exist on a nursing unit (or within any other hierarchical organization). If their education includes experiencing the role of nurse leader by directly stepping into that role, students can be asked to consciously implement some complexity science-inspired leadership principles and analyze the outcomes. Graduate students can be encouraged to explore leadership issues more deeply by engaging in a change project in which they use complexity science-inspired principles to engage unit or other staff in their project. The importance of exposure to these current understandings of leadership and management cannot be underestimated and is consistent with the IOM's vision for redesign of the health care system incorporating complexity science leadership principles (IOM, 2001). Students of today must learn these principles as preparation for becoming the leaders of tomorrow.

Strengthening the Web: Complexity-Inspired Classroom Management

The spider knows that it is not enough to create a minimalistic web structure but that, in order to maximize communication among the varying parts of the web and ensure the web's integrity and strength, each thread must be linked to multiple other threads. Similarly, integrating complexity science content into the curriculum is necessary but not sufficient to prepare tomorrow's nurses to practice in complex health care systems. Redesign of nursing education must include process as well as content. Benner et al. (2010) point out that as the amount of required content in nursing education programs increases, nurse educators increasingly rely on lectures presented with projected outlines and focus on teaching classification systems and taxonomies. These authors claim that this does not adequately support the development of an "attuned, response-based practice" (p. 43) and advocate for more of a focus on relational practices, which help students learn to process and use clinical information. Well-known educator Parker Palmer (1998) supports this position, stating that educators consistently underestimate students' thinking abilities and so focus on providing students with transfusions of data.

Nurse educators' tendency to act as if compelled to fill their students' brains with factual information may be related to the ultimate goal of student success on the nursing entry exam (or graduate level certification examinations). This is, of course, a valid goal because students must pass the NCLEX and other relevant exams for entry into or advancement in practice. Faculty may also want to boost the college's reputation, and thus their own self-esteem, through achievement of high "pass rates" on national examinations. Aside from issues of reputation and esteem, it is a fact that nursing schools are held hostage by regulations and accreditation standards that require schools to demonstrate achievement of certain levels on their aggregate NCLEX and/or certification scores in order to maintain program approval. The ultimate goal, however, should not be entry to practice alone. Curriculums that focus on patient safety, clinical reasoning, overall competence, and the ability and incentive to become lifelong learners should adequately prepare students for the NCLEX and/or other certification exams.

The principles of complexity science can be harnessed to foster exciting and different methods of teaching and learning in the classroom. Traditional education formats wherein educators "teach" from the front of a classroom and students are expected to sit passively and absorb material are contrary to teaching from a complexity science perspective. Table 13.2 illustrates the contrast between traditional educational philosophy and that inspired by complexity science principles.

Viewing a class through a complexity science lens, it is apparent that each class is a complex adaptive system (CAS). A CAS is composed of multiple interacting elements—in this case students and faculty members. Each of these elements is in itself a CAS. The class is itself, part of a larger whole, in this case the academic program that is part of the school. This "nested" structure illustrates a property of CAS called "embeddedness." Each level within the whole is part of and affects each other part and the whole. As a CAS, a class exhibits certain common properties. These properties include adaptable elements, nonlinearity, emergency, diversity, self-organization, the coexistence of order and disorder, and distributed control. Definitions of these and other complexity science terms may be found in Goldstein (2008). These properties should be capitalized upon to improve course procedures and classroom activities.

When speaking of students in a classroom or clinical setting, the term "adaptability" addresses the ability for learning and behavior change. Nursing

TABLE 13.2 Contrasting Approaches to Nursing Education: Traditional Versus Complexity-Inspired

TRADITIONAL EDUCATION APPROACH	COMPLEXITY-INSPIRED APPROACH
Subject-centered	Relationship-centered
Learning as linear process	Learning recognized as nonlinear
Learning is individual process	Learning is social/shared process
"One-size-fits-all" approach	Diverse methods recognize multiple learning styles and needs
Educator source of knowledge—students as recipients	Knowledge shared
Educator remains "objective"	Educator-student engagement encouraged
Educator maintains control over classroom	Control is shared
Experimentation with new techniques is risky	Experimentation is valued
Contradictory statements must be resolved	Paradoxes and contradictions are accepted

students must develop the ability to apply knowledge in both familiar and unfamiliar clinical situations in addition to learning new facts and ideas. Such methods as presenting clinical case studies and using focused questions to lead students through those cases fosters the development of clinical reasoning in a more effective way than does lecturing about a disease process or presenting incidence statistics. Contextualization of information to clinical situations supports development of clinical reasoning much more strongly than simply presenting students with facts, categorizations, and taxonomies.

Educators often speak of "building" upon knowledge from previous courses or classrooms. This implies that learning is a linear process and that "good students" are able to master a set of concepts or processes and quickly put them to use in practice. In linear learning models, mastery of a set of concepts is also assumed to prepare students for mastery of more complex related concepts. However, neither learning nor application always happens in a predictable sequence. Every experienced teacher can recall at least one student who was able to demonstrate mastery of complex material in class, as demonstrated by success on exams or graded assignments, but had greater than expected difficulty applying those concepts in a practice situation. Despite the fact that this student might be at the top of the class in grades, clinical faculty or preceptors express frustration with this student. Yet, at some point, usually when least expected, it "all comes together" and the student is able to demonstrate clinical application as well as theorctical mastery of the material. For this student, theoretical learning occurred ahead of the application skills. Another example of nonlinear learning is the student who has difficulty demonstrating mastery of the physiologic concepts related to blood sugar regulation in the classroom, yet a semester later is able to apply these concepts to a patient with diabetes, astutely recognizing hypoglycemia and articulating the appropriate nursing response. Each of these students, whose knowledge and ability to apply that knowledge appear to be "out of sync" are demonstrating nonlinear learning trajectories. Such nonlinear trajectories may be more common than is generally recognized.

Faculty and academic administration often speak of racial, religious, ethnic and/or cultural diversity among students, so diversity within a class may seem obvious to faculty members. Diversity also manifests itself in terms of such attributes as learning styles and abilities, aptitude, knowledge background, attitudes, morals, and interest/eagerness to learn. The presence of such diversity should affect an educator's presentation of course-related material, leading them to employ varied educational methods. Including a mixture of individual and group exercises and/or both verbal/auditory and kinesthetic classroom experiences is likely to result in more meaningful learning for a wider range of students. Unfortunately, educators, including nurse educators, demonstrate overreliance on lecture, or lecture–discussion format in their classrooms, with "lecture–discussion" often meaning lecture with time allowed for a few questions or comments (Benner et al., 2010; Palmer, 1998).

It is now even possible to utilize prepared slides with accompanying materials (lecture notes, assignments, exam questions, etc.) from textbook companies or as downloads from various Internet sites. Although educators may feel that this standardizes material across course sections, programs, or schools, it is clear that this "one size fits all" approach in the classroom fails to recognize students' diverse learning needs and does not adequately address development of clinical reasoning skills. Teachers employing such methods will surely notice that many students are sitting with confused looks, glazed eyes (or even closed eyes) after about 20 minutes. The effectiveness of educational methods wherein educators talk at students has been the subject of much criticism by expert educators. In the words of Parker Palmer (1998):

> When we teach by dripping information into the passive forms, students who arrive in the classroom alive and well become passive consumers of knowledge and are dead on departure when they graduate. (p. 42)

Nurse educators should thus experiment with more interactive approaches to learning, some of which are discussed in the paragraphs below.

Traditional views of education treat learning as an individual process wherein the teacher controls the learning environment and each individual student is responsible for absorbing information. In this perspective, the teacher is considered the ultimate expert and students are looked on as passive recipients of knowledge. This model fails to recognize learning's social nature. People are relational beings and their relationships with others influence their behavior and opinions. Students who are engaged relationally by their teachers are more likely to pay attention and remember the lessons. The teacher who asks a student to discuss health experiences within their own family or those of a patient for whom they have cared, and then uses those stories as teaching moments has engaged not only that student, but the entire class in a different and unique way as compared to the teacher who presents a lecture on the same topic. For example, Professor L. encourages Janet to describe her grandfather's struggle with diabetes, which led to amputation of his foot. She then asks the other students in the class to relate Grandpa's experience to the principles of diabetic care. It is easy to understand how personalizing the material in this way is likely to resonate more deeply with the students than a standardized lecture on the vascular complications of diabetes.

Role playing is an exercise sometimes used in the classroom, primarily to teach communication skills. This technique should be used more often and in more varied circumstances. Role playing is indeed very helpful in teaching communication skills and also history taking and how to deal with difficult personalities. When using role playing in the classroom, it is best to give the students minimum instructions and allow them to use their imaginations.

When students are given some control over the role play exercises, some surprising and delightful things are likely to occur. I once had two graduate students role playing taking a sexual history, something even experienced nurses often feel awkward and uncomfortable doing. The student playing the elderly patient began answering the questions in ways that were quite unexpected by the interviewer or the class, claiming that she had three lovers, one who was 30 years younger than herself and then stating she had both male and female lovers. It was a great lesson for all on how to deal with unanticipated information. The students had a great deal of fun during this class period, making it more likely that the material would be retained.

Although 1:1 relationships are important in learning, current understanding of human behavior places both learning and behavior change within the context of relationship networks (Christakis & Fowler, 2009). Such networks exist or will develop within any class, and teachers should learn to use the power of the group to facilitate learning. This goes beyond encouraging students to study together for exams, although the latter is helpful to many students. Educators should learn to utilize both group dynamics and knowledge held by class members in their educational processes. Carefully crafted group exercises during class time can capitalize on the network's power as well as shared knowledge. A simple example is asking a student to describe a patient they have cared for and then making the rest of the class responsible for asking questions about the case. Encourage everyone to present additional ideas and insights, in a process similar to in-hospital grand rounds. Such case discussions can be assigned as formal, graded assignments, or treated as a more casual "on-the-spot" class exercise. A bonus to such case discussions is that each student is exposed to many more patient cases than he or she could personally interact with. Benner et al. (2010) acknowledge that nurse educators are expert at using such methods in the clinical arena, such as at post-conferences, but exhort educators to connect classroom and clinical learning more effectively by increasing use of such exercises in the classroom. This is consistent with complexity science-inspired teaching and learning practices.

Group exercises can enrich nonclinical courses as well. Various discussion formats have been examined and used in college classes including whole class discussions, paired listening exercises, round-table discussions, and varying modalities of small group discussions. Brookfield and Preskill (2005) describe the use of "buzz groups," groups of four to five students who are charged simply with discussing a topic or issue. Relaxed buzz groups are completely unstructured, except for a designated topic, whereas structured buzz groups utilize a set of questions for the students to consider and require a report back to the teacher or larger group. "Snowball" or "Pyramid" groupings begin with individuals discussing an issue with a single other person. After a few minutes that pair joins another pair, making a quartet, and so on until the group comes together to share ideas and findings.

In organizing discussion groups, faculty members can allow the students to choose the peers they would like to gather together with. This embodies the complexity science principle of self-organization and allows students to work with those they are comfortable with. However, allowing students to select their own group members may decrease the diversity of opinion present in the groups, as students may consistently choose to talk with their friends or those who sit closest to them. Assigning students to groups while increasing diversity may decrease some students' comfort within the discussion group—which may sometimes be a good thing. Assigning students to groups can also be used to integrate new students into a class where many students already know each other and have formed cliques or subgroups.

Varying methods can be utilized to assign students to groups. Ideas for creating groups can be accomplished by such methods as coin tossing, random selection using a list of students last names, or by purposefully selecting groups based on students characteristics such as age or gender. One method of mixing students is to have each one choose a slip of colored paper from an envelope. For example, for four groups, use red, blue, white, and green paper strips. Students can be organized in like color groups. Some students will eventually try to manipulate this process by picking the same color paper slip as their buddies, so at times groups in which each student holds a different color can be used.

A type of group discussion that is helpful when topics are controversial is called a "conversation café" (http://www.conversationcafe.org/). This technique controls the conversation by allowing everyone a chance to speak for time-limited periods. A "talking object" is used to designate the speaker. Ground rules such as "be respectful," "don't interrupt," "listen openly," and "talk only when holding the talking object" are articulated and agreed upon at the beginning of the session. During the first few rounds the talking object is passed to each person in turn, going around the circle, after which people can request the talking object to add to what has already been said. A summary or concluding round is held last with each person speaking, again in turn. Examples of when to use conversation cafés include discussion of controversial topics, such as adoption by gay couples, and ethical issues, such as those that sometimes occur at end of life. Conversation cafés are also useful for limiting discussion monopolization. Brookfield and Preskill (2005) is a good source for additional discussion group ideas. Many teachers will develop their own ideas for discussion groups.

By using discussion groups, educators allow students some control over events in the classroom, thus the complexity science principle of "distributed control" is recognized. Some discussion groups consciously encourage self-organization, for example, snowball groups; however, most discussion groups include some self-organizing properties.

Another way to present interesting learning opportunities is through the use of film clips specifically chosen to stimulate thinking on a topic, followed by discussion. An example from this author's class on family nursing is to show a clip of a funny (or tense) family interaction and then ask the students to analyze the formal and informal family roles depicted by the characters in the movie. This idea can be adapted to teaching many different concepts.

Many nursing education programs are fortunate to have a simulation laboratory, sometimes with multiple simulation manikins. These high-tech facilities provide many opportunities to give students exposure to the complex aspects of patient care. A full discussion of simulation technology use is beyond the scope of this chapter, but a few ideas can be mentioned here. Manikins can be programmed to manifest complex and sometimes emergent conditions, providing opportunities for students to articulate and utilize their understanding of complexity science and pathophysiology as they develop clinical interventions. The manikins can also be programmed to respond in unexpected ways to interventions, helping students develop facility in reevaluating changing conditions and giving them experience in changing course midstream without jeopardizing any real people. The simulation lab is a good place for students to become exposed to varying types of health care technology including pumps, monitors, and use of electronic medical records. Caring for multiple simulated patients in a high-tech environment can teach students prioritization. Working as if on a "unit" that comprises multiple simulated patients and is staffed by multiple students can give students practice in communication and leadership. Although the possibilities are endless, the fact that simulation manikins are not human must be taken seriously. The downside is, of course, that there is no meaningful human interaction between the student and her "patient" in the simulation lab, so overreliance on simulation as a teaching method should be discouraged.

Changing from a standard lecture format to a more interactive format requires courage and openness to ambiguity. The educator must be able to surrender some control over the material, the classroom atmosphere, and other learning activities to the students. It is helpful to realize that even with the standard lecture or lecture–discussion format, the teacher does not really have as much control of the students' learning as might be assumed. Complexity science theorists state that every organization has both a formal, usually hierarchical, relationship structure, and an informal relationship structure termed a "shadow system." The shadow system operates outside of the official auspices of the organization and often has different goals (Zimmerman et al., 2001). Such shadow systems operate in even the most structured and controlled classroom environment. These unofficial and often unrecognized groupings can be quite powerful and may subvert the teacher's authority and disrupt the learning environment. Learning opportunities that engage students and create a more inclusive and welcoming atmosphere may temper negative influences that come from the shadow system.

Interactive educational methods improve learning through several mechanisms according to Brookfield and Preskill (2005). Most relevant to preparing nursing students for practice are increases in learning breadth and depth, improvements in intellectual ability, and development of key skills related to synthesis and integration of information. The latter skills are important tools for lifelong learning, which is critical for nurses who must remain up to date with advances in health care knowledge and technology. Interactive education also increases students' ability to listen carefully and communicate clearly, both key skills in working with patients, families, and in leading interdisciplinary care teams. Effective communication is also a key element in the prevention of medical errors. Additional benefits are that students learn to explore and tolerate diverse perspectives, live with ambiguity and complexity, and enjoy the self-respect that comes with participating in the cocreation of knowledge.

As it is necessary for the spider to use multiple threads and sophisticated techniques to craft her strong and beautiful web, so is it necessary to pay attention to both theoretical content and classroom process in order to create a true complexity science-inspired curriculum design. Each of these elements is necessary to instill an understanding of complexity science into nursing students and to impart to them the ability to use this knowledge to further patient safety, care quality, and nursing practice.

Complex System Barriers

Educational institutions are themselves complex systems that comprise parts or elements that are capable of acting independently as well as interdependently. These parts include students, faculty, administrators, departments, and committees, to name a few. According to Zimmerman et al. (2001), "creative self-organization" in complex adaptive systems can happen when there are the right amounts of certain characteristics such as information flow, diversity, anxiety, and connectivity. Change is most likely when there is a lot of creative energy. However, the opposite situation can exist where a system is stuck in a behavior pattern, creating multiple barriers to change. In nursing education programs there are some notable barriers that might be expected to slow or impede the incorporation of complexity science into the curriculum. One major anticipated barrier is faculty reluctance to engage in the work of integrating complexity science throughout the curriculum. This may be related to ignorance of complexity science or of its relevance to health care today or such factors as reluctance or inability to take on the work required for substantive curriculum change. However, despite overall curricular stasis in this regard, some small experiments are taking place across the United States. Some schools have begun to experiment with curricular change by inserting a single complexity science course into their curriculum or by including a short unit or two on this topic in another course. These efforts represent a good beginning but do not go far enough.

As noted earlier in this chapter, AACN has already set the standard for incorporation of complexity science into the curriculum. Academic deans and directors need to provide both impetus and support for curriculum change. In order to further this goal, faculty must be provided access to information and resources. To date, opportunities for this are scarce, but there have been some efforts in terms of conferences and publications. Plexus Institute (www.plexusinstitute.org), for example, has sponsored an annual conference on nursing and complexity science for the past several years. University of Kansas and Washburn University in Kansas have also participated in or sponsored nursing or health professions conferences, which included complexity science content. It is suggested that deans and directors consider funding faculty attendance at such conferences and perhaps supply faculty with necessary books and other literature. (Some important early books related to complexity science and health care are listed in the reference list for this chapter.) Bringing experts on complexity science and particularly on complexity science and nursing education to campus for conferences, colloquia, or curriculum consultations are cost-effective ways to provide education and other services to an entire faculty or program.

Barriers to implementation of experimental, innovative, and particularly interactive learning activities also lie within individual educators, deans, and directors. Some of these have already been mentioned in this article. However, physical facilities also play a role in either encouraging or impeding classroom innovation. Many nursing classes take place in rooms set up for showing overhead slides, wherein the teacher is enthroned behind a high podium with a computer screen between her and the students. Such physical environments discourage close interactions between teacher and students. Desks or tables are often bolted to the floor, making gathering into small groups, especially circles, challenging. Some classes share space with learning or simulation labs, again making space a rare commodity. Although physical barriers are difficult to overcome, they can be worked around. Portions of classes can be moved outside or to hallways or other common areas. Corners of rooms can be used for groups and most importantly, the faculty member can come out from behind the computer and podium and engage in close contact with the students. As the realization dawns that multiple, interactive methods are equally or more effective in preparing students for practice and for their licensing and certification examinations, new classroom designs that are more conducive to this type of education will be developed.

It is obvious that in order for full incorporation of complexity science into the curriculum to occur, many changes are needed. These changes are most likely to occur in steps. Indeed, complexity science recognizes the importance of incremental change and of "small experiments" that can provide results that fuel reflection and lead to more change. The danger is in failure to initiate change. Complexity science theory posits that healthy systems exist in a state where order and disorder coexist. It is in this state that learning and change are most likely to take place. Systems characterized by stability and control are likely to inhibit the conditions needed for growth and change. Unfortunately, the latter characteristics are often predominant in nursing education programs and among nursing education faculty and administrators. Nursing educators must cede some of their traditional, authoritarian methods of classroom interaction and become comfortable operating in the order–disorder zone, which is sometimes referred to as "the edge of chaos." Using educational methods inspired by complexity science in the classroom can help move classes and individual students to such a state, where learning is more likely to occur.

In order to prepare nursing students for practice in today's complex health care environment, experts recommend drastic changes in nursing education, including the incorporation of complexity science into the nursing curriculum. This chapter has discussed content and methods necessary to begin implementation of these recommendations. As nurses graduate with a foundation in complexity science, their ability to provide care in the complex systems where they will work after graduation should improve, as should their comfort in today's health care workplace. Long-term results should include increased retention of novice nurses, resulting in health care cost savings, and most importantly, improvements in patient safety and care quality.

REFERENCES

American Association of Colleges of Nursing. (2003). *Working paper on the clinical nurse leader*. Washington, DC: Author.

American Association of Colleges of Nursing. (2008). *The essentials of baccalaureate education for professional nursing practice*. Washington, DC: Author.

American Association of Colleges of Nursing. (2009). *Draft: The essentials of master's education in nursing*. Washington, DC: Author.

Begun, J. (2008). The challenge of change: Inspiring leadership. In C. Lindberg, S. Nash, & C. Lindberg (Eds.), *On the edge: Nursing in the age of complexity* (pp. 239–262). Bordentown, NJ: Plexus Press.

Benner, P., Sutphen, M., Leonary, V., & Day, L. (2010). *Educating nurses: A call for radical transformation*. San Francisco: Jossey-Bass.

Bowles, C., & Candela, L. (2005). First job experiences of recent RN graduates: Improving the work environment. *Journal of Nursing Administration, 35*(3), 130–137.

Brookfield, S., & Preskill, S. (2005). *Discussion as a way of teaching: Tools and techniques for democratic classrooms* (2nd ed.). San Francisco: Jossey-Bass.

Christakis, N., & Fowler, J. (2009). *Connected: The surprising power of our social networks and how they shape our lives*. New York: Little, Brown.

Ebright, P., Patterson, E., Chalko, B., & Render, M. (2003). Understanding the complexity of registered nurse work in acute care settings. *Journal of Nursing Administration, 33*(12), 630–638.

Goldstein, R. (2008). Resource guide and glossary for nonlinear/complex systems terms. In C. Lindberg, S. Nash, & C. Lindberg (Eds.), *On the edge: Nursing in the age of complexity* (pp. 263–294). Bordentown, NJ: Plexus Press.

Institute of Medicine. (2000). *To err is human: Building a safer health system.* Washington, DC: National Academy Press.

Institute of Medicine. (2001). *Crossing the quality chasm: A new health system for the 21st century.* Washington, DC: National Academy Press.

Institute of Medicine. (2004). *Keeping patients safe: Transforming the work environment of nurses.* Washington, DC: National Academy Press.

Kovner, C. T., Brewer, C. S., Fairchild, S., Poornima, S., Hongsoo, K., & Djukic, M. (2007). Newly licensed RNs: Characteristics, work attitudes and intentions to work. *AJN, 107*(9), 58–70.

Krichbaum, K., Diemert, C., Jacox, L., Jones, A., Koenig, P., Mueller, C., et al. (2007). Complexity compression: Nurses under fire. *Nursing Forum, 42*(2), 86–94.

Lindberg, C., Nash, S., & Lindberg, C. (Eds.). (2008). *On the edge: Nursing in the age of complexity.* Bordentown, NJ: Plexus Press.

Palmer, P. (1998). *The courage to teach: Exploring the inner landscape of a teacher's life.* San Francisco: Jossey-Bass.

Pellico, L. H., Brewer, C. S., & Kovner, C. T. (2009). What newly licensed registered nurses have to say about their first experiences. *Nursing Outlook, 57,* 194–203.

PricewaterhouseCooper's Health Research Institute. (2007). *What works: Healing the healthcare staffing shortage.* PricewaterhouseCooper. Retrieved from http://www.pwc.com/us/en/healthcare/publications/what-works-healing-the-healthcare-staffing-shortage.jhtml

Stacey, R. (2003). *Complexity and group processes: A radically social understanding of individuals.* New York: Brunner-Routledge.

West, B. J. (2006). *Where medicine went wrong: Rediscovering the path to complexity.* Hackensack, NJ: World Scientific.

Zimmerman, B., Lindberg, C., & Plsek, P. (2001). *Edgeware: Insights from complexity science for health care leaders* (2nd ed.). Irving, TX: VHA.

Response to Chapter 13

Patricia Welch Dittman

Complexity theory and its use in an academic setting have been utilized to bridge the gap between theory and practice. Complexity science has provided an understanding of how complex organizations work and survive. Mason (2009) describes complexity theory application to environments, organizations, and systems and recognizes that they all have large numbers of interconnected elements interacting with each other. Currently, there are complex systems that interact with students from health care professions on a daily basis. Some of these complex systems include the university and colleges where they study, to the health care organizations where they practice their professional skills. This practice is not just seen in the prescribed curriculum of the practice of nursing, but also in most health care professions.

Academic settings are starting to understand complexity science and its relationship to the interprofessional education (IPE) at both the under graduate and graduate levels. The interprofessional educational curriculum embraces the multidisciplinary student as well as faculty (Cooper & Spencer-Dawe, 2006; Cooper, Spencer-Dawe, & McLean, 2005). IPE educational offerings provide the building of multidisciplinary relationships that will provide a foundation for better understanding of scope of practice, teamwork, interdisciplinary collaboration and relationships, research and the use of communication for patient care, safety, and higher quality care.

A great example of the use of complexity theory and interprofessional education is a study by Cooper and Spencer-Dawe (2006). The study implemented the use of complexity theory concepts through IPE intervention at the University of Liverpool in the United Kingdom and measured its impact on 500 students from nursing, medicine, physiotherapy, occupational therapy, and social work. The intervention included interdisciplinary workshops, lectures, and e-learning packets that focused on narrowing the gap between theory and professional practice. The earlier this curriculum is introduced to the health care students, the greater the understanding and integration of these the principles of complexity theory are seen in practice. Health care research funding sources, IOM, and accreditation bodies all value an environment that fosters interdisciplinary relationships.

Nursing as a profession has also recognized the need for complexity science to be integrated into the curriculum for better student preparation. Nurses as key stakeholders in the complex health care system need to meet the system-level demands of change, problem solving, administration, technology, etc. The theory of bureaucratic caring (Ray, 1989, 2001) was formulated by the investigation of the meaning of spiritual–ethical caring based on the nurse–patient relationship, in the institutional culture of a complex health care system. The theory establishes the need for a health care system that proposes a synthesis of spiritual–ethical caring and bureaucratic processes (economic, technological, legal, political, educational, and physical), which should be open and interactive and respond to the nurse's decisions (Dittman, 1995). This theory of bureaucratic caring makes a connection to chaos and complexity theories (Coffman, 2006). Ray compares change in complex organizations as part of the living process and encourages nurses to embrace meaning in the ever-changing culture. Complexity, which values the wholeness, recognizes that the whole is in the part and the part in the whole (2001). Nurses who interact with health care or academic complex systems need to recognize that these systems are nonlinear and do not react proportionately.

As we start to address the needs of the profession of nursing we should begin with its academic preparation and entry to practice. Benner et al. (2010) in their new book titled *Educating Nurses: A Call for Radical Transformation* talk about apprenticeships of the nursing profession that are constantly interwoven in daily practice. The apprenticeships are knowledge, skill, and ethical comportment, which include caring. Balancing these three major responsibilities in a complex health care system calls for the nurse to be prepared academically in any or all areas of practice. Benner et al. (2010) make four suggestions for change in the preparation of nurses.

The nurse educator should:

1. Shift from a focus on covering decontextualized knowledge to an emphasis on teaching for a sense of salience, situated cognition, and action in particular situations.
2. Shift from a sharp separation of clinical and classroom teaching to integration of classroom and clinical teaching.
3. Shift from an emphasis on critical thinking to an emphasis on clinical reasoning and multiple ways of thinking that include critical thinking.
4. Shift from an emphasis on socialization and role taking to an emphasis on formation (p. 89).

This new paradigm of teaching and learning practices reinforces the need to prepare the nurse for a complex environment. The development of curriculum, instructional strategies, and evaluation and testing mechanisms with an emphasis on the complex caring practice of nursing will be based on the integration of knowledge, skill, and judgment. An instructional strategy that is increasing in its use is simulation in a lab environment that provides a safe place for the integration of classroom and clinical teaching. Simulation sessions also incorporate the use of group learning, as all members of the team participate in the scenario. In the debriefing sessions, where clinical and classroom teaching occurs, students are melded together for the benefit of student learning.

National nursing organizations and accreditation agencies are now addressing the need for reform in nursing education. The American Association of Colleges of Nursing (AACN) recognizes the need to include complexity science in the nursing curriculum. In both of their latest publications, "Clinical Nurse Leader Curriculum" (AACN, 2003) and the draft proposal for "The Essentials of Masters Education in Nursing" (AACN, 2009), complex health care systems and complexity science are discussed. The American Nurses Credentialing Center (ANCC, 2009) Magnet Nursing Designation is also grounded in the understanding of complex health care systems. The interrelationships that nurses create and grow only help to provide safe hospital environments. The use of Magnet criteria, that is, structured empowerment, exemplary professional practice, transformational leadership and new knowledge, innovation, and improvements have a complexity science component in that they value organizations that are dynamic and self-organizing.

It is at the edge of chaos where the concept of nursing education is positioned, and its actions related to the process of change should be addressed: "Modern science defines the edge of chaos as a dynamic, holistic and reciprocal process that drives change and creative reordering in systems" (Ray et al., 1995, p. 48). Complexity theory and its application to a complex academic environment will be more effective than the traditional linear theory of causality when dealing with development of educational projects, policies, or programs (Nordtviet, 2010). As the profession of nursing addresses its need for transformational change that values the characteristics of the profession and the complexity of a practice-based curriculum, we hope that the outcomes will minimize the practice–education gap (Benner, 2010).

REFERENCES

American Association of Colleges of Nursing. (2009). *Draft: The essentials of master's education in nursing*. Washington, DC: Author.

American Association of Colleges of Nursing. (2003). *Working paper on the clinical nurse leader*. Washington DC: Author.

Benner, P., Sutphen, M., Leonary, V., & Day, L. (2010). *Educating nurses: A call for radical transformation*. San Francisco: Jossey-Bass.

Coffman, S. (2006). Theory of Bureaucratic Caring. In A. Tomey & M. R. Alligood (Eds.), *Nursing theorists and their work* (6th ed.). St. Louis, MO: Mosby.

Cooper, H., & Spencer-Dawe, E. (2006). Involving service users in interprofessional education narrowing the gap between theory and practice. *Journal of Interprofessional Care, 20*(6), 603–617.

Cooper, H., Spencer-Dawe, E., & McLean, E. (2005). Beginning the process of teamwork: Design, implementation and evaluation of inter-professional education intervention for first year undergraduate students. *Journal of Interprofessional Care, 19*(5), 492–508.

Dittman, P. W. (1995). *The work-life views of the nurse manager during transition from primary care to patient-focused care*. Master of science thesis, Florida Atlantic University, Boca Raton, FL. Retrieved from Proquest, University Microforms International.

Mason, M. (2009). Making educational development and change sustainable: Insights from complexity theory. *Journal of Interprofessional Care, 29*, 117–124.

Nordtviet, B. H. (2010). Development as a complex process of change: Conception and analysis of projects, programs and policies. *Journal of Interprofessional Care, 30,* 110–117.

Ray. M. (2001). Marilyn Anne Ray, The Theory of Bureaucratic Caring. In M. Parker (Ed.), *Nursing theories & nursing practices* (pp. 421–431). Philadelphia, PA: F. A. Davis Company.

Ray, M. A. (1989). The theory of bureaucratic caring for nursing practice in the organizational culture. *Nursing Administration Quarterly, 13*(2), 31–42.

Ray, M. A., Didominic, V., Dittman, P. W., Hurst, P., Seaver, J., Sorbello, B., et al. (1995). The edge of chaos: Caring and the bottom line. *Nursing Management, 24*(9), 48–50.

CHAPTER 13
LESSONS IN COMPLEXITY SCIENCE: PREPARING STUDENT NURSES FOR PRACTICE IN COMPLEX HEALTH CARE SYSTEMS

- *Complexity science teaches us about solving problems in a complex world or "making things work" by knowledge of linear and nonlinear dynamics and complex adaptive systems in complex systems. How does this chapter illuminate the forces in today's health care system and what is necessary in the classroom for leadership in complex systems? How do teachers teach, whether in the classroom or in the health care environment, to solve problems based upon new thinking on the coexistence of order and disorder (chaos theory), transformational leadership, embeddedness, distributed control, and complex adaptive systems?*
- *Why do you think medical errors in health care have risen in the last decade?*
- *How do nursing educators teach students and other professionals in the practice environment about the complexity of infectious disease, such as MRSA, as complex systems in health care organizations?*
- *How do nurses educate for new approaches to leadership in organizations? What are the key concepts behind networks of relationship? Is a clinical nurse leader role (CNL) an answer to changing practice to being closer to patients at the bedside? What about leading in community health nursing?*
- *What are central curricular concepts from complexity that should be in nursing curricula? How should research be taught in today's health care arena? How do we understand patterning and design research to study patterns to solve complex health care and nursing problems?*
- *Dittman shared some examples of the integration of complexity sciences and nursing. She also shared Benner and other scholars' work on what is needed in nursing education today. How would you integrate complexity science/s into nursing education?*

FOURTEEN

Technological Change in Health Care Electronic Documentation as Facilitated Through the Science of Complexity

Todd Swinderman

The process of implementing new technology is often associated with turbulence. Nurse informaticians (experts in information technology) have a magnetic appeal, transforming nursing practice and information technology components of the Electronic Health Record (EHR) through their caring relationships. These caring relationships, at the edge of chaos, with the leadership of nurse informaticians make it possible for nursing and information technology to relationally self-organize or transform to reflect and support nursing and multidisciplinary health care practice. The respondent, Penprase, used the template to discuss complexity sciences, the theory of bureaucratic caring, nursing, medicine, and other professions in the practice of health care delivery and information technology. Penprase presented how information in the Swinderman chapter related to the micro, meso, and macro scaling. The macro scale related to complexity science tenets of multivariate patterns, Ray's Bureaucratic Caring Theory and readiness to adopt change in organizations and the "magnetic appeal" of the caring nurse in transforming nursing practice; the meso scale related to nursing information systems, the tension between disorder and order (chaos theory), and magnetic appeal in terms of creating the energy to positively attract the other through knowledge and practice of caring for change. The micro scale focused on the caring attractors (education, negotiation, translating, and acting as a liaison or magnetic attractor) for actual choice making by the nurse and the patient to enhance the electronic documentation system for change. Penprase supported the research on information technology and caring in nursing by projecting the emergence of new roles and information systems as individuals seek to understand the theories of complexity sciences and how they relate to change and self-organization within chaotic health care environments.

INTRODUCTION

Information technology is advancing rapidly and reshaping the look of health care. The Internet has enabled individuals to create personal health records and hospital systems have created the Electronic Health Record (EHR). Nursing plays an interactive role in both. Information technology is integral to everyday nursing practice as nurses increasingly rely upon information to support evidence-based practice. The pace at which information technology is changing is faster than nursing practice is changing. The practice changes associated with implementation of new technology is likely to create turbulence in the organization. The science of complexity provides a framework to make sense of the turbulence associated with transforming nursing practice with health care electronic documentation.

Nurses working in the field of information technology are often referred to as nurse informaticians. These nurses hold an ideal position for facilitating the coevolution of nursing practice and information technology. Swinderman's (2005) research demonstrated that nurse informaticians have a magnetic appeal in transforming nursing practice and information technology components of the EHR through their caring relationships. This chapter will illuminate how these caring relationships at the edge of chaos, with the leadership of nurse informaticians, make it possible for nursing and information technology to relationally self-organize or transform to reflect and support nursing and multidisciplinary health care practice.

Complexity Science

Chaos theory is a subset of complexity sciences (Briggs & Peat, 1989; Ray, 1998). Chaotic, far-from-equilibrium systems require great energy input from outside. They fluctuate between ordered behavior and patterns of randomness. These fluctuations increase randomly, with order and disorder sometimes existing simultaneously until they reach a bifurcation point (the edge of chaos) where one fluctuation becomes amplified and spreads to dominate the system. Thus, order develops out of chaos (Prigogine & Stengers, 1984), a transformation that is being studied in the context of emergence (Holland, 1998). As the system transforms, experiencing tension between order and disorder in the phase space, a choice point is reached where a new orderly system or disintegration may ensue (Davidson & Ray, 1991; Ray, 1994, 1998; Ray et al., 1995). The phase space is a snapshot in time used to facilitate visualization of the transformation. Oscillations between options may occur for an undetermined period of time. The choice point is where the system selects a trajectory moving away from other options. Attractors (trajectories toward which other nearby trajectories tend) can pull the transformation in a certain direction. More than one attractor may exist, the ones that attract many trajectories are called basins, and their strength can often be observed in the patterning of the system (Davidson, 2005).

Complexity science is relevant to nursing science because it provides a mechanism for modeling multivariate patterns of evolving order. Within chaos theory are other components that are applicable when applying chaos theory to nursing (Ray, 1994; Vicenzi, White, & Begun, 1997; Wheatley, 1999). "The recognition of practice complexity and the need for more appropriate models with which to understand and explain human reactions to illness are influencing nursing researchers and practitioners to turn to models of chaos and complexity" (Hamilton, Pollack, Mitchell, Vicenzi, & West, 1997, p. 237). "Chaos contains the stimulus to the human spirit to overcome obstacles, search the universe for answers, and undertake the painful process of universal and personal discovery" (Porter-O'Grady, 1990, p. 171). The nonlinear dynamics and emergent properties of complexity sciences are also inherent in health care organizations as systems (McDaniel & Driebe, 2001).

When a nursing information system for documentation is implemented, there is a departure from the handwritten method of documentation to computer documentation. This change is best viewed as the phase space within the chaos theory's framework. "For us humans, it is known as a moment of great fear, tinged perhaps, with a faint sense of expectations" (Wheatley, 1999, p. 88). This period of time is also known as dis-ease (Davidson & Ray, 1991; Ray, 1994). The new information system is then posed between death or disharmony and acceptance or transformation. The acceptance or lack of acceptance is a potential choice of the staff nurse. Transformation, on the other hand, is a process by which the staff nurses cocreate a new system of documentation that supports and reflects their nursing practice. Caring is the energy by which choice is facilitated to bring order (healing or well-being) out of chaos (dis-ease, need, pain, or crises) (Ray, 1994, p. 26). "The nurse in the caring relationship flows with the client [or other staff nurses] to enhance the depth and complexity of understanding of self—the nurse's and the client's selves" (Davidson & Ray, 1991, p. 83). Attractors can serve as a metaphor for creative activities in an organization in which innovation is possible, yet there is a boundary to the activities determined by the core competencies of the organization, as well as its resources and environmental factors affecting the organization (Zimmerman, Lindberg, & Plsek, 2001, p. 260).

The implementation and support of the new system that is replacing the handwritten or existing systems occurs in what complexity theorists call the phase space where there is tension between order and disorder. This is also the edge of chaos. It is possible that nurses might quit, become antagonized, and self-organize to find ways to abandon the electronic documentation system (Gardner & Lundsgaarde, 1994). The implementation might also enable nurses to transform their practice and documentation patterns into a new system with the electronic documentation system (Gardner & Lundsgaarde). Swinderman's (2005) research illuminates the meaning and patterns that occur in the phase space and the interconnectedness of the nurse informatician, the staff nurse, and the electronic nursing documentation. No clear path is predictable, a tenet of chaos theory. Caring attractors influence the transformation to electronic documentation or other unanticipated patterns or attractors may lead to the return of the handwritten documentation.

Complexity Science Research and Technological Change in Health Care

To understand the humanness of the interconnectedness between nurse informaticians and the implementation of electronic documentation, Swinderman conducted a hermeneutic–phenomenological research that illuminates the meaning of the lived experience of nurse informaticians. The method is a hermeneutic (interpretive) phenomenological (descriptive) method. "Hermeneutic phenomenology is grounded in the belief that the researcher of understanding and the participants come to the investigation with forestructures of understanding shaped by their respective backgrounds, and in the process of interaction and interpretation, the co-generate and understanding of the phenomenon being studied" (Wojnar & Swanson, 2007, p. 175). This research method is in keeping with themes of complexity sciences by acknowledging the interconnectedness of the researcher and phenomenon of interest, the researcher and the participant, the researcher and the text in finding meaning, and the researcher and the meaning of the lived experience as a whole. The Swinderman Conceptual Model For Nursing Informatics (see Figure 14.1) provides an illustration for understanding the meaning of implementing clinical information systems in complex health care environments.

Complexity captures the nonlinear essence of the organization and persons as a whole; the relationships and the processes define where the system started and coevolved in an unpredictable manner. Changes are irreversible and the process, although showing a sensitive dependence on previous conditions, is unpredictable as to future patterning. The input is not equal to the output. Little things may have a great impact, and big things may have little effect or none at all. Describing the concept of nonlocality seen in quantum mechanics, energy permeates the process and acts anywhere and at anytime. A small energy can provide the form to change the direction of many. The interpretive themes that support this metatheme are nonlinear nursing practice, holistic nursing practice, change, choice, and the nursing process.

The metaphor illuminating the universal whole or the meaning of the whole is the *magnetic appeal of nurse informaticians*. The magnetic appeal represents the attraction in the phase space of implementing an electronic documentation system that the nurse informatician brings to the edge of chaos. In the phase space, the nurse informatician influences the choice making at the choice point to facilitate repatterning or transforming

FIGURE 14.1 Swinderman Conceptual Model for Nursing Informatics

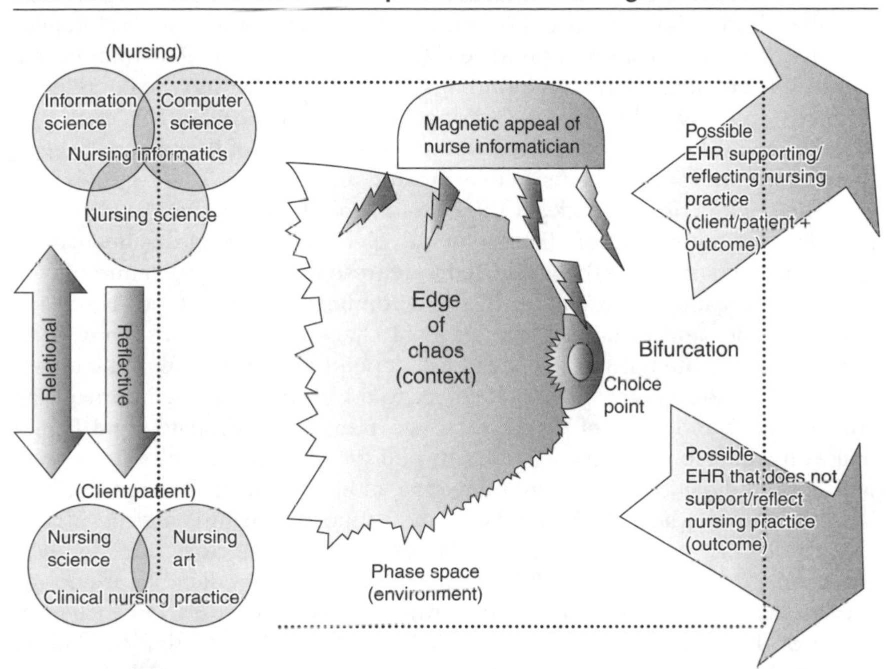

nursing practice to coexist with information technology. The nurse informatician exhibits magnetic attraction, creating an energy field with skills such as caring, educating, translating, negotiating, and even cheerleading at the appropriate times with different people within the organization such as leaders, staff in information technology, and clinical information software vendors to achieve transformation of nursing practice with the EHR.

At first, the art of handwritten documentation seems paradoxical to computerized data entry. However, art can be generated through the computer in the form of fractals. Fractals are ". . . complex by virtue of their infinite detail and unique mathematical properties (no two fractals are the same), yet they're simple because they can be generated through successive applications of simple iterations" (Briggs & Peat, 1989, p. 96). The same patterns are repeated from the microscopic to the macroscopic. Both the art of documentation and computer data entry can exist simultaneously as well, and still reflect the uniqueness of the patient and nursing interaction.

The Swinderman Conceptual Model (SCM) provides an illustration for understanding the meaning of implementing clinical information systems in complex health care environments. "Conceptual models also reflect sets of values and beliefs, as in philosophical statements as well as preferences for practice and research approaches" (Parker, 2001, p. 6). This model is a synthesis of qualitative research findings, chaos theory within the complexity sciences and Ray's Bureaucratic Caring Theory (Ray, 2001). This model is situated contextually and contains all four components of the metaparadigm, nursing, health (or well-being), environment, and person.

The context of this model is the cultural, social, and political aspects surrounding the transition from handwritten to electronic nursing documentation. Ray's bureaucratic caring theory is used to represent this integral nature of contemporary organizational culture. The environment is where there is tension between order and disorder at the edge of chaos and embodies many paradoxes. The ultimate trajectory selected will depend on the choices of the nurses made at the choice point, a heartfelt choice to do good. The nurse informatician with qualities and skills of caring, educator, translator, negotiator, and liaison makes the interconnections necessary to pull the transformation of the system in a desired direction. The desired direction, although unable to predict, might be possibly one where the EHR reflects and supports nursing practice.

The choice point is where the system selects a trajectory moving away from other options. The possibilities of options are unpredictable. If enough nurses make the choice to abandon the new system, cessation in use or death of the new system may ensue. It has been estimated that the failure rate of clinical information systems is 30% or more (Sittig, 2001). It is also possible that the EHR could transform into a medical model without the essence of nursing being incorporated. It is also possible for nursing to participate in the transformation of EHR where nursing continues to be the keeper of the care plan and the essence of nursing is incorporated.

The magnetic appeal of nurse informaticians represents the attraction in the phase space. The nurse informatician brings to the edge of chaos in the phase space gifts to influence the choice making where, at the choice point, the repatterning or transforming of nursing practice can coexist with information technology. The transformation of a system experiencing tension between order and disorder (the edge of chaos) is where the system is at a choice point in the phase space when a new orderly system or disintegration may ensue (Davidson & Ray, 1991; Ray, 1994, 1998; Ray et al., 1995). The phase space is a snapshot in time used to facilitate visualization of the transformation. Oscillations between options may occur for an undetermined period of time. The choice point is where the system selects a trajectory moving away from other options. Attractors along the way have the ability to pull the transformation in a desired direction (Ray, 1998).

The tension between order and disorder at the edge of chaos are due in part to the paradoxes: "The paradoxes of complexity are that both sides of many apparent contradictions are true" (Zimmerman et al., 2001, p. 17). In complex systems, it is no longer either–or and one or the other; rather it is both–and all at once. The first paradox in this research is that data emerged from a qualitative study of human interactions.

Nursing has historically been a hands-on profession. Struggles with nursing documentation have been recorded since the 1930s. Documentation provides the evidence for nurses to imbue the nursing situation with best practice. The evidence can be in the form of any of the four (empirical, aesthetic, ethical, and personal) fundamental patterns of knowing in nursing (Carper, 1978; Fawcett, Watson, Neuman, Walker, & Fitzpatrick, 2001). However, outcomes of nursing situations need to be documented to add to the database as evidence for future knowledge in nursing situations. This is an ongoing source of tension at the edge of chaos.

Tension and paradox are normal occurrences that transform organizations: "Complexity science is the study of living systems but living systems die" (Zimmerman et al., 2001, p. 17). Consider the same paradoxes in the following statement:

> For the person who says, "I don't want to learn about computers and how to use the Internet," the best response may be to say, "Die, it will be easier on you." Although facetious, that piece of advice reflects an element of truth. Technological advances and the challenge of adapting to them are not going to go away. (Porter-O'Grady & Malloch, 2003, p. 27)

Choice making occurs at the choice point. The choice point is where the system selects a trajectory moving away from other options. "The choice making of nurses occurs with the interest of humanity at heart, utilizing ethical principles as the compass in deliberation" (Coffman, 2005, p. 126). Although the choice may be initially forced upon the nurse by the organization to use a new computer system, if enough nurses are challenged by the system, that new system may be transformed or abandoned altogether. The possible transformation that Doug feared was an EHR that did not capture the essence of nursing. This notion is possible if nurses do not participate in the transformation of the EHR; therefore, nurses must make the choice with the interest of humanity at heart and enhance nursing practice while engaging with the EHR.

The metaphor *magnetic appeal of nurse informaticians* emerges as an apperception of the whole as the essence of the experience. The nurse informaticians have the ability to act like magnets or attractors and pull transformation in a desired direction. Magnetic attraction is both general and abstract. "Magnets function as transducers, transforming energy from one form to another, without any permanent loss of their own energy" (Magnetic Sales, 2000). Smith (1998) used the following example of how one magnet can be used in many different situations to attract a variety of objects. "I have just one (token) magnet in my workshop; but I use it on different occasions to pick up different things: yesterday, filings from the drill press; today, paper clips" (p. 170). The qualities and gifts of the nurse informaticians that promoted caring attractions were specifically identified as educating, negotiating, translating, and acting as liaison. These processes were applied on different occasions to attract different things, all in an effort to guide the transformation from handwritten documentation to the desired direction of the EHR.

The hysteresis loop, which characterizes each magnet material, describes the cycling of a magnet in a closed circuit as it is brought to saturation, demagnetized, saturated in the opposite direction, and then demagnetized again under the influence of an external magnetic field (Magnetic Sales, 2000). Hysteresis, in complex environments, describes patterns that reflect its past history and limit the unpredictability as to future patterning (Davidson, 2005). Moreover, when demagnetization occurs in complex health care organizations, there is a loss of security, cooperation, communication, community, and ultimately business (Ray et al., 2002).

Nursing practice is interrelational with information technology. Patients are creating and maintaining their own personal health records. Patients are accessing their own medical records and communicating incompleteness and potential errors in documentation. Patients are also ongoing contributors of information in the EHR. For example, some patients access their own EHR and enter daily weights or blood sugar recordings. These data entered by the patient are then monitored by the health care team.

Organizational culture for nursing practice will need to change so that time and education are devoted to teach nurses the skills to document nursing caring in the EHR. These processes could be enhanced by best practices where the nurse documents the therapeutic (caring) relationship between the nurse and the patient, the critical thinking and judgment process (based on empirical, aesthetic, ethical, and personal ways of knowing—evidence) of the nurse, and the patient preference (choice) (personal communication, V. Neely, 2005). This documentation captured in the EHR becomes cyclical in nature by adding to the database as evidenced in future decision making (choice) of both the nurse and patient. This creative, nonlocal process is representative of the nursing process.

Nurses work every day at the edge of chaos. As nursing and information technology coevolve, nurses will need to look inward and outward, listen to their better judgment, and then make the moral choice of what ought to be done with information technology. Nurse informaticians engaged in a caring relationship need to reorient the choice patterns of nurses and others to transform the way that documentation is entered into the EHR. Change must first occur with staff nurses. This process will further facilitate the caring relationship between the nurse and patient in relational self-organization.

Clearly, nursing administration must create a culture for change and acceptance in clinical information systems for nursing documentation. Nurse administrators must be revisionists, informed about leading-edge science and technology, and work with nurse informaticians in a collaborative relationship for implementing changes. Many nurse administrators are engaged in achieving American Nurses Credentialing Center (ANCC) Magnet Recognition (2005) designation in their health care organizations in an attempt to attract and retain nurses and demonstrate a level of excellence. The most successful implementations started with "top-down support" for the new clinical information system followed closely by "bottom-up" staff level participation in the designing process of information technology systems. This notion confirms Wheatley's (1999) statement, "[w]e need leaders to understand that we are best controlled by concepts that invite our participation, not policies and procedures that curtail our contributions" (p. 131). Lindberg, Herzog, Merry, and Goldstein (1998) stated "the leaders dealing in ambiguous environments might do well to foster conditions under which adaptability and creativity emerge from within the system" (p. 8).

Complexity sciences should be incorporated into nursing research to further understand the subtle flux and flow of dynamic patterning of interrelationships that exist in nursing. The clinical nurse leader is a graduate degree preparing practice level nurse generalist to deal with complex health issues in complex health care systems. The Plexus Institute, started by Curt Lindberg, is an agency that assists systems in understanding how to transform their organizations in terms of complexity sciences, but not in the traditional machinelike metaphor, however. The Swinderman Conceptual Model is one tool available for visualizing the meaning of implementing electronic documentation system in a complex health care environment.

Nurse informaticians have a *magnetic appeal* in transforming nursing practice and information technology components of the EHR through their caring relationships. The transformation is unpredictable but possible. The Swinderman Conceptual Model for Nursing Informatics illustrates the magnetic appeal through caring relationships of the nurse informatician working in complex health care environments. Through these caring relationships at the edge of chaos, with the leadership of nurse informaticians, it is possible that nursing and information technology will relationally self-organize or transform to reflect and support nursing caring and multidisciplinary health care information technology practice.

REFERENCES

American Nurses Credentialing Center. (2005). *Magnet recognition program*. Retrieved from www.nursingworld.org/ancc/magnet.html

Anderson, R. A., Crabtree, B. F., Steele, D. J., & McDaniel, R. R. (2005). Case study research: The view from complexity science. *Qualitative Health Research, 15*(5), 669–685.

Briggs, J., & Peat, F. D. (1989). *Turbulent mirror: An illustrated guide to chaos theory and the science of wholeness*. New York: Harper & Row.

Carper, B. (1978). Fundamental patterns of knowing in nursing. *Advances in Nursing Science, 1*(1), 13–23.

Coffman, S. (2006). Marilyn Anne Ray: Theory of Bureaucratic Caring. In A. Marriner-Tomey & M. R. Alligood (Eds.), *Nursing theorists and their work* (6th ed., pp. 116–139). St. Louis: Mosby.

Davidson, A. W. (2005, February). *Facilitating the mutual human-environment process.* Unpublished manuscript.

Davidson, A. W., & Ray, M. A. (1991). Studying the human-environment phenomenon using the science of complexity. *Advances in Nursing Science, 14*(2), 73–87.

Fawcett, J., Watson, J., Neuman, B., Walker, P. H., & Fitzpatrick, J. J. (2001). On nursing theories and evidence. *Journal of Nursing Scholarship, 32*(2), 115–119.

Gardner, R. M., & Lundsgaarde, H. P. (1994). Evaluation of user acceptance of clinical expert system. *Journal of the American Medical Informatics Association, 1*(6), 428–438.

Hamilton, P., Pollack, J. E., Mitchell, D. A., Vicenzi, A. E., & West, B. J. (1997). The application of nonlinear dynamics in nursing research. *Nonlinear Dynamics, Psychology, and Life Sciences, 1*(4), 237–261.

Holland, J. (1998). *Emergence: from chaos to order.* Reading, MA: Helix Books.

Lindberg, C., Herzog, A., Merry, M., & Goldstein, J. (1998). Life at the edge of chaos. *Physician Executive, 24*(1), 6–21.

Magnetic Sales. (2000). *Permanent magnet design guidelines.* Retrieved from www.magnetsales.com/Design/DesignG.htm

McDaniel, R. R., & Driebe, D. J. (2001). Complexity science and health care management. *Advances in Health Care Management, 2,* 11–36.

Parker, M. E. (2001). *Nursing theories and nursing practice.* Philadelphia: F. A. Davis.

Porter-O'Grady, T. (1990). *Reorganization of nursing practice: Creating the corporate venture.* Rockville, MD: Aspen.

Porter-O'Grady, T., & Malloch, K. (2003). *Quantum leadership: A textbook of new leadership.* Sudbury, MA: Jones and Bartlett.

Prigogine, I., & Stengers, I. (1984). *Order out of chaos.* Toronto, ON: Bantam Books.

Ray, M. A. (1994). Complex caring dynamics: A unifying model of nursing inquiry. *Theoretic and Applied Chaos in Nursing, 1*(1), 23–31. (Renamed *Complexity and Chaos in Nursing.*)

Ray, M. A. (1998). Complexity and nursing science. *Nursing Science Quarterly, 11*(3), 91–93.

Ray, M. A. (2001). The theory of bureaucratic caring. In M. Parker (Ed.), *Nursing theories and nursing practice* (pp. 421–431). Philadelphia, PA: F. A. Davis Company.

Ray, M. A., DiDominic, V. A., Dittman, P. W., Hurst, R. A., Seaver, J. B., Sorbello, B. C., et al. (1995). The edge of chaos: caring and the bottom line. *Nursing Management, 26*(9), 48–50.

Ray, M. A., Turkel, M. C., & Marino, F. (2002). The transformative process for nursing workforce redevelopment. *Nursing Administration Quaterly, 26*(2), 1–14.

Sittig, D. (2001, November 14). *The importance of leadership in clinical information system implementation process.* Retrieved from www.informatics-review.com/thoughts/leadership.html

Smith, B. C. (1998). *On the origin of objects.* Cambridge: The MIT Press.

Swinderman, T. D. (2005). The magnetic appeal of nurse informaticians: Caring attractor for emergence. (Doctoral dissertation, Florida Atlantic University, Christine E. Lynn College of Nursing, 2005). *ProQuest Digital Dissertations* (UMI No. 3162666).

Vicenzi, A. E., White, K. R., & Begun, J. W. (1997). Chaos in nursing: Make it work for you. *American Journal of Nursing, 97*(10), 26–31.

Wheatley, M. J. (1999). *Leadership and the new science: Discovering order in a chaotic world* (2nd ed.). San Francisco: Berrett-Koehler.

Wojnar, D. M., & Swanson, K. M. (2007). Phenomenology: An exploration. *Journal of Holistic Nursing, 25*(3), 172–180.

Zimmerman, B., Lindberg, C., & Plsek, P. (2001). *Edgeware: Insights from complexity science for health care leaders* (2nd ed.). Irving, TX: Voluntary Hospitals of America.

Response to Chapter 14

Barbara Penprase

CONCEPTS PRESENTED IN THIS CHAPTER THAT ARE RELEVANT TO THE COMPLEX SYSTEMS SEEN IN NURSING, MEDICINE, AND/OR OTHER EXAMPLES OF HEALTH CARE PRACTICE

Complexity	Magnetic attractors
Ray's Bureaucratic Caring Theory	Choice point
Informatics	Caring attractors
Phase space	Magnetic appeal to transformation

HOW INFORMATION CAN BE COLLECTED ON THESE CONCEPTS: INCLUDE INFORMATION SCALING FROM MICRO THROUGH MESO TO MACRO

SCALE	CONCEPT	RELATED TO	SEEN IN ASSESSMENT DATA
Macro	Complexity	Multivariate patterns of evolving order	Through adoption of nursing information system (the change)
Macro	Ray's Bureaucratic Caring Theory	The organizational culture and readiness to adopt change	Nurses' choices to adopt to change or not
Meso	Informatics	Nursing information systems	Magnetic appeal of nurse informatics; the strength of nursing practice to be attracted to the change
Meso	Phase space	Tension between order and disorder	Understanding the attraction in the phase space of implementing the change where tension and order compete

Meso	Magnetic attractors	Creating the energy to positively attract the change	Nursing skills such as caring, educating, translating, negotiating, and even cheerleading
Micro	Choice point	Choice making by the nurse	Nurses move toward one attractor over others
Micro	Caring attractors	Caring attractors already existing within nursing practice	Educating, negotiating, translating, and acting as liaison
Macro	Magnetic appeal	Transforming nursing practice	Caring relationships within nursing that serve as strong attractors and act as a magnetic appeal to adoption of change

CONCEPTS

Complexity theory offers a dynamic exciting approach to understanding the intricacy of nursing practice in a variety of areas such as nursing practice, nursing leadership, and nursing informatics.

Complexity theory proposes that organizations are best understood as complex systems comprised of dynamic networks of relationships. The theory suggests that all complex systems change to adapt to their environments, and one may understand these systems by examining patterns of relationships and interactions among the system's agents.

A single agent can influence many agents through a network of connection; therefore, interactions at the local level give rise to global patterns (Casti & Cilliers, as cited in Anderson, Crabtree, Steele, & McDaniel, 2005). Because agents act only locally, no single agent knows the system as a whole and complexity emerges from the patterns of interaction among the agents.

Science of complexity theory has a holistic approach with a focus on relationships and is congruent with the work of nursing.

Attractors are a force within the complex system that directs the shape and development of that system. Every complex system has attractors that help create boundaries of stability and instability (Haynes, 2003).

Edge of chaos is where dynamic energy exists, and creativity emerges linked from previous knowledge and the unknown (chaos).

EXAMPLES OF HOW THESE CONCEPTS AND/OR METHODS CAN BE APPLIED IN PRACTICE

Complexity theory proposes that organizations are best understood as complex systems that comprise dynamic networks of relationships. The theory suggests that all complex systems change to adapt to their environments and one may understand these systems by examining patterns of relationships and interactions among the system's agents. Agents are defined as the basic fundamental processing units of a system. Examples of agents may be people, human processes such as nursing processes, medical processes, administrative processes, and/or computer systems (Anderson et al., 2005).

Swinderman offers a forceful interpretation of the application of complexity theory to nursing practice integrated within information technology. He specifically focuses on the "caring relationships at the edge of chaos, with the leadership of nurse informaticians, make it possible for nursing and information technology to *relationally* self-organize or transform to reflect and support nursing and multidisciplinary health care practice." Although Swinderman's application to caring is unique when studying nursing behaviors during implementation of information systems, attractors remain a constant basic principle to complexity theory. An attractor is a force of some type to which a system is drawn, which influences the direction, state, or behavior of the system. Attractors direct the shape and development of that system. Every complex system has attractors that help create boundaries of instability (Haynes, 2003). All attractors emerge out of the adaptive processes of their setting and act as catalysts for information and beliefs about the environment (Penprase, 2003; Penprase & Norris, 2005). Swinderman's findings support that attractors serve as catalysts for change and those attractors that are based on nursing practice are key to how that change will be adopted. Swinderman's study supports that strong attractors already exist within nursing practice through caring principles that facilitate movement from chaos to order.

Swinderman Conceptual Model for Nursing Informatics outlines how nursing informatics can be adopted positively and successfully within nursing practice through the "magnetic appeal of nursing informaticians" within complex systems. Swinderman suggests that "skills such as caring, educating, translating, negotiating, and even cheerleading, at the appropriate times" will lead to transformation of nursing practice. However, Swinderman's study just as importantly notes the significance of the role of the nursing management team and culture in which the change takes place.

CONCLUSION: EVALUATION OF THE USEFULNESS OF THE CHAPTER CONCEPTS AND METHODS FOR NURSING AND HEALTH CARE PRACTICE

This study by Swinderman helps nurses identify how previous nursing practice tends to be strong attractors to the development and implementation of information systems. He also emphasizes the importance that nurse leaders within the change model are necessary to serve as strong attractors that result in positive change. Research related to complexity theory within the framework of nursing is in early stages and is limited. This chapter gives the reader an excellent illustration of the importance of complexity theory principles that can be used to understand and guide practice patterns related to change. Health care organizations, especially nursing departments, offer a rich environment for qualitative research and the application of complexity theory concepts, because health care is experiencing chaotic times with often unpredictable results.

A new understanding of nursing management roles will emerge as individuals seek to understand how the theoretical principles of complexity theory relate to adaptation to change within the chaotic hospital environment. Complexity theory alters the perception of unpredictable behavior and assists in understanding the deeper, hidden meaning of the environment as a guide to successful adaptation in the continuously changing health care environment. The findings of this research study suggest that understanding complexity theory concepts, such as attractors and phase transition, within an organization can be used to assist nursing management in understanding how change within health care institutions could occur. Also, by understanding the key concepts of complexity theory within institutions, communication strategies can be developed to better support project implementation.

The intricate and unpredictable nature of interactions occurring within a complex system leads to formation of new and unexpected patterns or properties (Holden, 2005). Information systems development will continue to evolve within health care. Nursing is integral in ensuring that these systems evolve to complement nursing practice in the future and be reflective of the nursing process.

REFERENCES

Anderson, R. A., Crabtree, B. F., Steele, D. J., & McDaniel, R. R. (2005). Case study research: The view from complexity science. *Qualitative Health Research, 15*(5), 669–685.

Haynes, P. (2003). *Managing complexity in the public services.* Maidenhead, UK: Open University Press.

Holden, L. (2005). Complex adaptive systems: Concept analysis. *Journal of Advanced Nursing, 52*, 651–657.

Penprase, B. (2003). Understanding hospitals' changing environments through the lens of complexity theory (Doctoral dissertation, Wayne State University). *Dissertation Abstracts International.* Retrieved from www.lib.umi.com/dissertations

Penprase, B., & Norris, D. (2005). What nurse leaders should know about complex adaptive systems theory. *Nursing Leadership Forum, 9*(3), 127–132.

CHAPTER 14
TECHNOLOGICAL CHANGE IN HEALTH CARE ELECTRONIC DOCUMENTATION AS FACILITATED THROUGH THE SCIENCE OF COMPLEXITY

- *Swinderman presented a model of nurse informaticians with respect to the use of chaos theory. How do you understand chaos theory as the science of change? What does this model mean to you?*
- *The caring nurse is a "magnetic attractor." What does the concept "magnetic attractor" mean to you?*
- *What does it mean to relationally self-organize?*
- *Penprase illuminated concepts that are relevant to complex health care systems. Explain the importance of understanding the micro, meso, and macro scaling variances in nursing and nursing informatics. What is the meaning of the choice point in the transformative process of relational self-organization or change using your understanding of chaos theory?*

FIFTEEN

Implementing Change in Nursing Informatics Practice*

Aric S. Campling

Marilyn A. Ray

Jacqueline M. Lopez-Devine

Implementing change to existing clinical information systems can provide a challenge for the nursing informatics specialist (NIS), as well as for nursing leadership in general. In the complex world of health care delivery, nursing leaders and informatics specialists can use an understanding of various change and implementation theories to promote acceptance of new systems and processes in nursing practice. In the future, nursing professionals such as informatics specialists and nursing managers must not forget their core mission of caring, such as how the language of the caring relationship, pattern, and holism is actuated. Choices and actions taken by these influential professionals must maintain a caring, humanistic focus. Nursing informatics specialists can provide opportunities to include the language of caring in electronic health records and stability and support in the face of chaos when technical change is imminent. This chapter describes select theories to assist with innovation and change management for information technological implementation and describes two cases where these theories were successfully used in nursing practice. Respondent Valentine highlights innovation-diffusion theory (a significant change theory) when major clinical information technological initiatives are introduced within health care organizations. She discusses how the change theories have to be tailored to fit the organizations' user population (i.e., for the most part, nurses). This response shows how the authors of this chapter implemented the change in nursing documentation by assessing organizational readiness, identifying

*The Department of Patient Services and the Department of Information Systems at Bethesda Memorial Hospital, Boynton Beach, Florida, supported the projects described in this chapter. The authors currently have no commercial, proprietary, or financial interest in the organizations or theories described in this article.

key stakeholders, and evaluating the implementation process. Nursing informatics must create new learning cultures where using the language of complexity, and the language of caring will be integrated, and nurses and others can recognize how technology as a tool can elevate practice and transform clinical practice toward improved caring and patient outcomes.

INTRODUCTION

The Challenge of Implementing Change

Nursing in the contemporary technological environment is complex (Coffman, in press; Ray, 1987). Implementing change to existing clinical information systems (electronic health records) can provide a challenge for the nursing informatics specialist (NIS), as well as for nursing leadership in general. Nursing leaders and informatics specialists need to understand various change and implementation theories to promote acceptance of new technical systems and processes. As nursing professionals, informatics specialists and nursing managers must not forget their core mission of caring, such as how the language of the caring relationship, pattern, and holism is actuated. These influential professionals' choices and actions must maintain a caring, humanistic focus. Nursing informatics specialists can provide opportunities to include the language of caring in electronic health records, stability, and support in the face of chaos when technical change is imminent (Barnard & Locsin, 2007; Ray et al., 1995; Swinderman, 2005, in press; Turkel, Ray, & Kornblatt, in press).

This chapter outlines select theories related to the implementation of change in clinical informatics systems, including innovation-diffusion, implementation, and change management within a complex caring dynamic. Practical examples describing strategies for using these theories will be presented.

CHANGE THEORY: INNOVATION-DIFFUSION

In implementing any kind of system change, the NIS must understand the social, organizational, and human caring undercurrents that drive the dynamics of the organizational system. Rogers' (2003) theory of innovation-diffusion provides a method of examining the processes of change and adoption of changes through the use of complex social channels within an organization. Social channels include change agents in relationship with professionals who have knowledge of the history of the diffusion of ideas and the potential outcomes of innovation. Thus, innovation-diffusion asserts that information regarding changes (or "innovations") is passed through these channels, wherein individuals follow a process of discernment to cope with the pending change. The informed NIS, as an agent for change, is one who understands and uses the theory's tenets to manage change and implementation.

Professionals first must have knowledge of complex systems that are always in a state of change or transformation, an ongoing emergent process. Change is based upon choices that are made relative to new innovations in a network of relationships that move between disorder and order through choice (Ray, 1998). Professionals acquire knowledge of the change, and they then have to incorporate this change into their current frameset, based on the perceived characteristics of the change (Rogers, 2003). They focus this integration process into a decision: whether to accept the change or to reject it. During the implementation, the individual participates in enacting the changes. Afterwards, they receive a confirmation of their decision—either negative or positive (Hilz, 2000).

Adopters of change are categorized in five ways: innovators, who stand at the forefront of change; early adopters, who accept the change right away, but are not the point-persons for implementation; early majority and late majority, who comprise more than 60% of those implementing the change; and finally the laggards who resist the change to the very end (Hilz, 2000). Despite the negative connotation of this latter term, these people may often have some insight into why an otherwise accepted system may ultimately fail, for example, if caring in nursing is not mentioned or integrated into the system of change.

The NIS must consider several factors—such as individual, organizational, social, and caring attributes—in estimating the potential for successful diffusion. At each stage of the innovation-diffusion process, there are tasks that can be performed to ease the implementation and acceptance of information systems change in an organization (Hilz, 2000; Rogers, 2003).

The organizational and social context into which change is introduced (as a means to the end of operations and management support to enhance organizational efficiency) must be considered when weighing the introduction's impact (Romano, 1996). System changes present challenges to currently held organizational and cultural beliefs regarding the individual's or group's work life. These challenges can be planned for, and managed, when considering any changes to or implementation of a system (Romano, 1996). As users find themselves challenged to accept new changes, they may attempt to cope by continuing to follow the old status quo. Even the most technically advanced, well-designed system can be crippled by users with low ownership of a project and who resist the implementation; however, knowledge of potential challenges and the development of strategies can help an organization overcome resistance and reach their future desired state (Lorenzi & Riley, 2002).

Change implementation, especially technical, within an organization can challenge the existing formal and informal system structures or cultures—despite or because of the organizational need for change. Major changes affect the network of human relationships and may increase anxiety and stress. Administrators should plan for some resistance because of this stress.

Resistance typically occurs when the users see no direct benefit from the change or when the change clashes with perceived norms and values. People may feel a threat of personal loss when facing change, whether that threat is job loss or just a challenge to a comfortable routine (Lorenzi & Riley, 2002). Resistance can be seen as a rejection of technological change, based on unawareness or perceived negative attributes (Hilz, 2000; Rogers, 2003; Romano, 1996). Managing change implementation requires an understanding of the factors surrounding potential resistance and various implementation methods that can be used to minimize the stress of implementation and thus enhance acceptance.

Benefits Management

More often than not, the clinical areas in health care organizations are held responsible for realizing the benefits of information systems change. A "benefit" is realized when an implementation results in a change that has a positive effect on the resources required to provide care or on quality of care itself (Hardiker, 2009; Hoehn, 1996; Hunter, 2002; Sensmeier, 2006). Difficulties in proving benefits from clinical information systems and changes to such systems typically stem from an inability to quantify the qualitative nature of nursing care (Hardiker, 2009). In addition, costs tend to be underestimated, and benefits are sometimes unrealistically estimated (Hoehn, 1996).

Two major models for benefit identification and realization exist. Cost justification describes a system where costs and priorities are defined, an appropriate implementation is selected, baseline data regarding system capability to accept the change are collected, benefits are concretely defined, management commitment to the change is obtained, the system is implemented, and the benefits are attained (Hoehn, 1996). This model has many shortcomings, the most severe of which is that this model is primarily cost driven in nature and seeks to justify a decision to make a change only after the change policy and process have been decided.

The second model is benefits management. In this process, the first step is to secure senior management sponsorship for a change. The NIS and the implementation team must evaluate the organization's culture, especially the nursing culture, current operations, and assess the potential for change, identify areas that need immediate operational improvement, and identify and prioritize the benefits attainable from any changes (Hoehn, 1996). NIS can select an implementation based on capability of the system to accept the change and the costs involved; validate the benefits of the implementation; then develop a plan and implement the system while seeking understanding and managing toward the validated benefits (Hoehn, 1996). After the implementation, the team should assess and fine-tune the system postimplementation; identify the actual attained benefits; and finally, reevaluate their current operations (Hoehn, 1996; Turkel & Ray, in press). This system creates a benefits-driven cycle of validation, monitoring, implementation, and confirmation. Understanding the process of benefits management, and the points at which the NIS and nursing administration can influence end-user acceptance of changes, requires a more focused understanding of the implementation process itself.

Implementation Framework

Mobilizing for major change involves three processes: identifying behavioral characteristics of change, understanding and influencing the decision-making process, and developing an implementation framework (Abbott, 1996). As described in the previous section, early and strong upper-management support is crucial to the success of any change implementation.

The first two processes described expand upon the role of the nurse as change agent and on the concepts found in Rogers' (2003) innovation-diffusion theory. The first step of the implementation framework, planning, calls for an implementation team to be formed. Efforts in this phase go toward understanding the current culture and environment and understanding the rationale for change (Abbott, 1996). Focus groups with key nursing personnel should be organized to develop knowledge of the nursing culture and caring practices. Resource organization and orientation is a key process early in the project organization, especially how nursing resources will be impacted. The project manager and core project team should develop the project plan. Realistic and flexible changes in process and workflow should be identified, planned for, and phased in. Any proposed issues and changes should involve early documentation and education far in advance of the proposed project deadline, also known as the "go-live" date. Implementation task forces are created to manage smaller identified tasks, and task force education and training is done (Abbott, 1996; Craig, 2002). This education emphasizes skills needed, aims to assist with understanding the change, and the necessity for change. Task forces or focus group leaders examine any areas that may require change prior to a go-live date, and are usually focused on specific areas, cultural norms, workflows, or application functionality. Several types of testing are done, including system testing and acceptance testing (Hoehn, 1996). This involves verification of the proposed implementation, and assures that users have a say in verifying the changes. When implementing a technical system involving nursing, nurses need to know that their voices have been heard. The implementation team develops system documentation, and then the end-users are trained and educated toward the same goals as the initial task forces and the larger project team. (With a new language and process for nurse caring practice on the horizon [Turkel et al., in press], NIS or information technologists must be aware and open to how a new language for assessing patients' needs can affect information system technological development.) At the education stage of the process, the prepared nurse manager or informatics specialist has the ability to influence the innovation-decision process (Abbott, 1996; Hilz, 2000; Rogers, 2003). Sensitivity to the core caring mission of nursing and nursing informatics is critical for successful acceptance and implementation. After testing and end-user training are complete, the implementation is set "live" and placed into production status, making the new system accessible and usable by the end users. The final step of the implementation process requires the committee to conduct a postimplementation review of the changes that were made and recommend further action, if necessary (Abbott, 1996). System evaluation should be carried out on an ongoing basis when patient caring is involved.

PROJECT IMPLEMENTATION

These theories and processes were put to use in implementing a number of priority-identified changes to computerized documentation procedures initiated at a regional community hospital. Two projects are detailed, describing how these projects benefited from innovation-diffusion theory, an understanding of the core mission of caring and resistance, and how they affect benefits management, and system/change implementation.

PROJECT I: REVISION OF NURSING CARE PLAN DOCUMENTATION

A representative of the hospital's medical staff asked for a meeting with the hospital president and the vice president of patient care services. Representatives from nursing and quality improvement were also present. At the meeting, the physician identified concerns regarding the relative usability of nursing care plan documentation. He explained, with samples from patients' charts, how the current design of the notes made them very difficult for the doctors to read, and filled the notes with information the doctors considered useless for their purposes. They did, however, find the nurses' narrative notes to be most useful because in these notes, the actual nursing care can be more easily identified. These free-text, typed narratives were interspersed throughout a sea of preformatted notes—notes that are generated from forms on a computer screen that have predefined selections of text from which nurses make selections.

With this problem identified by the physicians, management quickly took the lead in declaring a need for change to occur in the documentation process. A clinical documentation committee was formed with the intent of spearheading these changes. The committee comprised several vice presidents and nursing directors of clinical areas and information systems. As the committee continued to evaluate the current environment, the team found that nurses were also discontent with the charting processes. They voiced concerns about excessive time spent charting and about duplication of charting efforts. A number of areas for improvement were identified, and priorities were given to revising existing nursing care—plan pathways with a more interdisciplinary approach. With the full support of executive level management (in this case, including the support of the hospital president), a work plan was created to address the numerous issues, identified by both nurses and physicians, uncovered during the assessment phase. A task force was created, consisting of a nurse educator, director of clinical informatics, an informatics specialist, and an assistant nurse manager. This team worked to identify specifically how improvements could be made to the care plan documentation process. "Rebuild it from the ground up, if you have to," the president had advised. "Think out of the box." With each concrete milestone reached by the task force, a report was given back to the committee for advisement and a go-ahead to continue.

The task force met at regular intervals, at first to identify what needed change within the system. After much discussion and research on the existing information system's software capabilities, the team opted to create a new "problem list" based on common interdisciplinary activities surrounding specific conditions that a majority of the patient population presented with, conditions such as "cardiac dysfunction" or "electrolyte imbalance." None were so specific as to qualify as medical diagnoses, but they were concrete problems that couldn't be mistaken or confused. Serious consideration was given to the concurrent implementation of NANDA-based nursing diagnoses (North American Nursing Diagnosis Association [NANDA], 2001), but a decision was made to refrain from making too many significant workflow and process changes all at once, and maintain the "problem list" approach as it was a familiar bridge to the prior documentation method and met the needs of multiple disciplines utilizing the clinical computer system. The team then aimed to incorporate standard interventions and outcomes using a tested set of classifications known as the Nursing Interventions Classification (NIC) (McCloskey & Bulechek, 2000) and Nursing Outcomes Classification (NOC) (Johnson, Maas, & Moorhead, 2000).

Interventions are defined as treatments based upon clinical judgment and knowledge that a clinician performs to enhance patient outcomes (McCloskey & Bulechek, 2000). Each intervention is defined with a set of activities to follow. The NIC taxonomy enabled the ability to tie expected clinical interventions to specific problems. Each problem was given a specific intervention from NIC (e.g., "Electrolyte Imbalance" may have linked to it the intervention category, "Electrolyte Management"). Because of limitations in this system's software, particularly regarding the amount of text that could appear on the screen at one time, thought had to be given to how interventions would be documented. Because of this, a few select, critical activities associated with each intervention were identified and linked against the problem.

Outcomes change the focus from nursing activities to the patient's condition at a given point in time (Johnson et al., 2000). The NOC taxonomy designates a comprehensive list of desired patient outcomes. For each outcome, it defines a set of indicators by which to gauge an intervention's success toward attaining that outcome. For each set of problems and interventions identified, a key outcome was selected. Due to space limitations within the computer application, which were identified early in the assessment process, the team could not maintain a full list of indicators for each problem. The team, however, kept these referenced for future use.

These specifications were handed to two clinical informatics specialists who would be designing the computer screens and data/logic processing, which would become the new care plan pathway. In task force meetings, the team decided on a two-phase approach to implementation to ease the burden of change on the clinical users. The front-end system—the screens and displays where the users interact with the system—would first be aesthetically modified to reflect a shift in mind-set to a more interdisciplinary approach. "Care Plan" selections became "Problem List" items. (Concepts like problem list and care plan are continually being reevaluated in nursing as issues are revealed [Turkel et al., in press]).

This first stage was built and tested, and then verified by the task force. Task force members then brought these changes to clinical educators, who were actively involved in the process of bringing this change to the nurses. Educators' acceptance of the first-stage modifications was fairly widespread, which enhanced nursing staff acceptance. Very little resistance existed toward this change. The implementation was made, and post-evaluation indicated validation of acceptance, with no problems identified in the change.

At this point, the preliminary work of the phase two implementation was ready for testing. The task force decided that for each problem, a screen would be built to capture three items. The first was to capture implementation activities, based on the NIC. This was implemented as a specified selection list of three or four activities. When selected, these items would create a preformatted line on the patient note that read like a natural-language sentence. The second was to capture a general assessment of the patient's outcome status—whether they had met the outcome or not met the outcome—based on clinical judgment. Each screen contained the ability for clinicians to view a "help screen," which would detail for them the indicators that described each outcome. Third, in order to streamline the charting process, the team provided each problem screen with free-text space for users to type their narrative notes. This way, nurses and other clinicians would not have to click through other screens to free text their notes. The charting policies and process were reinforced and streamlined, but fundamentally unchanged. The new screens also offered space to chart free text notes along with their preformatted notes, which reduced time spent navigating through the system to reach the free-text charting area. The new pathways offer nurses' charting guidelines and assistance through "help screens," which also reflect the hospital's identified standards of care.

A testing environment was created in which the task force members could educate the educators and help them learn not only the changes coming but also the organizational necessity for the changes. The educators and nursing leadership solidly accepted the modifications. These team members then took the education initiatives to their staff, passed along the information regarding the pending changes, and listened for feedback and watched for resistance. Feedback from the testers and the educators allowed the informatics specialists to customize and refine the pathways as needed until they were ready for general use.

The task force returned to the committee with their final work and planned their "go-live" date. Educational initiatives were stepped up amongst end users with posters, booklets, pamphlets, and direct interaction between educators, management, and end users. Through this interaction, acceptance was increased hospital-wide. The team felt certain that medical, clinical, and administrative staff would all see realization of identified benefits. The planning team enabled the system changes on the specified date and performed continued testing of the changes on the live system, even as our end users were using the pathways successfully. Educators and administrators were readily available to assist staff specifically for this project, if needed, and the information systems helpdesk was on call as always.

Postimplementation evaluation was ongoing from the live date and involved discussion with clinical staff regarding perceived improvement in charting times, ease of use, and decrease in redundancy. Specific physicians were identified as point medical staff members to ask about the changes to the documentation and to see if the end result met their particular needs for more useful and informative nursing notes. Clinical nurses who were most influential in vocalizing the necessary changes were also tapped for feedback. These included end users, charge nurses, and clinical managers. Responses were all very positive, especially from those physicians and nurses who were having the most problems with the original process.

The task force continued to monitor the situation and make corrections and enhancements as needed. However, the overwhelming support from all areas on this project, and the universal perception of a definite need for major technical change to documentation, made acceptance of the change through the innovation-decision process very high. The nurse informatics specialist and nursing management involved were therefore able to deliver on expectations and validate those acceptances. They were able to meet the initial benefit analysis throughout the process of this implementation and to maintain a focus on caring processes. However, not all changes are successful on the first try.

PROJECT II: PAIN DOCUMENTATION, PHASE I

As part of the hospital's ongoing commitment to pain management (used as an example of major technical change implementation), a pain committee was formed with the purpose of examining current pain management methods within the hospital and recommending changes to be made to the processes and functions involved. The pain team developed a paper form for nurses, which they were required to use for each instance of pain documented in a given day. The form requested the incident date and time, the level of pain on a 10-point analog scale, location and description of pain, interventions used, and the outcome of those interventions measured after a specified time, again based on a 10-point pain scale. Prior to the housewide implementation of this form, the pain assessment pathway was included as part of the routine shift assessment charting in the computer-based record, but was found to have low utilization. The pathway was not conducive to charting multiple facets of a patient's pain experience and was thought to be confusing by many nurses. The switch to the paper form was driven primarily by changes to policy and process. In part, government or independent industry review organizations mandated these changes. Final implementation of the changes were recommended by the pain team and supported by upper management. Executives and managers later deemed the change to be a *backward* slide from the hospital's ongoing efforts to computerize as many aspects of the medical record as possible.

After successful implementation of the paper documentation process, the director of clinical information systems questioned nursing management and many clinicians regarding the new processes. A major concern voiced by nurses, nurse leaders, and educators was the fact that pain charting was no longer part of the computer-based record, creating an actual dissonance, an interruption on documentation workflow as users had to switch frequently between two documentation methods. They were also concerned about duplication of efforts, as many nurses also charted their pain assessment as part of a free-text note (as they no longer had access to pain charting in the assessment pathway). However, the pain team did not unanimously support automation of the pain pathways; some members insisted that it would cause a severe fall in compliance with newly established pain documentation requirements. These members also did not share the opinion that removing pain charting from the computer and placing it back on paper was a backslide. Because of this, the clinical informatics director had a difficult time securing executive-level support for a move to automated pathways.

The director was eventually able to obtain upper-level support from the vice president of patient services by utilizing the information from clinical users and educators, a good understanding of the joint commission pain initiatives, and the hospital's ongoing mission to automate clinical areas. With this backing, the pain team was asked to reevaluate the current process and assess the ability to migrate their documentation to the computerized systems. The committee determined that clinical nurses were willing to accept changes involved in moving the pain charting back to the computer. The process from systems assessment and needs assessment to implementation was a quick one for this change because the voices of especially nurse-users were taken into consideration. Also, a clinical informatics specialist spent time rebuilding existing pain charting pathways in the assessment application to mimic the paper charting pathways. Pain was added as a "body system" to the shift assessment pathways. The pain screens were designed to first ask a clinician user how many instances of pain they were charting and then load a sequential set of screens that prompted for the same information found on the paper document for each instance. This information was saved to the assessment flow sheet, which was then printed in a columnar format—each assessment in one column (up to four columns per page), with the assessment codes arranged by system down the left side. The building, testing, customizing, and rebuilding process ultimately took several weeks until the pain team members were satisfied that they had a product adequate to meet satisfactory charting standards and support nurse caring processes.

The clinical documentation committee decided to test this change out on a single unit, as a pilot study, to determine how compliance would fare. The initial pilot date was preceded with education and training of the clinical educators and charge nurses, who in turn brought that education to their direct care users. Initial acceptance of the change was high—nurses were eager to chart in the computer system as they always did. The pilot date came, and the new assessment pathways were set live for that single unit, who used it in place of the paper documentation for 2 weeks. At the end of 2 weeks, the clinical documentation committee conducted a post-pilot review.

The team discovered that predicted benefits of this pilot study were not realized. Clinical nurse users found the charting process to be slow and bulky, with too many screen flips to click through. In addition, logic built into the screen processing sometimes made it difficult to abandon the pain charting if a nurse found it necessary to do so (e.g., they might have been called away from the computer for an urgent issue). In general, the pilot users were very discontent with the changes and had reverted to using the paper documentation as a "lesser of two evils."

Physicians found the new paperwork generated by the assessment pathway to be confusing and had trouble finding the pain documentation details. Among the concerns voiced by the physicians came a suggestion: Doctors could read and understand the recently enabled problem list documents, and these were the primary documentation doctors checked on a regular basis to see how their patients were faring. The request came to move pain assessment into the problem list pathways and create a printed document that was clear, concise, and informative.

PROJECT II: PAIN DOCUMENTATION, PHASE II

The pain team returned to the table to revise the pain documentation policy and process. Fundamentally, the policies were sound, but the workflow the project team developed had created excessive work. The problem list application was a viable alternative and fit the concept of managing pain as a factor of a patient's care, not simply as an assessment item. The new pain "problem" documentation screen required either a pre- or postimplementation pain scale, but did not require both together (though both could be documented together, if applicable). If a pain intervention was done at the end of a shift, the leaving nurse could chart the preimplementation, and the oncoming nurse could complete the postintervention charting. Pain location was required, which could be typed in or picked from a list of sites. At least one of a set of pain qualities or characteristics had to be included and could be selected from a list or typed in. They must indicate if the pain was continuous or intermittent; if intermittent, nurses must indicate a frequency of recurrence. They then had a preformatted "pick-list" of expected interventions to select from based on hospital policy and regulatory standards. The screen contained a few lines on which the clinical nurses could type any free-text notes they deem necessary to ensure completeness of charting. This charting information then printed in a full-page-width format equivalent to other notes, with information entered on the screen being concatenated into simple, straightforward, natural-language sentences.

Clinical leaders then tested these screens and documents. The new pathway revisions were receiving acceptance, but this time with some mild hesitance related to the initial nonacceptance of pilot changes. However, in general, support for the revisions was good. The clinical documentation committee decided to go live with the problem-list-based pain pathways throughout the hospital. Clinical educators and resource staff managed the necessary staff education. The implementation was made and found easy acceptance in the nursing areas.

Postimplementation assessments were conducted. Initial feedback from and continued informal surveys of clinical areas indicated that users found the new process intuitive. Providers liked the new streamlined documentation and stated it was easier to understand. They were also pleased with the added benefit of seeing their pain documentation along with the other problem documentation and narrative notes. In general, there was a great validation of the second phase of implementation.

IMPLICATIONS FOR THE FUTURE

Innovation-diffusion theory (Rogers, 2003) provides a sound framework on which to build user support for changes identified as necessary by hospital nurses, physicians, or managers. This chapter presented the innovation-decision process, including implementation phases that a nurse informatics specialist within a hospital organization used to influence the acceptance of a major technical change. The idea of nurse caring was preserved as the NIS interacted with executives, managers, and practicing nurses and implemented the new system.

As nursing moves forward within the contemporary health care environment, nursing informatics specialists must remember that they are first and foremost members of a caring profession. They should never let the disorder introduced by required, organizational change of information technologies in hospital bureaucracies allow them or their staff to be distracted from their core caring mission (Coffman, in press; Crow, 2001; Ray, 1987; Swinderman, 2005; Turkel & Ray, 2004). The reality of economics, regulatory mandates, technology, and the mission of caring are integrated in any organization (Ray, 1987; Ray et al., 1995; Swinderman, 2005, in press; Turkel & Ray, 2000). However, a deeper understanding of the complexity of a hospital environment, professional norms, cultural values, and the meaning of caring in nursing, and the positive application of innovation and change theories can create a work environment conducive to enhancing relationship-centered care (Ray, 2010). The caring, ethical decisions and actions undertaken by clinical nurse specialists and nurse managers can foster a new order and stability from initial chaos associated with major change (Ray, 1998; Ray et al., 1995; Swinderman, in press). Technological caring is a necessity in a patient–nurse caring centered approach (Barnard & Locsin, 2007; Ray, 1987;

Swinderman, 2005). This new order also can assist in the examination of how the language of caring can be more fully integrated into nursing informatics science. For example, can pain experienced by patients be incorporated as part of the language of caring through an understanding of the actual suffering of patients and how nurses relate caringly to them? Turkel, Ray, and Kornblatt (2011) developed a new relational caring mnemonic of Recognizing, Connecting, Partnering, and Reflecting based upon a *nursing* rather than a medical paradigm to begin to answer some of the vexing questions. This new mnemonic offers challenges to the *nurse-directed* language of the nursing process. A nursing paradigm of relationship-centered caring in clinical practice enhances awareness of ongoing inquiry and the reevaluation of the nursing language used in nursing. Concepts like those advanced by the North American Diagnosis Association (NANDA), the NIC and NOC systems, and the language of the problem list or care plan are in need of open dialogue and research to determine their impact on the future of information or communication technologies (Turkel et al., in press). With a clear understanding of the complex organizational processes they will encounter, NIS professionals can continue to support an efficient technical system, but one that is grounded in caring, one that is socially and economically responsible, and one that reacts well to future requirements for innovation-diffusion investigation.

REFERENCES

Abbott, P. A. (1996). The nursing manager's role in successful system implementation. In M. E. Mills, C. A. Romano, & B. R. Heller (Eds.), *Information management in nursing and healthcare* (pp. 116–127). Spring House, PA: Springhouse.

Barnard, A., & Locsin, R. (Eds.). (2007). *Technology and nursing: Practice, concepts, and issues*. United Kingdom: Palgrave.

Coffman, S. (in press). Marilyn Anne Ray: Theory of bureaucratic caring. In M. Alligood (Ed.), *Nursing theorists and their work*. St. Louis, MO: Mosby/Elsevier.

Craig, J. (2002). The life cycle of a health care information system. In S. Englebardt & R. Nelson (Eds.), *Health care informatics: An interdisciplinary approach* (pp. 181–208). St. Louis, MO: Mosby.

Crow, G. (2001). Caring and professional practice settings: The impact of technology, change, and efficiency. *Nursing Administration Quarterly, 25*(3), 15–23.

Hardiker, N. (2009). Developing standardized terminologies in nursing informatics. In D. McGonigle & K. Mastrian (Eds.), *Nursing informatics and the foundation of knowledge* (pp. 97–106). Sudbury, MA: Jones and Bartlett.

Hilz, L. M. (2000). The informatics nurse specialist as change agent: Application of innovation-diffusion theory. *Computers in Nursing, 18*(6), 272–281.

Hoehn, B. J. (1996). Benefits management. In M. E. C. Mills, C. A. Romano, & B. R. Heller (Eds.), *Information management in nursing and healthcare* (pp. 99–107). Spring House, PA: Springhouse.

Hunter, K. M. (2002). Electronic health records. In S. Englebardt & R. Nelson (Eds.), *Health care informatics: An interdisciplinary approach* (pp. 209–230). St. Louis, MO: Mosby.

Johnson, M., Maas, M., & Moorhead, S. (Eds.). (2000). *Iowa outcomes project: Nursing outcomes classification (NOC)* (2nd ed.). St. Louis, MO: Mosby.

Lorenzi, N. M., & Riley, R. T. (2002). Managing change: An overview. *Journal of the American Medical Informatics Association, 7*(2), 116–124.

McCloskey, J., & Bulechek, G. (Eds.). (2000). *Iowa intervention project: Nursing interventions classification (NIC)* (3rd ed.). St. Louis, MO: Mosby.

North American Nursing Diagnosis Association. (2001). *NANDA Nursing diagnoses: Definitions and classification, 2001–2002.* Philadelphia: Author.

Ray, M. (1987). Technological caring: A new model in critical care. *Dimensions of Critical Care Nursing, 6,* 166–173.

Ray, M. (1998). Complexity and nursing science. *Nursing Science Quarterly, 11*(3), 91–93.

Ray, M. (2010). *Transcultural caring dynamics in nursing and health care.* Philadelphia: F. A. Davis.

Ray, M., Didominic, V., Dittman, P., Hurst, P., Seaver, J., Sorbello, B., et al. (1995). The edge of chaos: Caring and the bottom line. *Nursing Management, 26*(9), 48–50.

Rogers, E. (2003). *Diffusion of innovations* (5th ed.). New York: Free Press.

Romano, C. A. (1996). Organizational impact of information systems. In M. Mills, C. A. Romano, & B. Heller (Eds.), *Information management in nursing and healthcare* (pp. 83–91). Spring House, PA: Springhouse.

Sensmeier, J. (2006). Healthcare data standards. In V. Saba & K. McCormick (Eds.), *Essentials of nursing informatics* (pp. 217–228). New York: McGraw-Hill

Swinderman, T. (2005). *The magnetic appeal of nurse informaticians: Caring attractor for emergence.* Doctor of Nursing Science unpublished doctoral dissertation, Florida Atlantic University, Boca Raton, FL.

Swinderman, T. (2011). Technological change in health care electronic documentation as facilitated through the science of complexity. In A. Davidson, M. Ray, & M. Turkel (Eds.), *Nursing, caring, and complexity science: For human–environment well-being.* New York: Springer Publishing Company.

Turkel, M., & Ray, M. (2000). Relational complexity: A theory of the nurse-patient relationship within an economic context. *Nursing Science Quarterly, 13*(4), 306–313.

Turkel, M., & Ray, M. (2004). Creating a caring practice environment through self-renewal. *Nursing Administration Quarterly, 28*(4), 249–254.

Turkel, M., Ray, M., & Kornblatt, L. (2011). Instead of reconceptualizing the nursing process let's re-name it. *Nursing Science Quarterly.*

Response to Chapter 15

Kathleen Valentine

1. What is the relationship between theory and practice (clinical, administrative, or educational) highlighted in this chapter?

The authors provided an overview of Roger's Innovation-Diffusion Theory (2003), with references to theories of implementation and change management from (Barnard & Locsin, 2007; Hilz, 2000; Ray, 1998; Ray et al., 1995). As large-scale clinical technology initiatives are introduced within organizations, it is essential to use evidence-based implementation and change strategies that are specifically tailored to that organization's end users in order to realize the clinical transformation that is often cited as a benefit. The clinical benefits need to be clearly articulated at the outset of the project so that practice drives changes in technology rather than technology driving changes in practice. If benefits are relevant and evident, the clinician can then choose to engage in the personal changes necessary to adopt new behaviors and skills, which, when achieved, elevate practice. Organizations don't change; it is the people within organizations that change—or don't. The health care industry relies on knowledge workers who individually perform by their own self-guided vision of what is best and what is possible. Active change management welcomes resistance as a road map to success. Resistance helps to outline how to assist the end user prepare for adoption.

Silent conformity can be a sign of disengagement and disinterest. An example of this was evident in the second case the authors shared. Lack of overtly expressed resistance should not be mistaken for agreement. Resistance must be actively solicited and used to help transform early conflict into problem solving. By surfacing the fears, concerns, identification of system errors or flaws, projects can be improved and end users feel confident that their voices are heard and considered.

Therefore, the human side of any technological change requires detailed analysis related to clinician readiness; stakeholder identification, communication, and engagement; competency development, performance coaching, and reinforcement for the end user. The same type of detailed project management used with the technological installation must be interwoven with the human side of change (Fong & Valentine, 2007; Valentine, 2005).

2. How does a nurse, physician, health care administrator learn about complexity science/s and caring through the presentation of ideas or research?

What one learns about complexity from this case is that each case is unique. Though patterns can be gained and lessons learned from earlier change attempts, nothing can substitute for the due diligence of planning for both the human and technical interrelated dimensions of each change project.

The authors present an example of assessing organizational readiness and identification of key stakeholders, their relevance, and likelihood of support (Mitchell, Agle, & Wood, 1997). Through their example they were able to convey the importance of power, legitimacy, and urgency (Mitchell et al., 1997) in a change process. In the first example, the physician, senior leadership, and management had positional power. Legitimacy was established by the expressed physician value and need for clear communication in nursing notes. Urgency was established through a need to satisfy a perceived widespread dissatisfaction related to nurses' documentation, which ultimately affects patient care. Stakeholders with a combination of power, legitimacy, and urgency have the authority and impetus to influence the success of a change initiative (Mitchell et al., 1997). The sponsorship of the enterprise-wide committee provided the leadership and resources for a large-scale redesign of the electronic health record documentation, and a reframing of the nursing care plan to an interdisciplinary problem list.

As the NIS works with nurses around large-scale changes, seeing the nurse as a "caring person" informed through disciplinary knowledge of caring can serve to frame the dialogue and interactions about creating tools and resources to deliver "best care" to patients. In the second example, there was evidence that nursing was "hesitant" related to the changes in the pain record documentation. Recognizing resistance as indicative of frustration in a caring process can guide the NIS to inquire and seek to understand more about the resistance as an active indicator of what might need to be different. "Managing resistance" can be seen as an impersonal strategy to keep the project on track. An alternate frame of reference could be to view "resistance" as an opportunity for authentic engagement with those who care enough to give voice to their legitimate concerns and ultimately increase readiness for adoption.

In the second example, nurses voiced that they were unable to have a holistic view of the patients' stability and condition without having the "fifth vital sign" integrated into the electronic documentation tool. This is an example of the need for practice to shape technology rather than documentation tools shaping practice. Ultimately, a practice-based solution was discovered and implemented in part due to attending to the "hesitant" staff.

3. What meanings are illuminated in this chapter for nursing or other health care professions? Direct your answer to education, administration, research, or practice based upon the chapter's focus.

It is essential to have clear expectations and executive ownership of a large-scale change process. In the two examples provided, the first has clear sponsorship and visible reinforcement of the project's importance to the core business of caring for patients and communicating effectively across the interdisciplinary team. Initially, the second example did not have end-user input, an executive sponsor, systematic communication, detailed competency development for end users, and lacked a clear vision of the preferred future state. Without the clear and articulated vision the project became derailed into a suboptimal paper documentation alternative, which ended up costing time, effort, and end users' trust. Skepticism was higher after a failed process change attempt.

4. What is the relevance of this work to the future of the discipline and profession of nursing, health care professions, and health care in general?

Nursing disciplinary knowledge includes the specialty of nursing informatics. To advance the profession we will need to optimize our recognition and use of this specialty and spread core competencies about the use of technology into clinical practice that transforms practice to our preferred vision of the future. We have to collectively create learning cultures with continuous feedback about what is working and what can be improved. Technology is integrated as a decision support tool for nursing, as an administrative tool for system changes, and as a means to capture data for research and continuous improvement in clinical care. Though we will rely on the leadership of the nurse informatics specialist, each professional nurse must also choose to personally adopt necessary technological skills that promote excellence in patient care. In increasingly complex environments when "wisdom is applied with compassion" (American Nurses Association, 1998, 2008) by the NIS it is an expression of caring and helps perpetuate a learning culture. At the conclusion of an implementation, the NIS knows successful integration of technology, people, and caring priorities is achieved if the end user expresses the benefits of the change. For example, "I know that this organization cares about me, I am well prepared and confident to care for my patients well using this new technology" (Fong & Valentine, 2007). That preferred vision of the future keeps technology as a tool that elevates practice and transforms clinical practice toward improved quality patient care.

REFERENCES

American Nurses Association. (1998). *Nursing informatics: Scope and standards of practice*. Silver Spring, MD: Author.

American Nurses Association. (2008). *Nursing informatics: Scope and standards of practice*. Silver Spring, MD: Author.

Barnard, A., & Locsin, R. (Eds.). (2007). *Technology and nursing practice: Practice, concepts, and issues.* United Kingdom: Palgrave.

Fong, V., & Valentine, K. (2007, April 20). *Taxonomies define safe, effective practice or enable data retrieval?* Proceedings of the 6th Association for Common European Nursing Diagnoses, Interventions and Outcomes (ACENDIO) Conference, Amsterdam, The Netherlands, Session #39.

Hilz, L. M. (2000). The informatics nurse specialist as change agent: Application of innovation-diffusion theory. *Computers in Nursing, 18*(6), 272–281.

Mitchell, R., Agle, B., & Wood, D. (1997). Toward a theory of stakeholder identification and salience: Defining the principle of who and what really counts. *Academy of Management Review, 22*(4), 853–886.

Ray, M. (1998). Complexity and nursing science. *Nursing Science Quarterly, 11*(3), 91–93.

Ray, M., Didominic, V., Dittman, P., Hurst, P., Seaver, J., Sorbello, B., et al. (1995). The edge of chaos: Caring and the bottom line. *Nursing Management, 26*(9), 48–50.

Valentine, K. (2005). Electronic medical records promote caring and enhance professional vigilance. *International Journal for Human Caring, 9*(2), 121.

CHAPTER 15
CHANGING NURSING INFORMATICS PRACTICE

- *How can you integrate concepts from change-theory (innovation-diffusion) into changes that need to be made in your organization in terms of practice or organizational culture? Who in the organization would be an early adopter?*
- *Can caring practices be easily documented within your current electronic or paper documentation system?*
- *If you could make one change to your current documentation system, what would that be?*
- *Would you be able to implement in your organization the nursing care plan described in this chapter? What support would you receive? What barriers would need to be overcome?*

SIXTEEN

Human Rights and Humanoid Relationships in Nursing and Complexity Science

Rozzano Locsin
Marguerite Purnell
Tetsuya Tanioka
Kyoko Osaka

Probably robots or humanoids in nursing practice will be inevitable very soon, and we should be thinking about what the relationship will be between the machine and the nurse. How will the dynamics of human interaction be altered, and what will the "synthetic society" of humans comingled with humanoids look like? Embodiment in a living body is more than hearing, vision, touch, and also intestines, stomach, cells, and flowing fluids; embodiment is also the mind and spirit. All are involved in perception and influencing action, ethical or otherwise, in ways that we do not yet understand or appreciate. What is the most we can achieve in communicating with something that is not embodied? Nursing intuition is capable of exhibiting properties and principles that cannot be reduced to the cumulative effects of the properties. Nurses act and move in a manner not planned or anticipated, but perceived as being of value at that moment and in the situation of caring or authentic presence in relationship. How can a humanoid respond to the nurse's intuition through her directions to develop shared information? The answer is complex, but it will lie in an ethical caring response. Can a humanoid be ethically caring? The respondent, Barnard, reflects on the ideas presented in this chapter and explains how technology continues to develop and with its development, fundamental questions arise and need to be answered. Barnard explains the distinction of technology and technique for nursing, health care, and society. Technological competence demands that each nurse is not only able to use a technology but also have a critical and ethical understanding of how technology affects nurses and patients, other professionals, and society. Technology must be appropriate to care for the physical health of people, but at the same time, technology also must be cocreated to support people's emotional, ethical, and spiritual well-being.

INTRODUCTION

Observing Our Humanity

In a world that is complex, whose inhabitants are increasingly destitute, in poor health, and beset by environmental and societal calamities, it seems strange to be considering human–humanoid interactions and a vision of what might be "somewhere over the rainbow" in the provision of human care. Ideas of what may be considered "health" and the responsibility for its preservation are viewed through complex cultural lenses enmeshed within the issues of culture and access to care. As a human right, the provision of health care is connected to the availability of professional care in the form of goods and services. What is our collective obligation in the provision of care and what is the right of individuals as human beings to health care? Globally, the preservation of life at its most basic level depends upon the provision of care. However, in light of a decreasing pool of caregivers who, likewise, depend upon and are constrained by diminishing economic resources, what is the way to realize the idea of "somewhere over the rainbow?"

In health care and nursing in particular, shortages of caring personnel, qualified to respond expertly and humanely to individual and aggregate human needs, add their statistics to the burden of ever-increasing costs, delays, and filtered care. Differences experienced by persons in their ability to access sophisticated medical technology, such as machines—virtual or nanotechnology, pharmaceuticals, and cutting edge support equipment—further emphasize the financial divide between the "haves" and "have-nots." Amid this chaotic milieu await those who are ill or vulnerable, positioned between access to hope and disenfranchisement from care.

In considering the ineluctable situation where a vulnerable human being is set on one hand and the magnetic attraction of technological hope for cure or for continuing life is set on the other, the idea is evident that a commercial value has been created and sustained from out of the crucible of human frailty. *Access, availability*, and *economic value* of the services desired for health are innervated by the latest technologies and tacitly connected to the value of a human being; an unintended consequence of creativity and technological innovation that is genius in itself, but also alien to the genuine countenancing of human need.

However, in this era of unmatched technological exploration and innovation, do we dare or dare we dream of technology being routinely harnessed in a way that is startlingly different, but which may help overcome this chasm of human deprivation and need for health care? Can technology be employed to occupy the role of caregiver, ordinarily a human-to-human relationship, in cost-efficient ways that are able to provide sensitive care to those who are nursed, and to seamlessly interface with professional nurses? It is the nature of this interface between human and machine that is provocative and calls forth our deepest held values about what it means to be human. At risk for change are how we envisage and value ethically and economically the complex relations between humans and humans and between humans and nonhumans in contemporary health care delivery systems (Turkel & Ray, 2000, 2001, 2009). How do we regard human intentionality as a dynamic energy (Purnell, 2005) and its expression in machine programs, and how do we understand embodied human personality versus "designer" computer persona?

ROOTS OF MEDICAL TECHNOLOGY

When peeling back the layers in development and adoption of highly sophisticated technologies for medical use, the trajectory of modern medical technology that evolved amid great intellectual, social, and political upheavals of the 19th century (Baer, 1990) provided deep foundations for the present technological hegemony in health care. Prior to those early times of turbulence and rapid change, care was given in the home, usually by the women of the family, herbalists, midwives, or neighbors.

Within one generation, the Industrial Revolution that spurred worldwide technological development also changed the way in which human care was provided. New technologies and spheres of influence evolved and were adopted for use in health care, including the microscope and its concomitant study of bacteriology, the development of pharmaceuticals, and the remarkable discovery of the x-ray by Wilhelm Roentgen in 1895. Kalisch and Kalisch (1983) note succinctly that "The first use of the x-ray marked the beginning of medical care requiring equipment so elaborate that the average practitioner could not afford to install it. The natural result was the founding of more community hospitals in which local physicians could use such apparatus jointly" (pp. 141–142). These physician-owned *for profit* hospitals were vital for the ensconcing of medical technology and efficient provision of care. Patients were attracted to the hospitals as clients paying for hospital care and for nursing services provided by student nurses (Baer, 1990), gradually displacing nursing care in the home. The idea of hospital care thus became grounded in the *ability to pay for access* to the latest technological advancements of the day and was sold together with nursing as "services the public could only obtain in hospitals" (Sandelowski, 1999, p. 198). In contemporary health care, the promise and benefits of medicine continue to rest in the availability of cutting edge technology and not only in the conservation of life but in the potential cost savings in the reduction of health care personnel.

VALUE OF HUMAN LIFE

Although dependency on technology (Locsin, 2005; Sandelowski, 1993) is both implicit and explicit, the right, presumed or otherwise, of access to the latest health care technology is, in reality, created externally and separate from the individual by prevailing economic forces. It is, instead, based on the perceived and actual *value* of the technological services to the person. In health care and in the insurance systems that provide an actuarial framework for estimating cost of access to health care, human beings are valued by the potential cost of medical services and technology they *might* require for life. The difficulty in discerning where the calculated value of the technology ends and the value of the human person begins occurs at this intersection of technological hegemony and the human ability to participate in its accrued benefits. In the face of injury or imminent death, who will say that a person has no right to the technology that is essential for continuance of his or her life and well-being? We may well ask at what price comes the technology? Is there a solution set? What can contribute to the mitigation or alleviation of this precarious situation? For answers, we turn again to the ubiquitous computer and to a re-visioning of the future that ostensibly could be from the pages of a science fiction novel, but which is well within reach of this generation.

EQUIPMENT: PARTICIPANT OR COMPONENT?

What was once deemed futuristic and science fiction is now within the ordinary reference framework of modern society. Technologies that inform nursing and human–computer integrality may be understood in three broad ways:

a. *Technology as equipment*—These are more commonly understood as computers, gadgets, and instruments.
b. *Technology as participant*—This technology is more sophisticated and is used in assisting with human care. Examples of this technology are robots (Capek, 2001) and humanoid robots, which are distinguished by natural language processing and artificial intelligence in performing "care" functions. Artificial intelligence may be described as programmed intelligences such as "the capacity of a computer to perform operations analogous to learning and decision making in humans, including the perception and recognition of shapes in computer vision systems" (Artificial intelligence, n.d.).
c. *Technology as a critical component of those who receive care*—Patients with futuristic, life-sustaining technologies are characterized by the future post-humans, cyborgs, and techno sapiens (Locsin & Campling, 2004).

These descriptions are fluid and dependent upon their end users such as the one nursed, and the nurse, for their meaningful applicability. The impetus toward technological innovation in health care is such that the latest technologies in all forms are now taken for granted as critical to maintain human health and, in a larger sense, to ensure human continuance.

NURSING, HUMANOID ROBOTS, AND THE NURSING SITUATION

A nursing situation is described as the conceptual space where the intentional caring nurse and the one nursed are situated (Boykin & Schoenhofer, 2001). Consequent to the influential appreciation of the nurse–nursed relationship and its ultimate goal of human care, in describing this proposition consideration is needed of the behaviors expected from varying machine technologies, including humanoid robots. Should these technological marvels behave passively according to human care expectations or should they behave actively, given the natural evolution in development of programmed responses and enhanced artificial intelligence?

Passive Robots

First, let us consider passive robots and the functions needed for even simple functions, such as delivering medical materials. The following describes the physical and capacity features of many such passive robots widely used in health care institutions: A large body is used for storage and an arm may be designed for carrying special materials. Wheels, necessary for movement, with distance sensor and thermograph distinguish between humans and inanimate things. A sounding device or "voice" gives warning to people in proximity. Robots are able to perform these behavior patterns simply because they are programmed to do so. Such actions do not require any creativity, imagination, or interaction on their part.

Passive robots do not necessarily look like humans but politely respond to human stimuli with human sounding voices. Because its programmed aim is to safely carry medical materials such as pharmaceuticals, mutual communication between humans and robots is not necessary.

Active Robots

On the other hand, the description of an active robot is one in which a robot is programmed by humans to obey their particular instructions faithfully. It is hypothesized here that robots may conduct multiple nursing actions just like human nurses do, such as delivering medications, assisting in surgeries and performing select physical care. Are these robots to be only used as assistants in nursing care tasks such as carrying medical materials? Or do we need sophisticated robots that can call to humans and actively report the results of their active *judgment* about health and human care?

ROBOTS, EMPATHY, AND CARING: A CONTRADICTION?

The relationship between robots, empathy, and caring is integral to the appreciation of their harmonious coexistence. Empathy is described as "identification with and understanding of another's situation, feelings, and motives" (*American Heritage Dictionary*, 2006, p. 586). Equipping robots with empathic understanding as a nursing skill function sensitive to a multitude of human characteristics is necessary in this artificial simulated life form. To illustrate the subtleties in the development and expression of human empathy, the following hypothetical dialogue between an elderly patient and a nurse is presented:

The elderly patient held a photograph in one hand and tightly clasped the nurse's hand, reminiscing about her long-lost love. She said wistfully to the nurse, "My man had a family, so we couldn't marry. It was out of the question. But we did have so many rich memories together. One night, we strolled along a beach and enjoyed talking about so many things. I remember gazing at the starry heavens and thinking that it was all so perfect. It seems like yesterday" The nurse with a gentle smile, said "Oh, what a wonderful story! Now I understand how much this faded photograph means to you."

As portrayed in this simple narrative, nurse caring is about grasping the wholeness of patients and what is meaningful to them. Nurses are required to have interpersonal competency in understanding patients from their expressions, gestures, and words, in interpreting patients' experiences as meaningful and in conveying empathy. For robots to possess such human sensitivity, their repertoire of skills needs to include the idea of caring, in itself a challenging concept to capture, and the intention to care. If humans find it difficult to explicate what their caring encompasses, how much more difficult would it be to convey these attributes or characteristics in an essential array for the programming of a humanoid robot? In situations of caring, what would a robot–human relationship be like?

From a very narrow, superficial perspective, it is possible to make robots appear as if they are being kind to others by including in their programming the ability to enact what are generally construed to be gestures of kindness. It may be supremely difficult, if not impossible, to design a robot that has the capacity for human empathy, compassion, and kindness, in order to "understand" others' pain, especially when robots cannot "feel" pain as do humans. However, "computer-enhanced" persons with microelectric arrays inserted in neural plexuses are predicted to be able to efficiently transmit signals to robots so that boundaries disappear and thought and action become seamless between the created machine robot and the cybernetically altered human (Warwick, 2004). Thoughts from a human being could simply direct the robot.

Similarly, the problem of empathy and its associated nuanced and unique responses calls forth critical thinking about which behaviors actually can reveal empathy. What would be clemental requirements for empathetic expression? At a minimum, robots providing autonomous human care should be able to recognize and understand human language and make appropriate judgments in a situation. They should also be able to recognize and understand human nonverbal face, hand, and body posture expressions. However, recognition is only a prelude to the desired empathetic response that is complex, nuanced, and often ineffable. Nursing is a unique practice within a distinct culture of human health, and as such, it contains in its ontology a repertoire of embodied human responses that can be called forth in the creation of programs for integrating humanoid robot elements of understanding. In this way, the *gross estimation* of human feelings may be possible and interpretable, providing pathways for program inclusion in the artificial intelligence of a humanoid robot.

Although the development of highly sophisticated humanoid robots has been considered as one solution for provision of the "hands-on" bedside care offered now by nurses, the idea of viable, intentional, and satisfying relations between human and human robots has thus far been socially accepted only in science fiction and played out in the virtual world of movies. Who has not felt the pangs of compassion and empathy when watching a humanoid robot, Andrew, weeping and desiring to become human in *The Bicentennial Man* (Asimov, 1999), or the Tin Man in *The Wizard of Oz* (Baum, 1939) who desired a heart, or Data in the *Star Trek* series (Roddenberry, 1995), who evolved and desired to experience the laughter and emotions of being human? What is it that endears these characters to us? Is it because they bear physical attributes critical to our perception of human beings, or is it the story and culture of illusory experience in which they are embedded and which we ourselves create? Although these believable characters of fiction hold up a mirror to our frailties, in another turn toward being human they also reject their own synthetic existence in favor of the human experience, with all of its uncertainties, suffering, and finiteness of life. With all the pathos of an aging being, Andrew in *The Bicentennial Man* states, "but I tell you all today, I would rather die a man, than live for all eternity as a machine."

Contemporary science has brought the humanoid from the screen to tangible, artificial life in the similitude of humanoid faces that now embody aspectual human traits and express subtleties of emotion. For example, an *android,* the name for a machine or an artificial human life form, can mimic human appearance. An android is also the generic name for a *humanoid,* and a *bioroid,* both innovations of robot forms that emulate or simulate living biological–mechanical organisms. Humanoid robots are sophisticated, human-like creations with possible affective, responsive behavior and particular technological innovations such as sensory capabilities and sensitive hands, thought to be hallmarks of being human. A *cyborg* is a cybernetic organism, or a being with natural and artificial systems.

What truly constitutes being a human has been pondered over the ages by scientists, philosophers, and authors alike. Consider Mary Shelley's enduring novel *Frankenstein* (1994), the story of a grotesque "human" being created from human parts, complete with arms and legs and even a brain. To Frankenstein, being human was a matter of self-referential understanding, and even though he possessed the requisite number of human parts stitched together, the physical ugliness and incongruence of these parts were in opposition to notions of acceptability held by noncreated others. Although the character of Frankenstein may epitomize the completed human being as made up of the requisite number of parts, it also depicts the essential nature of being human from the view of being more than and different from the sum of the parts. Frankenstein has since remained an effigy of the insufficiency of our understanding of what it means to be human and, paradoxically, an example of the impetus to preserve what it is that we really do not understand—our humanity.

HUMAN OR HUMANOID? CHARACTERISTICS OF THE SYNTHETIC "LIFE"

Although it is common to compare the characteristics and likenesses of robots to human beings, it is also clear that visual, linguistic, and emotional essences expressed in the context of being human in this world are essential characteristics by which humanoid robots are adjudged as moving toward being and becoming "human." Various images from the visual arts, including paintings, films, and pictures, and words from stories and texts are analyzed to explicate the critical appreciation of computer–human becoming that is understood as essential to health and human care.

Contemporary development of humanoid robot technology brings the future to the present. Robots with increasingly human-like functions have been developed. For instance, a bipedal robot named *Asimo,* performing a difficult feat that was unsuccessful in the past, can now independently walk the streets (www.honda.co.jp), and a robot with lung mechanics similar to those in humans can blow a trumpet by vibrating its lips (http://www.toyota.co.jp). In addition to refining appearance and functions, many researchers have worked on the development of emotion-recognition systems for humanoids to render their "interpersonal" affective responses closer to those of actual human beings. Such systemic innovation enables robots to recognize emotions from biological information such as voices, facial expressions, and even externally gauged brain activities. By combining these technological innovations and mounting them within a robot with human appearance, the birth of a humanoid robot that can behave uncannily like a human being is no longer the realm of science fiction.

In the provision of bedside care, and in the sustained absence of human nurses to provide care, humanoid robots may well possess a "value-added" dimension in being able to offer tender "human" interactions, while always "knowing" how to render expert technical care. The initial cost of the humanoid would be rapidly recovered in the sustained cost savings in continuity of service "24/7" with no time needed for orientation, continuing education, meals, sleep, illness, payment for services, payroll taxes, or even retirement benefits. It is here that one must return to the implicit value of the human being and weigh it against the explicit value of the humanoid robot. Will institutional cost savings trump the desire of human persons for genuinely human touch in their care? The human patient's response to being provided with such humanoid health care robots is largely unpredictable, given the magnetic attraction to cutting edge health care technology and the hope that is associated with it. What might be the critical point at which they are accepted or rejected?

THE UNCANNY VALLEY: VIEWS FROM A SYNTHETIC LIFE

Over 40 years ago, Masahiro Mori studied the effects that the visual appearance of robots exerted on the human emotional attraction or repulsion to them. According to Mori's (1970) hypothesis, as a robot is increasingly made to resemble a human being in appearance and behavior, and our sense of familiarity increases, humans will feel more positive and empathic. However, when the robot becomes too human in appearance, a point is reached where humans suddenly start to feel repulsion; for example, when they see a hand with muscles, tendons, and veins and they realize it is a prosthesis, they feel a sense of strangeness, or "negative familiarity" (Mori, 1970, p. 34). Similarly, the feeling of negative familiarity occurs when seeing a dead person: The body of the person is there, for all intents and purposes whole and entire, but yet the person is gone. This feeling is the uncanny valley. When again the appearance and behavior of the robot approaches a verisimilitude of humans, humans may again start to feel positive and empathetic and possess the same sense of affinity to the robot as to humans themselves.

Despite mixed acceptance of Mori's hypothesis, the reaction of consumers who fuel the demand for health care technologies unquestionably molds the extent to which humanoid robots, in whichever guise they appear, will interact in their care.

FIGURE 16.1 Mori's (1970) Uncanny Valley
A simplified version of the figure appearing in the Energy article. Copyright © 2005 Karl F. MacDorman and Takashi Minato. Mori, M. (1970). Bukimi no tani [The uncanny valley] (K. F. MacDorman & T. Minato, Trans.). *Energy, 7*(4), 33–35. Available from http://www.android-science.com/theuncannyvalley/proceedings2005/uncannyvalley.html

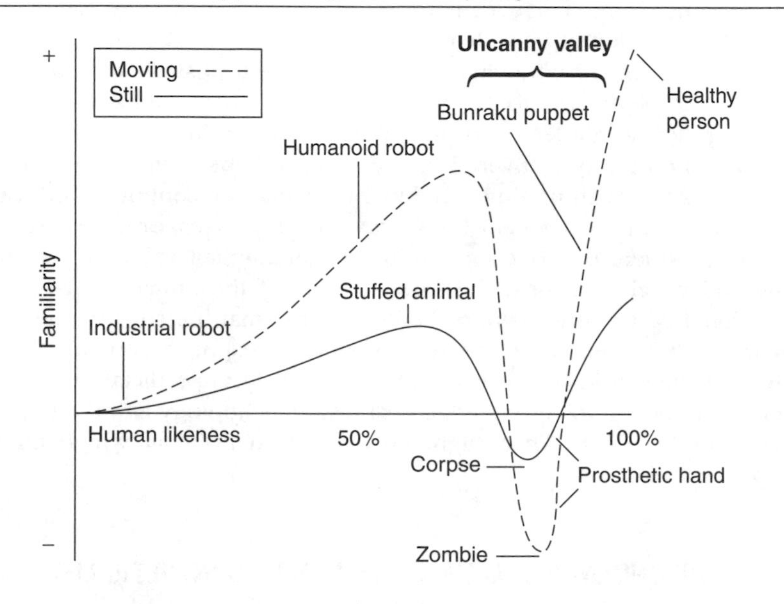

CROSSING THE LINE: CONSIDERATION OF ETHICAL ISSUES

Technology has crossed a line to where artificial intelligence and humanoid physicality, or our perception of it, have merged and consequently raised the complexity level of our appreciation. Kurzweil (1999) held this view in his discernment of spiritual machines in which the technological advancements of computers will exceed human intelligence toward a state of singularity. Life is no longer simple, with the age-old cycle of a person being born, living out the meaning of his or her life, becoming ill or old, and then dying. Babies are being born that can fit in the palm of a hand and are sustained by drugs and incubators; accident survivors live with artificial limbs, and those who are totally incapacitated may have their life prolonged for years sustained by a parenteral feeding tube. Technology mediates that which used to be a human process and which has now become a new reality in a human technology process.

This new reality that includes the acceptance of what amounts to the technological impersonation of a human being, raises uncomfortable ideas of ultimate hegemony over another, with selected human characteristics that may ensure servanthood, and indeed, even raise the specter of a type of slavery in which our power is enacted over another *perceived* as human being or vice versa. The awkward tension between the fusions of human–not human is palpable. At this point, we can ask what has happened to the human being who expresses the essential characteristics of humanity? Should this human being be the same, similar, or different?

It is uncertain how humans might react when as recipients of care, they receive nursing from humanoids—who or which, depending on how they are regarded, deliberately and physically touch them. Mori's (1970) hypothesis of the uncanny valley may hold true; however, the possibility exists that "familiar" robots equipped with skin farmed from humans, and with hands warmed by internal heaters and heat sensors, could immediately ascertain blood oxygenation levels on contact without arousing strangeness or negative familiarity in patients. Although this sophisticated robot's hand may emulate human touch, it also raises the notion of a fraudulent kind of touch that emulates caring but has as its end the cold, unbiased acquisition of empiric information through a programmed machine, rather than the compassionate caring for an embodied *person*. Conceivably, constant monitoring of the patient's internal physiology might be achieved with low-dose x-ray "vision" emitted from the humanoid robot's eyes or nano payload doses of pharmaceuticals administered through the hands, with or without permission—a science fiction comic book staple brought to life. This scenario could well prove to be an ethical nightmare, particularly in the care of persons who are indigent or unable to make decisions for themselves.

The foreseeable application of force by humanoids, particularly when patients resist "care" procedures, may turn what is programmed "intelligence" for gentle touch, into crude technologic force, no different from vice or hammer, against human flesh. In contemporary health care, delegation of care has also come to mean delegation of responsibility, and when created humanoids are involved, who is actually responsible?

When robots make mistakes or malfunction, a very human foible or technological nightmare, depending upon our lens, the possibility of offering forgiveness in the same way as to our fellow human beings presents a perplexing thought. How human are these humanoids? Will they possess a culture or will they be culture neutral? Will they adapt and evolve as do humans? Will they ever feel guilt or happiness and what is our responsibility to them? Moreover, when robots with human appearance and behavior malfunction or wear out after living with humans for an extended period of time, not unlike a family member, can they easily be disposed of, just like rubbish?

In contemplating this uneasy tension between human–humanoid relations that can be fraught with morality issues, what would one call the kinds of ethics that pervade this situation? The ethical morass for aberrations concerning humanistic creations in self-likeness is not accounted for in contemporary health care ethics and calls for consideration of a new ethics, one in which considerations of who and what are human, their benefits, and indeed rights, spring forth within the blended context of contemporary and futuristic technologies.

PRESERVING HUMANITY IN A TECHNOLOGICAL WORLD

We have come full circle from observing our humanity to preserving humanity in a world dominated by technology, in particular, appropriated technology that now dominates human health care. The concept of preservation has gained multiple meanings, although it is more popularly used in the political arena under the rubric of environmentalism, emphasizing the critical role it has garnered in supporting continuity of life on this planet. The future of human beings depends largely on the existence of an environment that is conducive to human living and human thriving. How do we ensure such an environment? As conducive to human living as our world needs to be, it is often the efficiency of things used, reused, replaced, or recycled that generates the complexity of preserving humanity, particularly within the realms of a sprawling technological world and other inter-world spaces that provided Rogers (1970) a typical view of what she called "homo spatialis." Thus, preserving humanity must become the mantra of contemporary and future nursing if it is to sustain and maintain its essential value, meaning, and practice in a technologically demanding environment.

What does it mean to be and live as a human being? What is it about human beings that needs preserving? There are many descriptions of human beings—each depending on the philosophical foundations of the description of being human, the functionality of being human, and the degree to which the human being contributes to society in general. Consider, for example, the human being as perceived from the philosophical view of positivism. Through this lens, a "whole" human being is understood as one who has a full complement of composite parts. Such a consideration begs the question for those human beings who have missing internal organs or missing visible parts such as arms or legs and those who have replacement body parts and prostheses. The question that is often asked is, "Are they less human?" Or are they, as Oscar Pistorius, a South African sprinter who runs with prosthetic legs, regarded as having an unfair advantage (Schwartz & Watermeyer, 2008)? When a person undergoes hemodialysis, is this person any less human because his kidneys are not fully functional? It is evident that the lack of a human part, essentially, does not render an individual less human or not human.

Another way of understanding and describing a human being is through function. An individual who is not able to independently function as a human without a technology, such as a comatose person on life support, may not be considered any longer as a human being. The idea of a "vegetative" state is a widely used pejorative, implying that the person is an object and therefore useless and disposable. Alternately, individuals who are appreciated as complete from the perspective of wholeness, regardless of the number of composite parts are understood as more than and different from the sum of the parts. This distinguishes human beings more clearly from created or copied versions, with the idea of the personal, human self as a unique, unitary human being (Rogers, 1970) and as essentially transformational (Newman, Sime, & Corcoran-Perry, 1991).

NURSING IN TECHNOLOGY-DENSE ENVIRONMENTS

Who are the persons that nurses nurse today? Robocop, bionic man and bionic woman, persons with artificial hearts, with pacemakers and implantable cardiac devices, mechanical valves, transplanted organs, titanium-enhanced long bones, genetically enhanced secretory organs, and "technoluxe" (Schwartz & Watermeyer, 2008) medical services such as reshaped noses, silicone-plumped lips, and botox-smooth faces—all receive nursing care. Importantly, however, what should the nurse know, or study, about these technology-enhanced persons?

Persons are living human beings with hopes, dreams, and aspirations. Human beings are unpredictable because of their originality and capacity to think, imagine, innovate, and create. As such, persons are not auto-matons or robots who can be programmed to perform, and it is here that the line is drawn between humans and created humanoids: Persons are always whole—a philosophical view that allows the practice of contemporary nursing in which persons are recognized as participants in their care rather than simply as objects of our care (Locsin, 2009). As such, nurses include their patients in the planning and implementation of their care: Instead of doing *for* their patients, nurses are now able to do *with* their patients. This sense of togetherness, a human attribute that is comforting and intrinsically unable to be replicated, describes in the intimacy of relationship, the complexity of what it means to be human and what it means to *not* be human. Our technological creations capture only an elusive glimpse of humanity through a very small window.

It behooves nurses to capture for their patients the utmost benefit of the *present* technology. Nursing occurs in the moment, and moment by moment. The implication of technology, wherever and in whatever environment it is presented, means mediating and transforming the technology for the person nursed. Technological competency as caring is the skilled demonstration of intentional, authentic caring activities by nurses who practice in environments requiring technological expertise, however that technology might be expressed (Locsin, 2009; Turkel & Ray, 2009). The practice model that is crucial to contemporary and future nursing is one in which the practice of caring in nursing can be expressed through technological competency. Will a new ethics evolve out of this view? In a study of technological caring in an intermediate care or step-down unit, nurses were confronted with issues related to caring for suffering "silent patients"—long-term care patients on ventilators

(Ray, 1998). Nurses faced one ethical crisis after another. Nurses experienced the same "agonizing vulnerability" to suffering as did their patients, resulting in moral chaos—moral blurring or moral blindness. Nurses were challenged to the last degree of the coping by their situations; they were compassionate; to cope with moment to moment experiences, they felt they may have to stop relating to their patients as persons. The paradox revealed the presence of compassion on the one hand because of feeling their own pain in the midst of feeling the pain of the patients, and the potential contradiction of unfeeling for patients as persons by developing moral blurring or moral blindness in order to cope within the ethical situation. In the process, however, nurses were disillusioned but not indifferent. A reflexive ethics emerged; the cry for help within the ethical situation forced a call to alleviate not only the patients' suffering but their own. They needed to understand the meaning of the use or abuse of technology—to grapple with their own values about technology, the moral status of the meaning of human being, and life and death. What may be paradoxical in the future is that although technological competency as ethical caring will be exhibited by the human nurse, human competency as technological caring may be exhibited by the humanoid, perhaps an ultimate expression of human–environment integrality. Given the expression of a call for help with nurses in intensive care patient situations, nurses will need to be aware of cocreating a type of reflexive ethics, an ethical intentionality and/or premeditation to deal with the ethics of the human–humanoid relationship. Can a humanoid be *ethically* caring?

The demand is great for a practice of nursing that is based on the authentic desire to know human beings fully as persons rather than as objects of care, a concept far removed from the idea of humanoid robots. Challenging the nurse to know persons in their wholeness means using every possible creative, imaginative, and innovative way to appreciate and celebrate their intentions to live fully and grow as a human being. Whether or not health care is regarded as a right, and regardless of the push for increasingly sophisticated technology to supplement or replace nursing care, we return to what our nursing practice must be; a uniquely *human* service for humans that is served by, not controlled by, our technological creations.

REFERENCES

American Heritage Dictionary of the English Language (4th ed.). (2006). New York: Houghton Mifflin Company.

Artificial intelligence. (n.d.). *Dictionary.com Unabridged*. Retrieved from http://dictionary.reference.com/browse/artificial intelligence

Asimov, I. (1999). *Bicentennial man* [DVD]. Columbus, C. (Director). Horner, J. (Composer). Travis, N., & DeToth, N. (Editors). Petersen, W., Columbus, C., Mark, L., Barnathan, M., Katz, G., Radcliffe, M., & Miller, N. (Producers). Buena Vista Home Entertainment, Burbank, CA. [Now Walt Disney Home Entertainment] (131 mins.).

Baer, E. D. (1990). *Editor's notes for "Nursing in America: A history of social reform": A video documentary.* New York: National League for Nursing.

Baum, F. L. (1939). *The wizard of Oz* [Movie]. Adapted from F. L. Baum's *The wonderful wizard of Oz*, published by N. Langley, F. Reyersen, & E. A. Wolf. Movie directed by V. Fleming. Produced by Metro-Goldwyn-Meyer.

Boykin, A., & Schoenhofer, S. O. (2001). *Nursing as caring: A model for transforming practice.* Sudbury, MA: Jones and Bartlett.

Capek, K. (2001). *RUR (Rossum's Universal Robots).* Mineola, NY: Dover Publications.

Coppola, F. F. (Producer), & Branagh, K. (Director). (1994). *Frankenstein* [Movie]. Adapted from Mary Shelley's *Frankenstein*. Distributed by TriStar Pictures. Retrieved from http://www.honda.co.jp/ASIMO

Kalisch, P. A., & Kalisch, B. J. (1983). *The advance of American nursing* (2nd ed.). Boston: Little, Brown.

Kurzweil, R. (1999). *The age of spiritual machines.* New York: Penguin-Putnam.

Locsin, R. C. (2009). "Painting a clear picture": Technological knowing as a contemporary process of nursing. In R. C. Locsin & M. J. Purnell (Eds.), *A contemporary nursing process: The (un)bearable weight of knowing in nursing* (pp. 377–393). New York: Springer.

Locsin, R., & Campling, A. (2004). Techno sapiens, neomorts, and cyborgs: Nursing, caring, and the posthuman. In R. Locsin (Ed.), *Technological competency as caring in nursing: A model for practice.* Indianapolis, IN: Center Nursing Press.

Mori, M. (1970). Bukimi no tani [The uncanny valley] (K. F. MacDorman & T. Minato, Trans.). *Energy, 7*(4), 33–35. (Originally in Japanese.)

Newman, M. A., Sime, A. M., & Corcoran-Perry, S. A. (1991). The focus of the discipline of nursing. *Advances in Nursing Science, 14*(1), 1–6.

Purnell, M. J. (2005). Inside a Trojan horse: Technology, intentionality, and metaparadigms of nursing. In R. Locsin (Ed.), *Technological competency as caring in nursing: A model for practice* (pp. 42–68). Indianapolis, IN: Sigma Theta Tau International Press.

Ray, M. (1998). A phenomenologic study of the interface of caring and technology: Toward a reflexive ethics for clinical practice. *Holistic Nursing Practice, 12*(4), 69–77.

Roddenberry, E. W. (1995). *Star trek* [Series]. Roddenberry, G. (Writer). Roddenberry, G., Lucas, J., Coon, G., & Freigurger, F. (Producers). Roddenberry Productions, Studio City, CA.

Rogers, M. (1970). *Theoretical basis of nursing.* New York: F. A. Davis.

Sandelowski, M. (1993). Toward a theory of technology dependency. *Nursing Outlook, 41*(1), 36–42.

Sandelowski, M. (1999). Troubling distinctions: A semiotics of the nursing/technology relationship. *Nursing Inquiry, 6*(3), 198–207.

Schwartz, L., & Watermeyer, B. (2008). Cyborg anxiety: Oscar Pistorius and the boundaries of what it means to be human. *Disability & Society, 23*(2), 187–190.

Turkel, M., & Ray, M. (2000). Relational complexity: A theory of the nurse-patient relationship within an economic context. *Nursing Science Quarterly, 13*(4), 307–313.

Turkel, M., & Ray, M. (2001). Relational complexity: From grounded theory to instrument development and theoretical testing. *Nursing Science Quarterly, 14*(4), 281–287.

Turkel, M., & Ray, M. (2009). Caring for "not-so-picture-perfect patients": Ethical caring in the moral community of nursing. In R. Locsin & M. Purnell (Eds.), *A contemporary nursing process: The (un)bearable weight of knowing in nursing* (pp. 225–249). New York: Springer.

Warwick, K. (2004). *I, cyborg*. Chicago, IL: University of Illinois Press.

Response to Chapter 16

Alan Barnard

Locsin, Purnell, Tanioka, and Osaka have written an interesting chapter that poses reflective ideas about a number of important questions for nursing practice, education, and health administration. The chapter explains that technology has a long association with nursing and remains the basis of health care action. Technology continues to develop and fundamental questions need to be answered in relation to technology and its influence and purpose related to person-focused care. Technology reflects sociocultural meanings and incorporates accepted assumptions and design. Robotics in health care is a recent phenomenon found primarily in Western countries that have an aging population and emphasizes technical efficiency. Technology is not simply a collection of things that we use to be efficient (Barnard, 1996). Locsin, Purnell, Tanioka, and Osaka emphasize three descriptions of technology and my reflection on their explanation causes me to ask if we have always used and integrated technology in nursing, what is so different now? Why, in more recent times, do we seem to be so concerned about it? Why do we feel a desire to contrast technology with human caring and person-focused nursing practice? The answer for many may lie with the sense that something is going on around us that is fundamentally unsettling. There is something emerging that cannot be seen, but is around us all the time, something we each experience yet can't put our finger on. In fact, the experience is not a thing but is a way of thinking and living. Technological development in postindustrial society is associated with a new system of efficiency and logical order greater than in any previous time. The system has been called technique (Barnard & Sandelowski, 2001; Ellul, 1964; Feenberg, 1999; Ihde, 1993) and can be defined as the creation of the kind of thinking and processes necessary for technology to develop and be applied in an efficient and rational manner in all activity. Importantly, there is nothing dangerous about efficiency as a goal, as inefficiency is not appropriate in health care. However technique is the creation of economic, human, and political systems that bring us increasingly into line with technology in order to obtain control over human and nonhuman phenomena. Technique reduces what should be intensely human-centric activities, such as nursing, to measurable and predictable processes that no doubt will openly embrace all forms of robotics as likely inclusions within health care. The importance of technique for nursing, health care, and society should not be underestimated. According to Lovekin (1991) "technique is a mentality, a system, a way of culture and society, and a way of life. Technique is metaphysics; a manifestation of the ultimately real; an addiction; and an obsession that nurses wish to do nothing about" (p. 65).

Technique is not a specific thing. It is a mentality that guides behavior, thinking, and outcomes. It is a system that seeks to reduce individual differences, emphasize specialization, enhance conformity, and bring sameness to product, processes, and thought. It is an attitude and a way of life. It is what Winner (1977) refers to as the gestalt of the modern man and woman. Examples of technique include economic rationalism, protocols, efficiency drives, diagnostic-related groups, and time-and-motion studies. Everything must be organized, planned, and reduced to a "one best way." It emerges, for example, that nursing care can be interpreted in terms of risks and a series of actions that can be reduced to performance measures as well as predefined protocols and language. Tradition, cultural difference, ethical/social values and sensitivity to individual difference fall away within a system of logical order. The participating midwife/nurse/patient/doctor/woman/man gains the pleasure of being part of a health care system and is reassured by the predictable nature of its organization, control, and efficiency. However, if participation is difficult because approaches to care restrict knowing the whole person, then there will be a feeling that care is excessively standardized. So what should we do as nurses, educators, and administrators? The first move is to ask yourself seriously, what gives meaning to your life and the profession? We each must be willing to be certain of the reasons we do things, the necessity and appropriateness of choices made, the suitability of care determined, and the necessity to seek out different practices, evidence, and perspectives where needed, even if it means swimming against the tide. The second move is to reflect upon the goals you have as a nurse, administrator, and educator, and to affirm the importance of genuinely determining care that not only addresses organizational needs, but the needs of individuals and families. Expertise in clinical practice is demonstrated through an ability to assess and influence decisions about the suitability of technology. It was encouraging to note that the U.S.A. Institute of Medicine 2001 report entitled *Crossing the Quality Chasm: A New Health System for the Twenty-first Century* emphasized the need for patient-centered and performance-based care that is devised around healing relationships and the provision of health care founded on needs and values. To this end, the values and questions expressed in the chapter by Locsin, Purnell, Tanioka, and Osaka are important for practice, administration, and education, and our responses to the many dilemmas before us should be directed by guiding principles such as:

1. Good health care matters because people matter, and this moral value must be continually emphasized especially within health care systems.
2. We each need to be involved in influencing the direction(s) of technological change and decision making because health care is political. Decisions about technology in care will impact directly on nursing care, the organization, and the experience of patients for whom we care.

3. Each person has an equal right to health care resources and quality nursing. The dignity and worth of each life supports the view that we are responsible to effectively utilize and integrate appropriate technology.
4. Technology use in practice must consider a patient's total physical, emotional condition such that treatment and intervention are appropriate and sensitive to needs.
5. Technological competence demands that each nurse not only be able to use technology safely but have a critical understanding of how technology must be appropriate and improve practice and caring.
6. It is true to claim that the majority of nurses desire to engage in person-focused care founded on compassion and concern. It is true also to claim that sometimes technology gets in the way of our ability to express that desire.

REFERENCES

Barnard, A. (1996). Technology and nursing: An anatomy of definition. *International Journal of Nursing Studies, 33*(4), 433–441.

Barnard, A., & Sandelowski, M. (2001). Technology and humane nursing care: A(n) (ir)reconcilable or invented difference? *Journal of Advanced Nursing, 34*(3), 367–375.

Ellul, J. (1964). *The technological society*. New York: Knopf.

Feenberg, A. (1999). *Questioning technology*. New York: Routledge.

Ihde, D. (1993). *Philosophy of technology: An introduction*. Bloomington, IN: Indiana University Press.

Lovekin, D. (1991). *Technique, discourse, and consciousness: An introduction to the philosophy of Jacques Ellul*. Cranbury, NJ: Associated University Presses.

Winner, L. (1977). *Autonomous technology*. Cambridge, MA: The MIT Press.

CHAPTER 16
HUMAN RIGHTS AND HUMANOID RELATIONSHIPS IN NURSING AND COMPLEXITY SCIENCE

- *How would you use concepts from caring or complexity science to support the idea that health care is a human right?*
- *Is a robot capable of caring expressions?*
- *Can a robot reflect on the meaning of an ethical caring dilemma?*
- *In the year 2025, will robots and technology replace humans in health care?*
- *How would robots self-organize?*
- *Could robots relationally self-organize?*

EPILOGUE

Emerging Ideas/Questions for the Future of Nurse Caring and Health Care in a Complex World

Marian C. Turkel
Marilyn A. Ray

> *Now we stand on the threshold of pause. Each one of us is going to look out at the world and into his or her heart. Out of this creative suspension will come a new impulse. Each one of us will be responsible for that impulse, for that which is going to carry us forward into this millennium. The combination of our impulses, thoughts, and new attitudes will create a new world. To do so we will not only consider our own hearts but we will begin to dialogue with others, with nature, and with the sacred. We have left the dream of absolute certainty behind. In its place each of us must now take responsibility for the uncertain future.*
>
> F. David Peat, 2002, p. 213.

The reading of the chapters, the responses, and the questions posed for reflection at the end of each chapter is finished now. As editors of this work on nursing, caring, and complexity sciences, we have been pleased to share the ideas, theories, models, and methods of diverse scientists, physicians, administrators, nursing scholars, and practitioners. Their ideas and research illuminate the knowledge related to the science and art of wholeness—the new science of networks of relationship (Mitchell, 2009; Newman, Smith, Dexheimer-Pharris, & Jones, 2008; Wheatley, 2006)—the unitary nature of mind and matter, and compassion and spirit. We extend an invitation to you, the readers, to begin the continual process of reflection and dialogue on the human–environment mutual caring process and request that you remain open to creative emergence, which makes possible the understanding of health, healing, well-being, dying, and death. The theoretical concepts in the chapters and responses are drawn from complexity science/s and nursing science, human–environment mutual process, the science of unitary human beings, caring science and caritas processes, entropy and chaos theory, complex caring dynamics and methods, relational caring complexity, the dynamics of story theory, complexity science, and research methods. Practice examples are innovative and creative and include hospital organizational program critique, complex organizational caring, administrative and leadership dynamics, complexity science applications in the practice of nursing, social conflict, the complexity of diabetes, complexity and nursing education, the language of caring in informatics and electronic health records, technological change in the hospital, and robotic technologies and ethical caring. Reflection and dialogue allow the reader to attain deeper levels of insight by learning through experience. Boyd and Fales (2003) described reflection as an internal learning process where an issue or idea is examined. Through this process, an individual comes to see the world differently, develops new insights, and acts differently. According to Johns (1995) reflection is the practitioner's ability to access and make sense of work experience in order to achieve more satisfying work. Moreover, according to Freshwater (1998, 2000, 2007) and Freshwater, Taylor, and Sherwood (2008), reflection involves a change in self allowing for the *transformational* potential of reflective practice to emerge, a focus on the significance of meaning within the relational caring experience in nursing. As editors, our vision is that we remain in a continual state of reflection and form a social network for dialogue among professionals to disseminate ideas of the sciences and art of holism, complexity, and caring, which will begin the transformation of the practice of health care, medicine, and nursing for mutual human–environment well-being.

You have experienced a heightened sense of awareness of how concepts from relational caring science and complexity science/s as networks of caring relationships have been or could be integrated in education, research, administration, and practice. Our challenge for the future as educators, researchers, leaders, and practitioners is how to illuminate these views in the current evidence-based practice or outcome and efficiency-oriented context of health care delivery today (Bar-Yam, 2004; Ray, 1998; Turkel, 2006). How do we make complexity and caring theories, models, or ideas captured in this book resonate in practice? As has been communicated, our new way of knowing holds that unitary knowledge not only 'stands for' or represents the complex world of science and art but also is a continual emerging reciprocal "dance of life" (Ho, 2003, p. 18) (and in practice, the force of death, which "wrestles with the force of life" [Buber, 1965, p. 64]). Thoma highlighted in her discussion of the science of wholeness that the essence of "[r]elationship itself is invisible yet knowable" (2003, p. 15) ". . . [T]he more carefully scientists look, the more complex phenomena are" (Mitchell, 2009, p. 299). We know, however, in the art of nursing, a nurse must *see, feel, and know* relationships to understand the dialogical principle and reflective process in the nursing situation or caring moment (Watson, 2008). Watson illuminated the caring moment as the 10 caritas processes (see the Prologue and D'Alfonso's Response to Chapter 7). Roach (2002) illuminated the caring moment as commitment (devotion), compassion (love), conscience (ethics), confidence (faith), competence (ability), and comportment (symbolic communicative interaction). The caring moment is similar to the philosopher Buber's (1965) notion of the "in-between," which can now be understood in science as the science of networks or relationship (Mitchell, 2009). When nurses speak about presence or "being there" with the patient (or others) they are engaged in a complex miracle unfolding, applying the knowledge of holistic science and seeing and feeling the "dance of life" experienced through loving kindness. Thus, we can see that a theory, model or an idea is necessarily simpler than the actual happening or event that is shared (Osberg, Biesta, & Cilliers, 2008). Nursing's reality represents something much more—our living relationships.

As a complex caring dynamic (Ray, 2010), nursing is an evolving system that calls for understanding what is deepest in ourselves—seeking understanding of the human–environment mutual process and seeking understanding of what is emerging in our spiritual and ethical imagination and creativity. Within the sphere of the integration of Nightingale's commitment to nature as a reparative process, Rogers' illumination of the science of unitary human beings as unitary and holistic, complexity science/s as the science of interconnectedness, nonlinear dynamics, self-organization and change, and relational caring we are aware of the reciprocal patterns of transforming spiritual and ethical energy for healing, health, well-being, or a peaceful death. Through reflection on the contents of this book we understand that it is imperative to exchange an objective, mechanistic, and reductionistic view of nursing science with a view of nursing as a holistic spiritual and ethical caring science. Nursing is the continual unfolding of the intimate human–environment mutual caring process. Thus, in science and nursing, the *integration of science and human experience* is the changing vision. The scientist Mitchell (2009) calls complexity the new science of networks. It is a process of "complexification" (Mitchell, 2009). We have come to learn that the theories or ideas, and indeed the practice in the dynamic quantum realm are a process of the *flow of living energy in networks of relationship*. This coherent energy captures values, principles and ideologies, such as unitary, belongingness, interconnectedness, relationship, uncertainty, unpredictability, increasing complexity and diversity, nonlinearity, pandimensionality, chaos (disorder and order), hysteresis (sensitivity to initial conditions), moral choice, nondestructive technology, spirituality, loving kindness, and emergence and much more into experience. We have learned that the slightest change in our interactions can produce the unexpected, even something miraculous. But as we engage in this dynamic "quantum" process of gaining new knowledge, we become more conscious of how these insights help us to change, and change in ways that allow us to see how we must behave responsibly and respectfully in our engagement with others in the cocreativity of nursing practice *and* the world. This is where the true essence of caring comes into being. As Maturana and Varela (1998; Goodwin, 2003, p. 14) stated, "We have only the world we can bring forth with others, and only love helps bring it forth. . . This is the biological foundation of social phenomena: without love [caring], there is no social process and, therefore, no humanness." Intellectual knowing is deeply connected with experiential understanding, loving kindness, ethical commitment, spiritual openness, and a sense of reverence for the "dance of life" that is continually unfolding.

REFERENCES

Bar-Yam, Y. (2004). *Making things work: Solving complex problems in a complex world*. Boston: Knowledge Press.

Boyd, E., & Fales, W. (1983). Reflective learning key to learning from experience. *Journal of Humanistic Psychology, 23*(2), 99–117.

Buber, M. (1965). *Between man and man*. New York: Macmillan.

Freshwater, D. (1998). The philosopher's stone. In C. Johns & D. Freshwater (Eds.), *Transforming nursing through reflective practice*. Oxford, UK: Blackwell Science.

Freshwater, D. (2000). *Transformatory learning in nursing education*. Unpublished, PhD thesis, University of Nottingham, Nottingham, England.

Freshwater, D. (2007). Reflective practice and clinical supervision: Two sides of the same coin? In V. Bishop (Ed.), *Clinical supervision* (2nd ed.). Basingstoke, UK: Palgrave.

Freshwater, D., Taylor, B., & Sherwood, G. (2008). *International textbook of reflective practice in nursing*. Oxford, UK: Blackwell Publishing and Indianapolis, IN: Sigma Theta Tau International.

Goodwin, B. (2003). Patterns of wholeness: Holistic science. *Resurgence, 1*(216), 12–14.

Ho, M. (2003). Dance of life: Holistic science. *Resurgence, 1*(216), 18–19.

Johns, C. (1995). Framing learning through reflection within Carper's fundamental ways of knowing in nursing. *Journal of Advanced Nursing, 22*, 226–234.

Maturana, H., & Varela, F. (1998). *The tree of knowledge: The biological roots of human understanding* (Rev. ed.). Boston: Shambhala.

Mitchell, M. (2009). *Complexity: A guided tour*. Oxford, UK: Oxford University Press.

Newman, M., Smith, M., Dexheimer-Pharris, M., & Jones, D. (2008). The focus of the discipline of nursing. *Advances in Nursing Science, 14*(1), E.16–E.27.

Osberg, D., Biesta, G., & Cilliers, P. (2008). From representation to emergence: Complexity's challenge to the epistemology of schooling. In M. Mason (Ed.), *Complexity theory and the philosophy of education*. Malden, MA: Wiley-Blackwell.

Peat, F. D. (2002). *From certainty to uncertainty: The story of science and ideas in the twentieth century*. Washington, DC: Joseph Henry Press.

Ray, M. (1998). Complexity and nursing science. *Nursing Science Quarterly, 11*(3), 91–93.

Ray, M. (2010). *Transcultural caring dynamics in nursing and health care*. Philadelphia: F. A. Davis.

Roach, M. (2002). *Caring, the human mode of being* (2nd ed., Rev.). Ottawa, ON: Canadian Hospital Association Press.

Thoma, H. (2003). All at the same time: Holistic science. *Resurgence, 1*(216), 15–17.

Turkel, M. (2006). Integration of evidence-based practice in nursing. In S. Beyea & M. Slattery (Eds.), *Evidence-based practice in nursing: A guide to successful implementation.* Marblehead, MA: HCPro.

Watson, J. (2008). *The philosophy and science of caring* (Rev. ed.). Boulder, CO: University Press of Colorado.

Wheatley, M. (2006). *Leadership and the new science: Discovering order in a chaotic world* (3rd ed.). San Francisco: Berrett-Koehler.

Addendum

Competency Bingo!

Warning: Playing This Game May Make You Aware of Industrial Influences in Your Environment

Throughput	Targets	Noncompliant	Fix It	Just Get It Done
Light a Fire Under Them	No Margin, No Mission	Bottom Line	Competency	Heads in Beds
Benchmark	Make It Happen	Let's Be Objective	Management Operations Reviews (MORs)	Productivity
Check Lists	Scripts	Competition	That's How It's Done Here	Manage It
Politically Correct	Job/Role	Service Initiative	No Budget	Nursing Tasks

Literacy Bingo!

Warning: Playing This Game May Raise Your Awareness of Caring Science

Patient/Family Journey	Vision of Excellence	Honoring Beliefs	Understand It	Being Fully Engaged and Present
Stoke the Fire Within	No Mission, No Margin	Caring Economics—Ethic	Literacy	Covenant
Self & Systems Actualization	Engage Others	Let's See the Humanity—The Subjective	Vital Signs Dialogue	Foster Harmony Between Caring & Cost
Presence/ Intentionality	Authentic Voice	Collaboration	Open to Creative Emergence	Coach "em"
Alignment With Humanistic Values	Life's Work	Authentic Transpersonal Moments of Being	Explore Infinite Possibilities Together	Caritas Processes

How to Play: Check off each phrase you hear during a meeting or phone call. When you get five blocks horizontally, vertically, or diagonally, you have "Bingo!" Save your games to inform your culture change.

TABLE 1 Education Template

COMPLEXITY SCIENCE CONCEPTS	APPLICATION TO EDUCATION
Relationship-centered: Every class fosters caring relationships where respect for person is honored and loving kindness (caritas) is transmitted	• Caring contract versus rules • Honoring gifts that each other brings • Integrating multiple ways of knowing (aesthetic, empirical, ethical, and personal)
Nursing is a caring, holistic process of recognizing, connecting, partnering, and reflecting	• Assist students in understanding that this involves knowing the wholeness of the patient, understanding what is most important to the patient, involving the patient/family in the plan of care, and understanding the patient's response to nursing from their perspective
Every class is a complex system wherein students are involved in continuous interaction within a network or web of relationships	• Great changes can emerge from small interactions; everything is open to spiritual–ethical choice making and everything is open to possibilities; even the impossible can happen; it can be the mystery unfolding-increasing awareness, learning, understanding, and emergence
Complex responsive processes of relating in the classroom	• Focuses on human beings as relational, caring human beings • Teaching–learning is a caritas process (compassion, loving attention to, and loving kindness) • Teaching is an ethical choice-making process • Learning is a social process • Need for increased teacher–student engagement • Teaching and learning experience is nonlinear: Diverse interactions regarding nursing (theory, research, leadership, and practice) can lead to wider patterns of understanding • All phenomena interrelate; people are in intersubjective relationships • Student and teacher are involved in the "information" relating to their own inquiry; information is unitary; and there are no objective questions • Patterns have the potential for continuity and change • Small changes can lead to new patterns of interacting and relating • Diversity in the classroom leads to new patterns of communication and learning • Paradoxes and opposite statements are accepted • Emergence: New ideas are continually emerging

(*continued*)

TABLE 1 (*continued*)

COMPLEXITY SCIENCE CONCEPTS	APPLICATION TO EDUCATION
Complexity science in the classroom	• Foster interdependence • Cocreate a caring-based learning environment (honor the mutual-human-environment relational process for learning and well-being) • Encourage spiritual–ethical caring in choice making • Recognize the embeddedness of people within a group, organization, and society • Appreciate diverse cultural values, beliefs, and attitudes that are embedded within teachers, students, and administrators and how they inform us about complex individual, group, and organizational system interactions and meaning systems • Encourage transcultural caring dynamics and competency • Plan opportunities for collaboration • Use diverse methods and learning styles depending on needs and cocreative processes • Encourage games and role playing that facilitate new patterns of learning and teaching • Understand that learning is nonlinear • Invite learning through conversation • Invite learning through story and discovery • Embrace new ways for problem-solving in transformative exchanges • Invite paradox in the classroom • Become familiar with unpredictability • Develop ways for pattern seeing, pattern recognizing, and pattern transformation • Capitalize on existing patterns of interactions to cocreate new patterns of interaction and change

(*continued*)

TABLE 1 *(continued)*

COMPLEXITY SCIENCE CONCEPTS	APPLICATION TO EDUCATION
	• Foster student and teacher reflecting and intuition • Understand ideas at the "edge of chaos" using chaos theory in complexity sciences (disorder and order are present at the same time, and with ethical and spiritual choice-making lead to transformation [change]) • Encourage relational self-organization (transformation through caring and caritas processes)

Boykin, A. (Ed.). (1993). *Living a caring-based program.* New York: National League for Nursing.

Boykin, A., & Schoenhofer, S. (2001). *Nursing as caring: A model for transforming practice.* Boston: Jones and Bartlett. (Re-release)

Davidson, A., & Ray, M. (2008). *Complexity for human-environment well-being.* Creating Jazz: Transforming Exchanges in Education & Practice. 35th Annual National Conference on Professional Nursing Education and Development, University of Kansas, School of Nursing, Kansas City, MO.

Lindberg, C. (2008). *Classroom jazz: Using the principles of complexity science to foster learning in the classroom.* Creating Jazz: Transforming Exchanges in Education & Practice. 35th Annual National Conference on Professional Nursing Education and Development, University of Kansas, School of Nursing, Kansas City, MO.

Lindberg, C., Nash, S., & Lindberg, C. (2008). *On the edge: Nursing in the age of complexity.* Bordentown, NJ: Plexus Press.

Ray, M. (2010). *Transcultural caring dynamics in nursing and health care.* Philadelphia: F. A. Davis Company.

Turkel, M., Ray, M., & Kornblatt, L. (2011). Instead of reconceptualizing the nursing process, let's rename it. *Nursing Science Quarterly.*

Watson, J. (2008). *Nursing: The philosophy and science of caring* (2nd Rev. ed.). Boulder, CO: University Press of Colorado.

TABLE 2 Leadership Template

COMPLEXITY SCIENCE CONCEPTS	APPLICATION TO LEADERSHIP
Emergence of new through connections	• Discovery action dialogues. Allowing for openness during meetings, withholding judgment—you are surprised by where you wind up given where you started.
Paradoxes are normal	• Situations that seem to be paradoxes are opportunities for creative solutions rather than compromise. When staff say, "We cannot extend visiting hours, we have to allow our patients to rest and receive treatment," embrace the apparent paradox by saying, "How can we create extended visiting hours in a way that patients receive all the rest and treatment they require?"
Self-similarity (patterns that repeat)	• Humans are good at identifying patterns, and they largely relate to connections within the system, but you have to use your senses. Ask questions, such as, "What do you see?" and "What strikes you?"
Patterns are connected	• Through networks—by role, by friendships, by working relationship, by shift, and by geography.
Embeddedness (located within a larger entity) patterns on a macro-level that are repeated at a meso- and micro-level	• How does the way an intern and nursing assistant interact relate to how our organization interacts with the local community?
Hysteresis (sensitive to initial conditions)	• Outcomes to complex problems are dramatically influenced by starting conditions; who is in the room, what kind of day are they having, and what sort of issues they are facing at home.
Systems theory—things/individuals are connected and influence each other	• Health care associated infection prevention system includes staff, hand hygiene dispensers, environmental services, microorganisms, visitors, patient transport, etc.
Cocreation	• The process by which stakeholders who appear to have conflicting interests reach mutually advantageous conclusions, using complexity-based approaches, such as dialogue, brainstorming, trial and error, etc.
Mutual dependence	• Our own experiences influence and are influenced by others. What is good for the frog needs to be good for the fly.

(*continued*)

TABLE 2 ***(continued)***

COMPLEXITY SCIENCE CONCEPTS	APPLICATION TO LEADERSHIP
Nonlinearity	• The result of your interventions does not relate to the amount of resources you put into your intervention—small changes may have large and sustainable results and big campaigns may flop.
Self-organization	• Groups of people/things seek out what is best for the whole if left to their own devices, allowing them to exceed what they might have imagined as limitations. A team of caregivers who have never worked together assembling around a patient who has suddenly deteriorated.

Lindberg, C., Herzog, A., Merry, M., & Goldstein, J. (1998). Healthcare applications of complexity science. Life at the edge of chaos. *Physician Executive, 24*(1), 6–20.

Lindberg, C., Nash, S., & Lindberg, C. (2008). *On the edge: Nursing in the age of complexity.* Bordentown, NJ: Plexus Press.

TABLE 3 Practice Template

COMPLEXITY SCIENCE CONCEPTS	APPLICATION TO PRACTICE
Unpredictability	• Nurses'/Physicians' workload • Time involved to develop a caring relationship with a patient • Patients' individual response to nursing interventions • Family response to changes in the patients' condition • Patients' choices with respect to the plan of care
Self-similarity (patterns that repeat)	• Increased number of admissions to trauma unit every Friday and Saturday • Patients readmitted because of inability to pay for medications • Patients response to chronic illness
Paradoxes are normal	• Need to administer pain medication to a patient with "drug-seeking behavior" • Ethical caring dilemmas (finding the caring under the uncaring patient family expressions or behaviors) • Caring for the not so picture perfect patient (Turkel and Ray)
Nonlinear	• Caring moment between a nurse and patient • Nursing intuition • The art of nursing • Spirituality • Healing touch • Patients' energy fields
Hysteresis (sensitive to initial conditions)	• Patients' preconceptions based on previous experiences may impact perception of care during the current experience. • Patients' experience in the ER may affect their perception of care in other units. • Nurse, physician, health care team, and the patient create the healing environment together.

(*continued*)

TABLE 3 ***(continued)***

COMPLEXITY SCIENCE CONCEPTS	APPLICATION TO PRACTICE
Mutual dependence	• The relationships among the nurses, the physician, the health care team, and the patients influence the healing.
Emergence	• Being open to miracles and the unexpected—a patient who survives a 45-minute cardiac arrest and a patient who survives her cancer diagnosis by believing in hope and miracles • Being open to the patients' expressions of both positive and negative feelings
Embeddedness (located within a larger entity)	• How does the relationship between the registered nurse and the patient care assistant influence the patients' perception of care on the nursing unit? • Registered nurses from the perioperative areas influence patient satisfaction at the organization level by visiting patients on the nursing units after surgery
Self-organization	• Registered nurses' involvement in evidence-based practice and research transforms care at the bedside • Response of health care providers to community needs in a disaster (e.g., hurricanes, blizzards, and floods)
Cocreation	• Registered nurses and patient care assistants develop an innovative care delivery model • Registered nurses and physicians work in collaboration to improve patient safety • Registered nurses and administrators work together to find a balance between caring and economics

Lindberg, C., Herzog, A., Merry, M., & Goldstein, J. (1998). Healthcare applications of complexity science. Life at the edge of chaos. *Physician Executive, 24*(1), 6–20.

Lindberg, C., Nash, S., & Lindberg, C. (2008). *On the edge: Nursing in the age of complexity.* Bordentown, NJ: Plexus Press.

Turkel, M., & Ray, M. (2000). Relational caring complexity: A theory of the nurse-patient relationship within an economic context. *Nursing Science Quarterly, 13*(4), 307–313.

Turkel, M., & Ray, M. (2009). Caring for "not-so-picture-perfect patients." In R. Locsin & M. Purnell (Eds.), *A contemporary nursing process: The (un)bearable weight of knowing in nursing* (pp. 225–249). New York: Springer Publishing Company.

TABLE 4 Research Template

COMPLEXITY SCIENCE CONCEPTS	APPLICATION TO RESEARCH
Entropy and complexity (2nd law of thermodynamics) increasing disorder in the absence of intervention choice making	• Describe and interpret the influence of the quality of the nurse–patient caring relationship in terms of choices for healing, improved health, and well-being, that is, using Complex Caring Dynamic Model of inquiry (Chapter 2) (see also Chapters 3 and 6).
Pattern seeing Pattern mapping Pattern recognizing	• Use of holistic approaches to study self & life pattern forms of health, healing, well-being, or dying through the dynamics of the nurse–patient caring relationship (i.e., see chaos theory, appreciative inquiry, and nonlinear hierarchical modeling, cosigner analysis, fractal pattern analysis [pattern seeing, mapping and recognizing sets of objects, events, elements, or patterns of self-similarity], time series, and phenomenology [description and hermeneutics (interpretation)]). (See Chapters 2, 9, and 10).
Pattern interconnection	• Use mixed method design (fractal analysis, qualitative including story-telling and imaginative, phenomenological and hermeneutic, and critical theory approaches) to see, map, and recognize patterns in the integration of human, spiritual and sociocultural experiences, genetics (kinship structure), laboratory results, x-rays or other technologies, electronic health records, etc. in terms of health, healing, dying, and well-being.
Complex caring dynamics	• Use creative research approaches to the study complex patterning (e.g., of "opposing things happening at the same time–*paradox*"); explore a *qualitative* approach to fractal analysis or complex caring inquiry to study caring and networks of relationship and spiritual–ethical choices or the caring moment. Use appreciative inquiry to study life patterns and transformative processes.
Hysteresis	• Identify initial conditions (hysteresis) of the researcher in relation to his/her influence on research itself (i.e., pay attention to the concept of "bracketing" presuppositions in qualitative researcher, and ways of knowing of the researcher in relation to interventions in quantitative research methods.

(*continued*)

TABLE 4 ***(continued)***

COMPLEXITY SCIENCE CONCEPTS	APPLICATION TO RESEARCH
Modeling	• Identify theoretical/conceptual models that guide research or discover conceptual/theoretical models *from* qualitative research methods, that is, ethnography, grounded theory, phenomenology, hermeneutics, and critical theory/action research. Examine how theory influences quantitative and qualitative research approaches (see Chapter 10). Identify the significance of mathematical modeling (see Chapter 9).
Scale (micro, meso, and macro)/ Embeddedness	• Examine scaling variances and processes, for example, for patients (Micro level), collect data (information) from a micro cellular level (chemical/insulin reactions, blood, sugar, and other lab results), to (Meso level) patient reactions to diet and nutrition, and caring interactions, to Macro level (community level and/or health care policy) (see Chapter 12, the story of patients with diabetes)
Nonlocality	• Research how improvement/change in health or well-being relates to healing modalities used from a distance (the study of prayer, therapeutic touch, Reiki, or other alternative healing modalities).
Emergence (the emergence of new patterns/interconnections, relational self-organization, and possibilities)	• Explore subtleties of forms of patterning (expressive, aesthetic, intuitive, imaginative, mysterious, and miraculous) through all forms of inquiry (quantitative, qualitative, critical, appreciative, creative, and mixed-method design approaches); identify how the study of the caring nurse as a "magnetic attractor" is significant to patient care, choices of health and well being, and organizational leadership, informatics, and the ethics of human-humanoid interrelationships (see all innovative chapters in this book).

Brown, C., & Liebovitch, L. (2010). *Fractal analysis.* Los Angeles, CA: Sage.

Cowling, R., & Repede, E. (2010). Unitary appreciative inquiry: Evolution and refinement. *Advances in Nursing Science, 33*(1), 64–77.

Davidson, A., & Ray, M. (1991). Studying the human-environment phenomenon using the science of complexity. *Advances in Nursing Science, 42*(7), 73–87.

Davidson, A., Cortes, S., Conboy, L., Ray, M., & Norman, M. (2006). *Complexity for human-environment well-being.* Retrieved December 2, 2009, from http://necsi.org/events/ICCS6/newpaper?id=216

Morse, J., & Niehaus, L. (2009). *Mixed method design: Principles and procedures.* Walnut Creek, CA: Left Coast Press.

Ray, M. (1994). Complex caring dynamics: A unifying model of nursing inquiry. *Theoretic and Applied Chaos in Nursing, 1*(1), 23–32.

Reed, P. (1997). Nursing: The ontology of the discipline. *Nursing Science Quarterly, 19*(1), 76–79.

Reed, P. (2006). The practice turn in nursing epistemology. *Nursing Science Quartely, 19*(1), 51–60.

Index